CHEMODYNAMICS

Chemodynamics

ENVIRONMENTAL MOVEMENT OF CHEMICALS IN AIR, WATER, AND SOIL

LOUIS J. THIBODEAUX

College of Engineering
University of Arkansas
Fayetteville, Arkansas

A WILEY-INTERSCIENCE PUBLICATION

JOHN WILEY & SONS, New York • Chichester • Brisbane • Toronto

Library of Congress Cataloging in Publication Data

Thibodeaux, Louis J
    Chemodynamics, environmental movement of chemicals in
air, water, and soil.

    "A Wiley-Interscience publication."
    Includes indexes.
    1. Environmental chemistry.   I.   Title.

QD31.2.T47   574.5′2   78-31637
ISBN 0-471-04720-1

Printed in the United States of America

10 9 8 7 6 5 4 3 2 1

To Elwana, Scott, and Shelly

PREFACE

Chemodynamics* is an applied science activity concerned primarily with the movement and fate of synthetic chemicals within the three geospheres of the environment (i.e., air, water, and earthen solids). The goal of this book is to present and evaluate existing methods, commonly referred to as models, for studying the movement of substances from the site of entry into the environment to the various geospheres for the purpose of estimating exposure along the way.

In a typical situation the engineer or environmental scientist is faced with a host of questions: what information is needed to attack a problem, how can it best be obtained, and how can a reasonable model be selected from the available alternatives. The purpose of this book is to teach how to answer these questions reliably and wisely. To do this, emphasis is placed on the qualitative description of simple models and mechanisms. This is accompanied by a quantitative description of the problem. This approach assesses the current ability of the existing knowledge in engineering science in determining the rates, lifetimes, routes, and reservoirs of chemical substances moving through the environment and attempts to estimate the resulting level of exposure to susceptible targets, both living and nonliving. This is a teaching book; therefore, the approach is to use simple models and ideas. These help develop a strong intuitive sense and a guide in constructing integrated environmental and ecosystem models for simulating chemical movement and fate.

---

*Word coined by Virgil Freed at 176th American Chemical Society Meeting, Los Angeles, California, 1974.

This book is prepared for an audience consisting of engineers (chemical and environmental), environmental chemists, public health scientists, and other scientists concerned with chemicals in the geospheres. The subject matter undoubtedly demands increased attention from both industrial and governmental organizations. The burden of responsibility for the ultimate fate and effects of man-made chemicals is being placed on the manufacturer. More legislation will be forthcoming and governmental regulatory agencies will likely increase monitoring activities in this area. A working conference of the National Academy of Sciences, National Academy of Engineering,* was convened

...to study the principles of protocols for evaluating chemicals in the environment; to provide us with the benefit of your insight and knowledge as to the interactions and effects that take place when chemicals enter the environment: and thus to help us to test for the adverse effects that may result when chemicals enter the environment...

The timeliness of chemical transport and related topics in the subject area of chemodynamics is readily apparent. The timelessness of the subject is also assured until the world community decides that the net social gain from man-made chemical substances is negative.

This is an introductory book. The subject matter has been developed for and used as a technical elective for chemical engineering seniors and graduate students. The next largest group of users has been graduate students in environmental engineering (a subdivision of civil engineering). The emphasis, instead of being on the distinctive and traditional chemical engineering activity of mass transfer, is on the movement of chemicals within the environment. Although the mathematical level is not particularly difficult (elementary calculus and linear first-order differential equations are all that is needed), this does not mean that the ideas and concepts being taught are particularly simple. A large number of problems have been included. Problem solving, that process of applying the concepts to new situations, is essential to learning.

The book is organized so that it can be effectively used by the student and the practicing professional. The student will discover that the subject matter is developed in a logical manner, commencing with simple concepts, progressing through model development, and finally discussing application to real world chemodynamic situations. The professional,

---

*William D. Ruckelshaus, "Message to the Working Conference," in N. Nelson, Chairman, *Principles for Evaluating Chemicals in the Environment*, National Academy of Sciences, Washington, D.C., 1975, p. 1.

familiar with the fundamentals of chemodynamics, will find that the book contains a large number of useful models and formulas, easy-to-follow example problems, and tables of chemical, physical, and environmental data directly applicable in assisting the formulation of a quantitative assessment of the impact of almost any chemical on the exterior environment. A brief review of the organization of the book follows.

Chapter 1 defines precisely the scope and content of the book. This is done by discussing the meaning of the word *chemodynamics* and presenting a single, specific example. The familiar chemodynamics example, reaeration of natural streams, serves to illustrate the format, depth and utility of all the models presented in the book. Chapter 1 also contains an introduction to the nomenclature and the International System of Units (SI).

The next two chapters are brief presentations of important fundamental concepts. Chapter 2 is concerned with the equilibrium condition for chemicals across environmental interfaces. The basic concept of equilibrium, useful theoretical developments, and some data are given for chemicals on either side of the air–water, water–soil, and air–soil interfaces. Chapter 3 reviews the fundamental mechanistic ideas and gives the working model formulas for chemical transport in the region near environmental interfaces. These two chapters contain very brief introductions to the important concepts of equilibria and transport. A basic knowledge of these concepts is necessary to understanding and using the more practical results contained in the following four chapters.

Chapters 4, 5, and 6 are concerned with the movement of chemicals across the air–water, water–soil and air–soil interfaces, respectively. These are highly applied chapters concerned with interphase transport in both directions across the respective interfaces. Many useful models are presented and the results of some model predictions of chemical movement are discussed.

Chapter 7 is concerned with intraphase movement of chemicals and is also an applied chapter. Intraphase movement is concerned mainly with the dispersion of a chemical within a single phase of the environment. The Gaussian plume model for air pollutant dispersion is a familiar example of the subject content of this chapter. Emphasis is placed on the vertical and horizontal movement of chemicals in water bodies. The effect of stratification is accounted for in several models.

Chapter 8 contains a brief review of other transport processes responsible for the movement of chemicals in the exterior environment. Chapter 8 concludes with a list of books and significant reports relevant to and directly related to the subject of chemodynamics. This material is suggested for use as supplemental reading and/or reference in a one-semester course.

The appendixes contain tables of chemical, physical, and environmental data. These data are useful to the student working the exercise problems and to the professional making model calculations of chemical lifetime, movement rate, and concentration level.

Louis J. Thibodeaux

Fayetteville, Arkansas
March 1979

ACKNOWLEDGMENTS

Many persons have contributed directly or indirectly to this book and it is fitting to mention some of them by name:

Professors J. A. Havens and C. S. Springer of the Chemical Engineering Department and Professor D. G. Parker of the Civil Engineering Department, University of Arkansas, deserve many thanks for their continued interest in the field of chemodynamics and their willingness to discuss both general and specific topics of the subject area.

Professor C. M. Thatcher of the Chemical Engineering Department, University of Arkansas, is thanked for his encouragement that I produce a manuscript. His help in attempting to develop a consistent system of mnemonic nomenclature was invaluable.

Professor J. R. Cooper, Chairman, Chemical Engineering Department, University of Arkansas, is thanked for approving of the subject matter and allowing it to be offered on a yearly basis as a technical elective course for senior students.

Students have made a tremendous, indirect contribution to the development of this book by displaying enthusiasm for the subject matter. A small number of students have made direct contributions by suggesting subject areas and exercise problems and by proofing sections of the manuscript. These students are Randy Bayliss, James A. Reinhardt, Gregory D. Reed, Albert L. Hood, Duane J. Lewis, Michael Mourot, Li-Kow Chang, Howell Heck, Davey Stallings, Karen Kuhn and Wayne Bequette.

Most of the manuscript was typed by Vicki Asfahl. Lynn Hachenberger is also acknowledged for assisting with typing the manuscript. Joyce Thibodeaux contributed by typing corrections, additions, and accessory material but she contributed mostly by insisting that I go upstairs and write.

L. J. T.

CONTENTS

LIST OF SYMBOLS

Symbols that appear infrequently or in one section only are not listed. Dimensions are given in terms of mass (M), amount of substance (mol), length (L), time (t), and temperature (T). Boldface symbols are vectors or tensors.

A = area, L^2.

a = absorptivity, dimensionless.

a = acceleration, L/t^2.

a = interfacial area per unit volume, L^{-1}.

C_d = drag coefficient, dimensionless.

$\hat{C}_p$ = heat capacity at constant pressure, per unit mass, L^2/t^2T.

$\hat{C}_v$ = heat capacity at constant volume, per unit mass, L^2/t^2T.

c = total molar concentration, mol/L^3.

c_{ij} = molar concentration of species i in phase j, mol/L^3.

D = characteristic length, L.

D_p = particle diameter, L.

$\mathcal{D}_{ij}$ = molecular diffusivity of species i in phase j, L^2/t.

$\mathcal{D}_{ij}^{(t)}$ = turbulent or eddy diffusivity of species i in phase j, L^2/t.

D_{ij} = diffusion coefficient of species i in phase j, L^2/t.

D_{A31} = diffusion coefficient of chemical A in soil phase pore spaces filled with air, L^2/t.

D_{A32}^T = thermal diffusion coefficient of chemical A in soil phase pore spaces filled with water, L^2/t.

d = diameter, L.

d = zero-plane displacement, L.

e = 2.71828...

e = emissivity, dimensionless.

F = a fraction denoting efficiency, removal efficiency, and so on, dimensionless.

f_{ij} = fugacity of species i in phase j, M/Lt^2.

f_{ij}^0 = pure component fugacity of species i in phase j, M/Lt^2.

G = mass velocity, M/tL^2.

g = gravitational acceleration, L/t^2.

H = diffusion hindrance factor, dimensionless.

H_A = Henry's law constant, M/Lt^2.

H_R = relative humidity, dimensionless.

h = heat transfer coefficient, M/t^3T.

h = Plank's constant, ML^2/t.

h = elevation, L.

$\mathbf{J_i}$ = molar flux of species i relative to mass average velocity, mol/tL^2.

$\mathbf{j}$ = mass flux of species i relative to mass average velocity, M/tL^2.

j_D = Chilton-Colburn j factor for mass transfer, dimensionless.

j_H = Chilton-Colburn j factor for heat transfer, dimensionless.

$^1K_{i2}$ = overall liquid phase mass transfer coefficient for species A across a gas–liquid interface, mole fraction driving force, mol/tL^2.

$^1K_{i2}'$ = overall liquid phase mass transfer coefficient for species A across a gas–liquid interface, concentration driving force, L/t.

$\mathcal{K}_{A12}^*$ = equilibrium partition or distribution coefficient for species A between air and water, mole fraction ratio, mol/mol.

$\mathcal{K}_{A12}^*$ = equilibrium partition coefficient or distribution coefficient for species A between air and water, concentration ratio, L^3/L^3.

k = thermal conductivity, ML/t^3T.

k_i'' = heterogeneous chemical reaction rate constant, n is order of reaction, $\text{mol}^{1-n}/L^{2-3n}t$.

k_i''' = homogeneous chemical reaction rate constant, n is order of reaction, $\text{mol}^{1-n}/L^{3-3n}t$.

$^3k_{i1}$ = individual gas phase mass transfer coefficient for species i across the air–soil interface, mole fraction driving force, mol/tL^2.

$^3k'_{i1}$ = individual gas phase mass transfer coefficient for species i across the air–soil interface, concentration driving force, L/t.

L = characteristic length, L.

l = Prandtl mixing length.

M = molar mean molecular weight, M/mol.

M_i = molecular weight of species i, M/mol.

$\mathfrak{M}$ = moles of material, mol.

$\mathfrak{M}_i$ = moles of component i, mol.

m = relative soil moisture content, dimensionless.

m = mass of material, M.

m_i = mass of component i, M.

N = Avogadro's number, mol^{-1}.

$\mathbf{N}$ = total molar flux with respect to stationary coordinates, mol/tL^2.

$\mathbf{N_i}$ = molar flux of species i with respect to stationary coordinates, mol/tL^2.

$\mathbf{n}$ = total mass flux with respect to stationary coordinates, M/tL^2.

$\mathbf{n_i}$ = mass flux of species i with respect to stationary coordinates, M/tL^2.

p = fluid pressure, M/Lt^2.

p_i^0 = pure component vapor pressure of species i, M/Lt^2.

Q = volumetric flow rate, L^3/t.

$\mathbf{Q}$ = energy flow rate across a surface, ML^2/t^3.

$\mathbf{Q_{12}}$ = radiant energy flow from phase 1 to phase 2, ML^2/t^3.

$\mathbf{q}$ = energy flux rates relative to mass average velocity, M/t^3.

R = gas constant, $ML^2/t^2T\text{mol}$.

R = radius of sphere or cylinder, L.

R_i = molar rate of production of species i, mol/tL^3.

r = radial distance in both cylindrical and spherical coordinates, L.

r_i = mass rate of production of species i, M/tL^3.

r_h = hydraulic radius, L.

T = *absolute* temperature, T.

t = time, t.

V = volume, L^3.

$\mathbf{v_i}$ = velocity of species i, L/t.

$\bar{\mathbf{v}}_j$ = mass average velocity of phase j, L/t.

$\mathbf{v_j}$ = time averaged, point velocity of phase j, L/t.

v_j' = fluctuating velocity of phase j, L/t.

$\mathbf{v}_\infty$ = approach velocity or velocity for removed from interface, L/t.

$v_* = \sqrt{\tau_0/\rho_j}$ = reference or friction velocity, L/t.

W = molar flow rate, mol/t.

W_i = molar flow rate of species i, mol/t.

w = mass flow rate, M/t.

w_i = mass flow rate of species i, M/t.

x = rectangular coordinate, longitudinal direction, L.

x_i = mole fraction of species i in water, dimensionless.

y = rectangular coordinate, vertical direction, L.

y_i = mole fraction of species i in air, dimensionless.

z = rectangular coordinate, lateral direction, L.

z_i = mole fraction of species i in solid, dimensionless.

$\alpha_j = k/\rho C_p =$
thermal diffusivity for phase j, L^2/t.

$\alpha_j^{(t)}$ = turbulent or eddy thermal diffusivity for phase j, L^2/t.

α_i^* = relative volatility (gas–liquid) of species i, dimensionless.

β = cloud cover factor, dimensionless.

$\Gamma(x)$ = gamma function of x.

Γ = mass flow rate of flowing liquid film per unit width of wetted surface, M/Lt.

Γ_a = dry adiabatic lapse rate, T/L.

γ = existing, in general diabatic, lapse rate in the surrounding air, T/L.

γ_{ij} = chemical activity coefficient of species i in phase j, dimensionless.

Δ_i = deficit mole fraction of species i, dimensionless.

δ = boundary layer thickness, L.

ϵ_j = fraction void space occupied by phase j in a solid, dimensionless.

ζ_{ij} = concentration coefficient of volumetric expansion for species i in phase j, dimensionless.

θ = angle in cylindrical or spherical coordinates, radians.

θ = wave period, t.

κ_1 = Prandtl's constant, dimensionless.

κ_2 = von Karman's constant, dimensionless.

λ_i = latent heat of vaporization of species i, L^2/t^2.

μ_j = viscosity of phase j, M/Lt.

μ_{ij} = chemical potential of species i in phase j, dimensionless.

ν = frequency, Brunt-Vaisala, t^{-1}.

$\nu_j = \mu_j/\rho_j =$
    kinematic viscosity of phase j, L^2/t.

$\pi = 3.14159\ldots$

ρ = reflectivity, dimensionless.

ρ_j = mass density of phase j, M/L^3.

σ = mean displacement or dispersion, L.

σ_{A1} = collision diameter of species A in a gas, L.

τ = residence time, t.

τ_o = shear stress at fluid–solid interface, M/t^2L.

ϕ = porosity, dimensionless.

ϕ_i = mass fraction of species i in water, dimensionless.

ψ_i = mass fraction of species i in air, dimensionless.

Ω = density gradient, L^{-1}.

ω_i = mass fraction of species i in solid, dimensionless.

ω_j = frequency of oscillation, t^{-1}.

OVERLINES

·   local value.

~   per mole.

∧   per unit mass.

−   time smoothed.

BRACKETS

$\langle a \rangle$ average value of a over a flow cross section.

$[=]$ has dimensions of.

SUPERSCRIPTS

o initial value.

*   equilibrium condition, value or solubility.

′   deviation from time-smoothed value.

′   denotes mass transfer coefficient has dimensions of L/t.

(*t*) turbulent.

(*l*) laminar.

T   thermal diffusivity.

1   gas interface.

2   liquid interface.

3   solid interface.

SUBSCRIPTS

A, *B*, *C*, etc.

  species in multicomponent system.

a   arithmetic mean driving force or associated transfer coefficient.

b   bulk or "cup-mixing" value.

i   interface.

i   arbitrary chemical species (*A* for chemical *A*, *B* for chemical *B*, etc.).

j   arbitrary phase (1 for air, 2 for water, 3 for soil, 4 for second liquid, etc.)

m   completely mixed system.

o   at origin of space dimension.

o   quantity evaluated at a surface or interface.

p   plug flow system.

x   variable has movement in *x* direction

y   variable has movement in *y* direction

z   variable has movement in *z* direction

1, 2, 3, 4

  see subscript *j* above.

COMMONLY USED DIMENSIONLESS GROUPS

Fr = Froude number.

Gr = Grashoff number for heat transfer,

Gr_{ij} = Grashoff number for mass transfer.

Nu = Nusselt number for heat transfer.

Nu_{ij} = Nusselt number for mass transfer.

Sh_{ij} = Sherwood number for mass transfer.

Pr = Prandtl number.

Re = Reynolds number.

Sc_{ij} = Schmidt number.

MATHEMATICAL OPERATIONS

$$\text{erf}(x) = \frac{2}{\sqrt{\pi}} \int_0^x e^{-t^2} dt =$$
    error function of x.

$$\exp(x) = e^x =$$
    the exponential function of x.

$\ln(x)$ = the logarithm x to the base e.

$\log(x)$ = the logarithm x to the base 10.

$$\Gamma(x, u) = \int_0^u t^{x-1} e^{-t} dt =$$
    the incomplete gamma function.

$$\Gamma(x) = \int_0^\infty t^{x-1} e^{-1} dt =$$
    the complete gamma function.

∇ = the "del" or "nabla" operator.

$I_0(x)$ and $K_0(x)$ =
    modified Bessel functions, of x, of the first and second kinds.

CHEMODYNAMICS

CHAPTER 1

INTRODUCTION

1.1. INTRODUCTION TO ENVIRONMENTAL CHEMISTRY AND ENGINEERING

Chemodynamics

With an unparalleled surge in creativeness, the human race has produced literally hundreds of thousands of "unnatural" chemicals. This ever-increasing array of chemical agents is regarded as a potential threat to humanity and its living environment. Many of these xenobiotics or anthropogenic substances have found their way into the biosphere and have been classified as toxic or potentially harmful chemicals. Figure 1.1-1 illustrates the pathways by which pesticides are transported between environmental compartments. Since we are going to continue using chemicals, it is important to be able to trace their transport in the natural environment.

Environmental chemodynamics is the name given to a subject that deals with the transport of chemicals (intra- and interphase) in the environment, the relationship of their physical-chemical properties to transport, their persistence in the biosphere, their partitioning in the biota, and toxicological and epidemiological forecasting based on physical-chemical properties. A comprehensive and systematic study of chemical movements in the environment is an interdisciplinary undertaking and must utilize the principles of such disciplines as chemistry, physics, systems analysis, modeling, engineering, and medical and biological sciences.

This book is concerned with several topics of environmental chemodynamics. Specifically the subject is the interphase transport of chemicals

Pesticide cycling in the environment involves complex processes

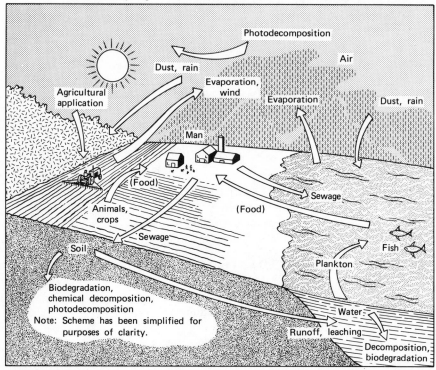

Figure 1.1-1. Pesticide pathways between environmental compartments. Reprinted, with permission of the copyright owner, AMERICAN CHEMICAL SOCIETY, from the April 22, 1974 issue of CHEMICAL AND ENGINEERING NEWS.

and energy between the air, soil, and water phases of the environment. We focus on the mechanisms and rates of movement of chemicals across the three interfaces: air–soil, soil–water, and water–air. Our focus is on the region near the interfaces and the natural forces that affect and control transport in that region.

Once an anthropogenic substance enters the natural environment human-initiated forces aimed at controlling, manipulating, modifying, and attenuating are usually secondary to the existing natural forces. These natural forces derive their energy from the sun and are manifest mainly in the form of fluid movements (air and water) and solar radiation. Observables that quantify aspects of these natural forces include temperature, incident radiation, flow velocity, pressure, relative humidity, and concentration.

Chemical is used here in the broad sense and includes water, oxygen, carbon dioxide, sulfur dioxide, and DDT. The movements between the four ecosystems—atmosphere, hydrosphere, lithosphere, and biosphere— that constitute the environment are referred to as *interphase transfers*. The interphase transfer of water and oxygen is desirable, whereas that of sulfur dioxide and DDT is undesirable for the most part. The movement of chemicals within the environment has a profound effect upon its livability. The rates of transfer are important and the magnitude can affect environmental livability. The natural processes that promote these exchanges are ever present and are responsible for the magnitude and direction of the exchanges, be they desirable or undesirable.

Once a chemical enters one of the mobile phases (i.e., air or water), it becomes dispersed rapidly because of fluid movements. Movement within a phase is termed intraphase mass transfer, diffusion, or dispersion. Interphase mass transfer is important to the movement of synthetic chemicals between the various phases of the ecosystem. People and the other organisms that constitute the biosphere reside, to varying degrees, within the other three spheres. Figure 1.1-2 traces the direct and indirect routes of

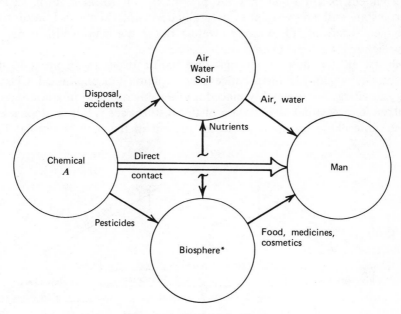

*Plants and animals excluding man.

Figure 1.1-2. Routes of synthetic chemicals to man.

synthetic chemicals through the environment and eventually to the human organism.

People encounter potentially harmful substances by direct contact with chemicals contained in food, food additives, medicines, cosmetics, the workplace, home, and so on. There are, however, several indirect contact modes of a more subtle nature. The continual intake of air and water is an indirect source of many chemical substances because of residuals in those phases of the environment. The pathways for entry of chemicals into these necessary elements of the ecosystem are shown in Fig. 1.1-3. The biosphere, in the form of natural foodstuffs, is another indirect source made even more critical by "bioamplification" processes. A large portion of our food consists of organisms at or near the ends of food chains all of which reside in some portion of the atmosphere, hydrosphere, or lithosphere. Medicines, cosmetics, and other similar personal goods made of living matter constitute another indirect contact source.

This book is concerned with natural interphase transfer processes in general and is specifically concerned with the movement of chemicals between the three portions of the ecosystems in which people find themselves. In general, chemicals are not always unnatural entities in the environment except when they are present in concentrated form. In isolated instances the movement of naturally occurring chemicals is addressed here (i.e., absorption of oxygen in water), but in the main we focus on the movement of synthetic chemicals.

Almost all activities of a progressive nature result in an upset in the natural status quo. The disturbance may be planned or unplanned. Chemicals and energy are inevitably placed into the environment in a nonnatural manner. The eventual assimilation of this disturbance is largely dependent on the interphase transport processes present in the natural setting. The

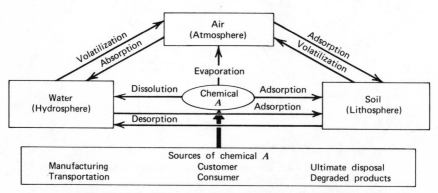

Figure 1.1-3. Movement of chemicals in the environment.

discussion in this book provides: (1) a qualitative understanding of these natural processes, and (2) quantitative tools by which to assess the response and/or recovery of the physical environment to chemical and energy stresses.

Chemical and energy stresses are eventually relieved but the rate is controlled by the natural exchange process. Interphase transport is usually the critical step. The phases involved are the atmosphere, hydrosphere, and lithosphere. We emphasize here the role of natural environmental transfer rates in assimilating stress. The question in many instances is what the rate of assimilation is and not whether the environment can assimilate the disturbance.

Intensity and lifetime are two important pieces of information with regard to environmental insults. Intensity is measured as a concentration or temperature and lifetime is measured in real time. Both of these variables are intimately related to the interphase transport phenomena and their rates.

The emphasis in this book is on application of chemical engineering transport concepts (i.e., heat, momentum, and mass transfer) to situations involving the natural environment. Chemical engineers readily deal with interphase transport; however, the application is typically inside process equipment (i.e., reactors, mass transfer columns, heat exchangers, etc.). Much of the body of chemical engineering knowledge can be applied to situations in a natural setting (i.e., out-of-doors). Here we redirect chemical engineering and transport science principles to address problems in the natural environment as opposed to the artificial environment.

Transport science is quite advanced. The emphasis here is mainly on particular applications of this science to the natural environment, focusing on the environmental impact of humanity's activities on the natural state.

International System of Units

Public laws declare that the policy of the United States shall be to coordinate and plan the increasing use of the metric system and to encourage educational agencies and institutions to prepare students to use the metric system of measurement as part of the regular education program. Under the acts, the "metric system of measurement" is defined as the International System of Units as established by the General Conference of Weights and Measures in 1960 and interpreted or modified for the United States by the Secretary of Commerce. The Secretary has delegated this authority to the Assistant Secretary for Science and Technology. Appendix A contains an interpretation and modification of the International System of Units (hereafter SI) for the United States.

Table 1.1-1. SI Base and Supplementary Units

Quantity	Name	Symbol
SI base units		
Length	meter	m
Mass	kilogram	kg
Time	second	s
Electric current	ampere	A
Thermodynamic temperature	kelvin	K
Amount of substance	mole	mol
Luminous intensity	candela	cd
SI supplementary units		
Plane angle	radian	rad
Solid angle	steradian	sr

The SI is based on seven fundamental or base units, as they are called, and the entire system is built from them. These units, their definitions, and their symbols have complete international agreement in every nation and in international organizations. In rapid review the SI base unit names and symbols are given in Table 1.1-1. These units are combined by dividing, multiplying, raising to negative or positive powers, to produce the needed derived units. There are 17 of them, which have been given special names. Any one of the base or derived units can be made any size, very large or very small, by the addition of a prefix. The student is urged to consult Appendix A for a list of the derived units, prefixes, and additional information regarding SI. Note that the unit of mass is the only one defined with a prefix. The gram (g) is then a valid SI unit.

A USEFUL INTERPRETATION OF THE MOLE AND MOLECULAR WEIGHT

Most of the SI units should be familiar to the student; however, some units important to the subject of chemistry need additional interpretation. The unit for the amount of substance, the mole, is defined as the amount of substance in a system that has the same number of entities, that is, molecules, atoms, ions, electrons, and particles, as there are atoms in 0.012 kg of carbon-12. There are 6.022 E 23 (Avogadro's number) atoms in 12 g of carbon-12. Since chemicals move about, react, exchange phase, and diffuse as well-defined entities it is more useful and often easier to quantify the number of these entities rather than the mass. The most useful entity for the purposes of chemodynamics is the molecule.

Atomic weight is the relative mass of an atom based on a scale in which a specific carbon atom (carbon-12) is assigned a mass value of 12. A table of relative atomic weights for the chemical elements appears in Appendix C. Molecular weight is the sum of the atomic weights of all the atoms in a molecule. Atomic weight and molecular weight can be interpreted to have dimensions of g/mol, kg/k mol, mg/m mol, and so on. The atomic weight of carbon is 12.011 g/mol. The molecular weight of carbon dioxide is 43.999 kg/k mol. The molecular weight (also atomic and ionic) weight of species i is denoted by M_i. Review the example problems on the utility of the mole in environmental chemistry.

SCIENTIFIC NOTATION AND MACHINE COMPUTATION

With the advent of the computer and the electronic calculator, a short-cut form of scientific notation has been adopted. The essence of the change is to represent the power-of-ten multiplier with the letter E. For example the standard scientific notation for Avogadro's number is 6.023×10^{23}; in "electronic machine" notation it is $6.023E23$. Scientific notation for the numerical value of the natural log of $\frac{1}{2}$ is -6.93×10^{-1} and in machine notation it is $-6.93E-01$. The advantages of using machine notation, besides compatibility with electronic computing devices, are the absence of powers and a shortening of the written statement. The electronic computation machine form of scientific notation is used throughout this book.

Multicomponent and Multiphase Notation

It is necessary to introduce an intricate and descriptive system of notation in dealing with the multicomponent and multiphase nature of environmental chemistry. Chemical transport within the environment necessitates movement of the particular chemical species with respect to the other species. If the movement takes place within a single phase it is termed *intraphase transport* and if it takes place between phases it is termed *interphase transport*. The following is a brief introduction to the notation to be employed throughout the textbook.

SPACE VARIABLES AND DIRECTIONS

Chemical movement implies a species change of position with respect to a fixed point in space. Since most movements of concern in this book are at or near the surface of the Earth the space variables and positive directions are defined with respect to that frame of reference. The dimension of the space variables is length (L).

$0 \equiv (0,0,0)$, the origin is a fixed, defined point in three-dimensional space.

x direction is horizontal and positive to the right.

y direction is vertical and positive upward.

z direction is horizontal and positive toward the reader.

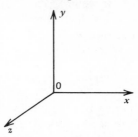

The surface of the land, sea, seabed, and so on, normally coincides with the x-z plane so that the positive y direction is pointing away from the center of the Earth. Normally, fluid flow (i.e., water or air) is from left to right in the x direction. The "natural" space orientation is used whenever possible, but special orientations may be desirable during the analysis of some problems. When this occurs the specific orientation is defined.

CONCENTRATION AND PHASE DENSITY

The quantitative intensity of existence of a chemical species within any phase of the environment is an important piece of information. It is defined as concentration, expressed as an intensive measure, and has many forms of notational representation. It was recognized earlier when the mole was introduced, that chemical species and such substances exist as well-defined entities. The most important entity with respect to environmental chemistry is the molecule; however, atomic and ionic entities are important also. The concentrations of chemicals are "naturally" expressed in molar units. Mole fraction of chemical species A represents the moles of A with respect to the total molar quantity of the phase. The mole fractions should represent the particular phase; specific definitions and notation are

$x_i \equiv$ mole fraction of chemical species i in water   (mol/mol)
$y_i \equiv$ mole fraction of chemical species i in air   (mol/mol)
$z_i \equiv$ mole fraction of chemical species i in soil   (mol/mol)

Other mole fractions such as species i in oil are defined when needed. Note that mole fractions must always be subscripted and the space variables are not, so that no confusion can result in the notation.

The simplest multicomponent system is a binary system consisting of species A and B. For this system the law of the whole (i.e., sum of the parts) requires

$$x_A + x_B = 1, \qquad y_A + y_B = 1, \qquad z_A + z_B = 1$$

In general for an N-component system consisting of $A, B, \ldots, N$

$$\sum_{i=A}^{N} x_i = 1, \qquad \sum_{i=A}^{N} y_i = 1, \qquad \text{and so on}$$

Molar concentration is also a convenient quantitative measure of chemical species intensity. A c is used for concentration:

$c_{i1} \equiv$ molar concentration of species i in air $\quad (\text{mol}/L^3)$

$c_{i2} \equiv$ molar concentration of species i in water $\quad (\text{mol}/L^3)$

$c_{i3} \equiv$ molar concentration of species i in soil $\quad (\text{mol}/L^3)$

The second subscript, when it is an integer, denotes phase j with 1 for air, 2 for water, and 3 for soil. Species concentrations in other phases are defined by employing additional integers in a similar fashion when they are needed. The sum of the molar concentrations in a phase is the molar phase density. In the case of air

$$c_1 = c_{A1} + c_{B1} + c_{C1} + \cdots + c_{Nj} \tag{1.1-1}$$

Similar molar densities exist for water and soil.

Mass units are also useful and convenient for expressing concentration of environmentally important chemical species. The mass fraction of species A represents the mass of A with respect to the total mass quantity of the phase. The mass fraction should also represent a particular phase. Specific definitions and notations are

$\phi_i \equiv$ mass fraction of species i in water $\quad (M/M)$

$\psi_i \equiv$ mass fraction of species i in air $\quad (M/M)$

$\omega_i \equiv$ mass fraction of species i in soil $\quad (M/M)$

Other mass fractions can be defined if needed.

Mass concentration is denoted by ρ; specifically,

$\rho_{i1} \equiv$ mass concentration of species i in air $\quad (M/L^3)$

$\rho_{i2} \equiv$ mass concentration of species i in water $\quad (M/L^3)$

$\rho_{i3} \equiv$ mass concentration of species i in soil $\quad (M/L^3)$

Mass and molar concentrations are related by $c_i = \rho_i / M_i$. Just as for mole fraction and molar concentrations the "law of the whole" demands

$$\sum_{i=A}^{N} \phi_i = \sum_{i=A}^{N} \psi_i = \sum_{i=A}^{N} \omega_i = 1$$

The sum of the mass concentration in a phase is the mass phase density. In

Table 1.1-2. Notation for Concentration in Multicomponent Systems

Basic definitions of molar

$$y_A \equiv \frac{c_{A1}}{c_1}, \qquad x_A \equiv \frac{c_{A2}}{c_2}, \qquad z_A \equiv \frac{c_{A3}}{c_3}, \qquad \text{etc.} \tag{A}$$

$$c_{A1} = \frac{\rho_{A1}}{M_A}, \qquad c_{A2} = \frac{\rho_{A2}}{M_A}, \qquad c_{A3} = \frac{\rho_{A3}}{M_A}, \qquad \text{etc.} \tag{B}$$

$$y_A + y_B + y_C + \cdots + y_N = 1, \qquad x_A + x_B + x_C + \cdots + x_N = 1, \qquad \text{etc.} \tag{C}$$

$$c_1 = c_{A1} + c_{B1} + \cdots + c_{N1}, \qquad c_2 = c_{A2} + c_{B2} + \cdots + c_{N2}, \qquad \text{etc.} \tag{D}$$

Basic definitions of mass

$$\psi_A \equiv \frac{\rho_{A1}}{\rho_1}, \qquad \phi_A \equiv \frac{\rho_{A2}}{\rho_2}, \qquad \omega_A = \frac{\rho_{A3}}{\rho_3} \tag{E}$$

$$\rho_{A1} = c_{A1}M_A, \qquad \rho_{A2} = c_{A2}M_A, \qquad \rho_{A3} = c_{A3}M_3, \qquad \text{etc.} \tag{F}$$

$$\psi_A + \psi_B + \psi_C + \cdots + \psi_N = 1, \qquad \phi_A + \phi_B + \phi_C + \cdots + \phi_N = 1, \qquad \text{etc.} \tag{G}$$

$$\rho_1 = \rho_{A1} + \rho_{B1} + \cdots + \rho_{N1}, \qquad \rho_2 = \rho_{A2} + \rho_{B2} + \cdots + \rho_{N2}, \qquad \text{etc.} \tag{H}$$

Additional relations

$$x_A M_A + x_B M_B + \cdots + x_N M_N = M_2, \qquad M_1 = \frac{\rho_1}{c_1}, \qquad \text{etc.} \tag{I}$$

$$\frac{\psi_A}{M_A} + \frac{\psi_B}{M_B} + \cdots + \frac{\psi_N}{M_N} = \frac{1}{M_1}, \qquad \text{etc.} \tag{J}$$

$$x_A = \frac{\phi_A / M_A}{\phi_A / M_A + \phi_B / M_B + \cdots + \phi_N / M_N}, \qquad \text{etc.} \tag{K}$$

$$\psi_A = \frac{y_A M_A}{y_A M_A + y_B M_B + \cdots + y_N M_N}, \qquad \text{etc.} \tag{L}$$

the case of air

$$\rho_1 = \rho_{A1} + \rho_{B1} + \rho_{C1} + \cdots + \rho_{N1} \tag{1.1-2}$$

Similar mass densities exist for water and soil. Table 1.1-2 contains a summary of the preceding concentrations and their interrelation.

TRACE QUANTITY ENGINEERING AND VERY DILUTE SOLUTIONS

We may define *trace quantity engineering* as the segment of engineering that covers the identification, control, and handling of trace components.[2]

The limitation on what constitutes a trace quantity is somewhat arbitrary at this point. For trace quantities, parts per million (ppm) and parts per billion (ppb) have been conventional terms for expressing concentrations. Although continued use of these terms is anticipated, their inexactness should be recognized, especially when the trace quantity is not completely soluble in the diluent material.

A trace quantity of chemical species A in water $(B \equiv H_2O)$ is a dilute binary mixture. The mole fraction A is

$$x_A \equiv \frac{\mathfrak{M}_A}{\mathfrak{M}_A + \mathfrak{M}_B} \qquad (1.1\text{-}3)$$

where $\mathfrak{M}_A$ = moles of A and $\mathfrak{M}_B$ = moles of B. Consider a fixed volume of the mixture $V(L^3)$ and assume A and B form ideal solutions. The volume V does not change when we add molecules of A and remove molecules of B so that the mole fraction can also be expressed as

$$x_A = \frac{c_{A2}}{c_{A2} + c_{B2}} \qquad (1.1\text{-}4)$$

For very dilute aqueous solutions (i.e., $c_{A2} \ll c_{B2}$) the mole fraction is proportioned to molar concentration. Figure 1.1-4 shows the relation between x_A and c_{A2} for an ideal aqueous mixture. The relationship is linear, with a slope of unity for $c_{A2} \leqslant 0.05 c_{B2}$. Above the 5% "breakpoint" the x_A versus c_{A2} curve begins to deviate from the linear relation between the variables. For water as a liquid at 4°C the molar density $c_{B2} = 55.56$ mol/L. It is then convenient to define *dilute aqueous chemical solutions* of species A as those of molar concentration 2.773 mol/L or less.

The preceding ideas can be extended to gaseous and solid solutions. Trace quantities of chemical species in air are dilute gaseous solutions or mixtures and trace quantities of chemicals within soil or earthen solids form dilute solid mixtures. Solid mixtures of interest in the area of chemodynamics differ drastically from fluid mixtures in that they are not homogeneous and are not usually true solutions. An example of this is a trace chemical A absorbed onto the surface of lumps of soil. The result is a heterogeneous dilute mixture. Figure 1.1-4 shows the relation between y_A and c_{A1} for air solutions. Table 1.1-3 contains a convenient summary of the phase densities and definitions of molar concentration ranges for dilute chemical solutions and mixtures in the three phases of the environment.

The convenience of the preceding definition has other advantages for the quantitative treatment of dilute solutions. One advantage is that the

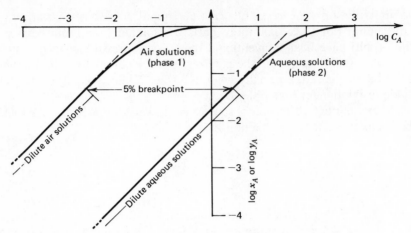

Figure 1.1-4. Mole fraction and molar concentration for chemical solutions.

relation between mole fraction and molar concentration is simplified. These relations are

$$y_A = \frac{c_{A1}}{c_1}, \qquad x_A = \frac{c_{A2}}{c_2}, \qquad z_A = \frac{c_{A3}}{c_3} \qquad (1.1\text{-}5A,B,C)$$

where c_1, c_2, and c_3 are respectively the molar densities of air, water, and soil. The molar density of air is a strong function of temperature and pressure. From the ideal gas law: $c_1 = p/RT$. The molar density of water is a weak function of temperature (see Appendix D). The density of most earthen soils is constant over a relatively large range of temperature and pressure.

In a similar fashion the relation between mass fraction and mass concentration is simplified:

$$\psi_A = \frac{\rho_{A1}}{\rho_1}, \qquad \phi_A = \frac{\rho_{A2}}{\rho_2}, \qquad \omega_A = \frac{\rho_{A3}}{\rho_3} \qquad (1.1\text{-}6A, B, C)$$

where ρ_1, ρ_2, and ρ_3 are the corresponding mass densities. All the equations in Table 1.1-2 remain valid for dilute solutions, but several important ones are simplified drastically. The relations between mole fraction and mass

Table 1.1-3. Phase Density and Dilute Solutions

Phase	Pure Phase Density		Definition of Dilute Solution[e] as a Maximum Concentration of Chemical A	
	Molar (mol/L)	Mass (g/L)	Molar (mol/L)	Mass (g/L)
Air[a]	$c_1 = 0.0446$	$\rho_1 = 1.293$	$c_{A1} = 0.00223$	$\rho_{A1} = 0.06465$
Water				
fresh[b]	$c_2 = 55.56$	$\rho_2 = 1000$	$c_{A2} = 2.778$	$\rho_{A2} = 50.00$
sea[c]	$c_2 = 55.25$	$\rho_2 = 1019$	$c_{A2} = 2.763$	$\rho_{A2} = 50.95$
Soil[d]	$c_2 = 44.10$	$\rho_3 = 2650$	$c_{A3} = 2.205$	$\rho_{A3} = 132.5$

[a] At (0°C, 760 mmHg).
[b] At liquid at 4°C.
[c] Of 3.5% salinity as NaCl.
[d] Silica (100% SiO_2).
[e] Definitions are based on $0.05 \geqslant x_A, y_A, z_A > 0$.

fraction (i.e., equations (K) and (L) in Table 1.1-2) become

$$x_A = \frac{\phi_A M_B}{M_A}, \qquad y_A = \frac{\psi_A M_B}{M_A}, \qquad z_A = \frac{\omega_A M_B}{M_A} \quad \text{(1.1-7A, B, C)}$$

so the conversion involves only a ratio of molecular weights.

PARTS PER MILLION AND PARTS PER BILLION

As noted earlier, parts per million (ppm) and parts per billion (ppb) are loose, conventional terms for expressing concentrations, although the more appropriate unit for pollutant concentrations is weight per unit volume. For gases, ppm is used on a volume basis, with temperature and pressure usually unspecified. For liquids, both weight and volume are used—it is to be hoped—on a consistent basis. For solids (i.e., particles not in solution, such as dust) in liquids, a weight basis is probably rational. Clearly confusion can set in if parts per million (or billion or trillion) is used other than for a general qualitative indication of concentration.

It is best if the vague ppm, ppb, and so on, usages are abandoned and chemical concentrations in fluids are placed on a mass (or mole) per unit volume basis. For gases (including air) temperature and pressure must be specified precisely so that the volume represents a known quantity of substance. The familiar standard conditions of chemistry and physics (i.e., STP) are 0°C and 760 mm Hg and the molar volume of any gas is 22.4

L/mol. All gas concentrations given either as c_{A1} or ρ_{A1} should be referenced to STP or some other standard. Although most liquid volumes including water change only slightly with temperature and pressure a precise definition is needed. Water at 1.0 atm pressure and 4°C has a mass volume of 1.0 L/kg and this is an excellent reference.

The use of ppm, ppb, and so on, for specifying trace quantity chemicals associated with soil or other solids is ideal provided weight ratios are used (i.e., g A/E6 g soil). This measure is imprecise unless the soil is on a water-free (unbound) basis. Reporting and measuring concentrations with the soil dried in air for a specified time at 1 atm and 100°C will establish a fairly precise standard for this case. Simple conversions exist between ppm, ppb, and ppt and concentration measured in SI units, provided ppm, ppb, and ppt are precisely defined for each phase. See Example 1.1-3 for useful conversion factors.

More on Notation

Multicomponent and multiphase notation was introduced in defining species concentration within various phases. A mnemonic system is used throughout the book. This technique of notation makes reading and interpreting equations easier and is a direct aid in computer programming; however, writing them is somewhat more tedious. As an example c_{A2} can be read as the "molar concentration of species A in water."

By the use of overlines, superscripts, and subscripts along with the latin (print and script) and greek alphabets, a fairly self-consistent system of notation can be developed. It works well most of the time but is not perfect. A complete list of notation appears in the front matter of the book. Examples of notation interpretation are as follows:

c_{A2i}^{*} = equilibrium molar concentration of species A in water at the interface

y_A = mole fraction of species A in air

$\mathcal{D}_{A3}^{(t)}$ = turbulent diffusivity of A in a solid phase

z = a distance from the origin in the z direction

N_{Az} = flux rate of species A in the z direction

$\dot{k}_A$ = a local value of the mass transfer coefficient of component A

$^{k}k_{A2}$ = gas interface mass transfer coefficient for species A in water

The normal procedure is to read the superscripts and overlines first, then the basic symbol followed by the reading of the subscripts. No more than three subscripts are used. The first subscript position denotes the species and is an uppercase latin letter. The second subscript position denotes the phase ($1 \equiv$ gas, $2 \equiv$ liquid, and $3 \equiv$ soil) and is an integer. The third subscript usually specifies position within the phase (i.e., $i \equiv$ interface). When mole fraction and mass fraction symbols are used, a maximum of two subscript positions are employed. The phase subscript notation is superfluous since x implies water, y gas, and z solid. If the notation system seems confusing at this point its usefulness and clarity should become evident as the text material unfolds.

Example 1.1-1. Equal Molar Trace Quantity Concentrations of Chemicals in Water and Air

We desire to prepare standard solutions of low molecular weight chlorinated hydrocarbons for laboratory use.

(a) Calculate the mass (kg) of chloroform ($CHCl_3$) and vinyl chloride (CH_2CHCl) in a 1 liter (L) aqueous solution in which the concentration of each is 1.0 mol/m^3.

(b) Repeat the calculation for a 1 liter air mixture at standard temperature and pressure (STP).

SOLUTION

(a) Since the molar concentration of each chemical is equal, each liter of water contains

$$6.022E\,20 = \left(6.022E\,23\,\frac{\text{molecules}}{\text{mol}}\right)\left(\frac{1\ \text{mol}}{\text{m}^3}\right)\left(\frac{1E-3\ \text{m}^3}{\text{L}}\right)(1.0\ \text{L})$$

molecules of chloroform and $6.022E\,20$ molecules of vinyl chloride. The mass of chloroform ($A \equiv CHCl_3$), molecular weight 119.39, is

$$m_A = \left(\frac{1.0\ \text{mol}}{\text{m}^3}\right)\left(\frac{119.39\ \text{g}}{\text{mol}}\right)(1.0\ \text{L})\left(\frac{1E-3\ \text{m}^3}{\text{L}}\right)\left(\frac{1E-3\ \text{kg}}{\text{g}}\right)$$

$$m_A = 119.39E - 6\,\text{kg}$$

The mass of vinyl chloride ($B \equiv CH_2CHCl$), molecular weight 62.5, is

$$m_B = (1.0)(62.5)(1.0)(1E-3)(1E-3) = 62.5E - 6\ \text{kg}$$

(b) Since the molar concentration of each chemical is equal, each liter of air contains $6.022E\,20$ molecules of each chemical. The mass of chloroform

$A = C_2H_5OH, M_A = 46.06$
$B = O_2, \quad M_B = 32.00$
$C = CO_2, \quad M_C = 44.01$
$D = H_2O, \quad M_D = 18.02$

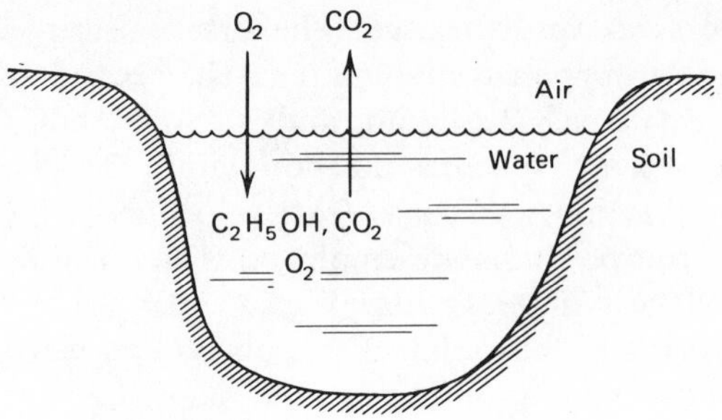

Figure E1.1-2.

and vinyl chloride in air is 119.39E−6 and 62.5E−6 kg, respectively.

Example 1.1-2. Chemical Reaction and Movement in Molar Units

At a certain point in a river ethanol in the water is being oxidized by a microbial enzyme reaction. The source of oxygen is from the air above the river surface. Assume that the process is occurring at steady state. Atmospheric oxygen moves across the air–water interface at a constant rate. Once the oxygen is in the water the oxidation of ethanol occurs at a constant rate. The stoichiometry of the reaction is

$$C_2H_5OH + 3O_2 = 2CO_2 + 3H_2O$$

The evolution of carbon dioxide from the water also occurs at a constant rate. Although ethanol is volatile assume it remains in the water and that the rate of ethanol disappearance $(-r_A)$ is 5E−4 mg/L·s. Additional information is given in Fig. E1.1-2.

(a) Calculate the reaction rate of oxygen consumption (k mol/s) within a cubic meter of water located at the surface.

(b) Calculate the molar flux rate of carbon dioxide (k mol/s·m^2) as this trace chemical desorbs through a square meter of the surface.

(c) Convert the numerical results of parts (a) and (b) to mass (kg) units.

SOLUTION

(a) Reaction rate of oxygen: First convert the ethanol reaction rate to molar units.

$$-R_A = 5E-4 \frac{mg}{L \cdot s} \left| \frac{kg}{E6\ mg} \right| \frac{1000\ L}{m^3} \left| \frac{k\ mol}{46.06\ kg} \right.$$

$$-R_A = 1.09E-8\ k\ mol/m^3 \cdot s$$

For each mole of ethanol three moles of oxygen are required; the rate of oxygen disappearance is

$$-R_B = 3(-R_A) = 3.27E-8 \ k \ \text{mol/m}^3 \cdot s$$

The oxygen consumption rate in 1 m³ of water is $3.27E-8 \ k$ mol/s.

(b) Molar flux rate of carbon dioxide: The stoichiometry demands that 2 mol of CO_2 be formed for each mol of C_2H_5OH reacted; therefore,

$$R_C = 2(-R_A) = 2.18E-8 \ k \ \text{mol/m}^3 \cdot s$$

A cubic meter of water will have 1 m² of exposed interface in contact with air to yield

$$N_{Cz} = 2.18E-8 \ k \ \text{mol/m}^2 \cdot s$$

(c) Converting from molar to mass units: Molecular weight is used in the conversion

$$-r_B = 3.27E-8 \ \frac{k \ \text{mol}}{\text{m}^3 \cdot s} \left| \frac{32.00 \ \text{kg}}{k \ \text{mol}} \right| = 1.05E-6 \ \text{kg/m}^3 \cdot s$$

$$n_{Cz} = 2.18E-8 \frac{k \ \text{mol}}{\text{m}^2 \cdot s} \left| \frac{44.01 \ \text{kg}}{k \ \text{mol}} \right| = 9.59E-7 \ \text{kg/m}^2 \cdot s$$

Example 1.1-3.   PPM, PPB, and PPT Converted to Concentrations Expressed in SI Units

Derive useful conversion factors for converting the following conventional concentrations to SI units:

(a) Parts per million (volume ratio) of chemical A in air.

(b) Parts per million (mass ratio) of chemical A in water.

(c) Parts per million (mass ratio) of chemical A within a soil.

Give the ppb and ppt result for each case also. The molecular weight of species A is M_A (g/mol).

SOLUTION

(a) 1 ppm A in air. Assume all components behave as ideal gases and STP is used as the reference state.

$$1 \ \text{ppm} = \frac{1 \ \text{m}^3 \ A}{E+6 \ \text{m}^3 \ \text{air}} \left| \frac{1000 \ \text{L}}{\text{m}^3} \right| \frac{\text{mol}}{22.4 \ \text{L}} \left| \frac{1000 \ \text{m mol}}{\text{mol}} \right| = \frac{1}{22.4} \text{m mol} \ A/\text{m}^3$$

$$1 \ \text{ppm} = \frac{1}{22.4} \frac{\text{m mol} \ A}{\text{m}^3} \left| \frac{M_A \ \text{mg}}{\text{m mol}} \right| = 1 \times \left(\frac{M_A}{22.4} \right) \text{mg} \ A/\text{m}^3 \qquad \text{(E1.1-3A)}$$

In similar fashion

$$1 \text{ ppb} = 1 \times \left(\frac{M_A}{22.4} \right) \mu g \, A/m^3 \tag{E1.1-3B}$$

$$1 \text{ ppt} = 1 \times \left(\frac{M_A}{22.4} \right) ng \, A/m^3 \qquad \text{all at STP conditions} \tag{E1.1-3C}$$

(b)   1 ppm A in water. Use liquid water at 4°C, 1 atm as the reference state of volume.

$$1 \text{ ppm} = \frac{1 \text{ g } A}{E+6 \text{ g H}_2\text{O}} \left| \frac{1 \text{ g}}{\text{cm}^3} \right| \left(\frac{100 \text{ cm}}{\text{m}} \right)^3 = 1 \text{ g } A/m^3 (= 1 \text{ mg/L}^*)$$

$$\tag{E1.1-3D}$$

$$1 \text{ ppb} = 1 \text{ mg/m}^3 \tag{E1.1-3E}$$

$$1 \text{ ppt} = 1 \ \mu g/m^3 \tag{E1.1-3F}$$

(c)   1 ppm A within soil. Since soil is not normally measured on a volume basis mass is used.

$$1 \text{ ppm} = \frac{1 \text{ g } A}{E+6 \text{ g soil}} \left| \frac{1000 \text{ mg}}{\text{g}} \right| \frac{1000 \text{ g}}{\text{kg}} \right| = 1 \text{ mg } A/\text{kg soil}$$

$$\tag{E1.1-3G}$$

$$1 \text{ ppb} = 1 \ \mu g \, A/\text{kg soil} \tag{E1.1-3H}$$

$$1 \text{ ppt} = 1 \text{ ng } A/\text{kg soil} \tag{E1.1-3I}$$

The Material Balance

This section may be skipped now. It will take on more concrete meaning when specific applications of chemical movement are addressed. A general form of the component material balance is presented for future use. The equation is simply a statement of the law of conservation of mass. Consider the volume element of arbitrary shape illustrated in Fig. 1.1-5. Where the composition within the system under study is uniform throughout (i.e., independent of position), the element may be defined as the whole system V and the accounting made for chemical A. Where the composition within the system is not uniform a differential element of

---

* Liter (L) is not a valid SI unit but widely used and impractical to abandon.

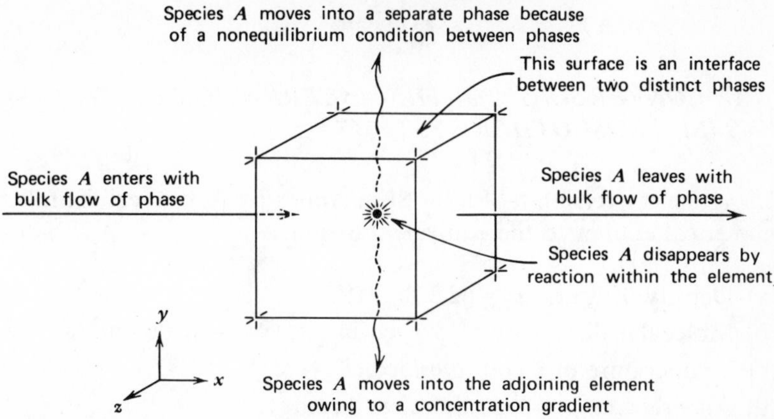

Species A moves into a separate phase because
of a nonequilibrium condition between phases

This surface is an interface
between two distinct phases

Species A enters with
bulk flow of phase

Species A leaves with
bulk flow of phase

Species A disappears by
reaction within the element

y

x

z

Species A moves into the adjoining element
owing to a concentration gradient

Figure 1.1-5. Material balance for an element of volume fixed in space through which a fluid is flowing.

volume, dV, must be defined and the accounting of species $\overset{.}{A}$ made for it. Thus, as illustrated in Fig. 1.1-5, we have

$$
\begin{Bmatrix} \text{rate of} \\ \text{mass } A \\ \text{flow into} \\ \text{element of} \\ \text{volume} \end{Bmatrix} = \begin{Bmatrix} \text{rate of} \\ \text{mass } A \\ \text{flow out} \\ \text{of element} \\ \text{of volume} \end{Bmatrix} + \begin{Bmatrix} \text{rate of mass } A \\ \text{loss due to} \\ \text{mass transfer*} \\ \text{from the element} \\ \text{of volume} \end{Bmatrix}
$$

$$
+ \begin{Bmatrix} \text{rate of mass } A \\ \text{loss due to} \\ \text{chemical reaction} \\ \text{within the element} \\ \text{of volume} \end{Bmatrix} + \begin{Bmatrix} \text{rate of ac-} \\ \text{cumulation of} \\ \text{mass } A \\ \text{in element} \\ \text{of volume} \end{Bmatrix} \qquad (1.1\text{-}8)
$$

A special form of Eq. 1.1-8 results if the volume element is assumed to move as a slug with the mean velocity of the fluid. In this case the flow into and flow out of terms are omitted. Additional assumptions can and will be placed on this general material balance as the needs of the problem dictate.

*Mass transfer includes interphase and intraphase movements.

Problems

1.1A CONVERTING TO THE INTERNATIONAL SYSTEM OF UNITS (SI) FROM OTHER SYSTEMS

1. Using the conversion tables for SI in Appendix A, convert the following numerical entities to the equivalent SI units:

   (a) Density of water, $\rho_2 = 62.3$ lb_m/ft^3.
   (b) Molecular diffusivity of oxygen in air, $\mathcal{D}_{0,1} = 0.206$ cm^2/s.
   (c) Temperature of a soil interface, $T_{3i} = 75°F$.
   (d) Air pressure, $p_1 = 1$ atm (14.696 $lb_f/in.^2$).
   (e) Latent heat of vaporization of water ($H_2O \equiv A$), $\lambda_A = 1051.5$ Btu/lb_m.
   (f) Mass flux rate of chemical A, $n_A = 4.1E+8$ $lb_m/s \cdot ft^2$.
   (g) Stefan-Boltzmann constant, $\sigma = 0.1712E-8$ $Btu/hr \cdot ft^2 \cdot R^4$.
   (h) Water viscosity, $\mu_2 = 1$ cp.
   (i) Water heat capacity of $\hat{c}_{p2} = 1$ $cal/g \cdot °C$.
   (j) The gas constant, $R = 1.987$ $cal/mol \cdot °K$.

2. Using only base units, verify the dimensions of the following variables, the notation for which appears in the list of Symbols.

   (a) $\hat{c}_p$ = heat capacity, L^2/t^2T.
   (b) $\mathbf{Q}$ = rate of energy flow across a surface, ML^2/t^3.
   (c) p = fluid pressure, M/Lt^2.
   (d) F = force of a fluid on an adjacent solid, ML/t^2.
   (e) W = rate of doing work on surroundings ML^2/t^3.

1.2. ILLUSTRATION OF OBJECTIVES AND CONTENT—REAERATION OF NATURAL STREAMS

The objectives of chemodynamics were stated in Section 1.1. The scope and concept of this book were also described. In the absence of technical detail this discussion was necessarily somewhat abstract. In this section the objectives and content are redescribed in terms of a single, specific example—the movement of oxygen from air into the water of a natural stream. This is usually the first "chemodynamic problem" encountered by students commencing studies in environmental pollution.

Introduction

The discharge of organic impurities, such as municipal sewage and industrial waste, into natural streams presents a problem of primary importance in the field of environmental engineering. The decomposition of this organic matter by waterborne bacteria for metabolic processes results in the utilization of the dissolved oxygen in the stream.

$$C_xH_yO_z + nO_2 \xrightarrow[\text{enzymes}]{\text{microbial}} \text{products of } CO_2, H_2O, \text{etc.}$$

The organic impurities denoted by the general formula $C_xH_yO_z$ are usually measured as an "oxygen demand". This simplifies an otherwise complex stoichiometry. The biochemical conversion of 1 g of "oxygen-demanding" organic matter requires 1 g of molecular oxygen. The replacement of this molecular oxygen (i.e., O_2) by reaeration occurs through the water surfaces exposed to the atmosphere. The concentration of organic matter can be so great that there results a condition in which the receiving stream is completely devoid of dissolved oxygen. Because every stream has a limited capacity to assimilate organic wastes, evaluation of the natural purification capacity of a stream is of fundamental and practical value.

Stream reaeration is concerned with the physical and chemical transport of oxygen from the air into water and is the classical problem in chemodynamics. Stream reaeration, also called reoxygenation, is a very old but still a very important problem, attracting the interest of environmental scientists. The mechanism of natural stream reaeration is still undergoing investigation.

Oxygen Equilibrium between Air and Water

Molecular oxygen exists in the Earth's atmosphere and comprises 20.95% (volume). Oxygen is soluble in water to a small degree, reaching a maximum of 14.16 mg/L at 0.0°C and 760 mm Hg. A table of dissolved oxygen solubility in equilibrium with dry air appears in Appendix C. To represent this equilibrium condition in mathematical notation let $A \equiv O_2$. The entries in the table are then represented by

$$\rho_{A2}^* = f(T) \tag{1.2-0}$$

where ρ_{A2}^* is the mass concentration of oxygen in water in equilibrium with air. The f represents the table and T represents temperature, noting that the concentration is a function of temperature. For a specified temperature ρ_{A2}^* is the maximum (stable) concentration of oxygen in clean water.

Deoxygenation and Reoxygenation

The classical work of Streeter and Phelps in 1925[3] presented a mathematical analysis of the organic waste and oxygen content in water known as the *dissolved oxygen sag*. Consider the idealized stream shown in Fig. 1.2-1. A volume element of water V, moving at the mean velocity v_2, contains oxygen of concentration ρ_{A2} and organic material ($\equiv B$) of concentration ρ_{B2}. This volume element, located downstream from the organic waste point of entry a distance L, is assumed not to mix with elements upstream or downstream. Assume only the two mechanisms biochemical oxidation and interphase mass transfer are occurring within the element. A material balance, according to Eq. 1.1-8, for component B in the volume yields the simple equation

$$0 = 0 + 0 + V(-r_B) + \frac{d}{dt}(V\rho_{B2}) \tag{1.2-1}$$

The rate of disappearance of B is usually assumed to be a first-order rate equation:

$$-r_B = k_B''' \rho_{B2} \tag{1.2-2}$$

where k_B''' is the rate constant. The negative sign appearing in front of the rate symbol denotes the disappearance of species B. The rate is normally defined positive for appearance of some reaction product.

Oxygen, component A, enters the volume element through the air–water interface. As is shown in a later section of the book the major controlling resistance to the absorption of oxygen resides in the water phase. The

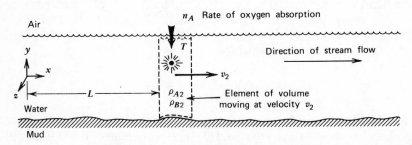

Figure 1.2-1. Natural stream reaeration.

interphase flux rate through the interface is represented by

$$n_A = {}^1k'_{A2}(\rho^*_{A2} - \rho_{A2}) \tag{1.2-3}$$

where ${}^1k'_{A2}$ is the liquid phase mass transfer coefficient of component A associated with the air interface and $(\rho^*_{A2} - \rho_{A2})$ is the concentration departure from equilibrium for species A in water. It is common practice to measure organic concentrations in water as "oxygen demands." The gross pollutants measures such as chemical oxygen demand (COD) and biochemical oxygen demand (BOD) are measured in mg O_2/L. This makes the stoichiometric ratio, in an otherwise complex biochemical reaction, unity. A material balance on component A according to Eq. 1.1-8 results in

$$0 = 0 - n_A A_{xz} + \left(\frac{1}{1}\right) V(-r_B) + \frac{d}{dt}(V\rho_{A2}) \tag{1.2-4}$$

A_{xz} is the interfacial area in the x-z plane of air–water contact for the volume V. Using Eqs. 1.2-2 and 1.2-3 and constant V gives

$${}^1k'_{A2} \frac{A_{xz}}{V}(\rho^*_{A2} - \rho_{A2}) - k'''_B \rho_{B2} = \frac{d\rho_{A2}}{dt} \tag{1.2-5}$$

In the interest of simplifying Eq. 1.2-5 the oxygen deficit is defined

$$\Delta_A \equiv \rho^*_{A2} - \rho_{A2} \tag{1.2-6}$$

With this definition, and the fact that ρ^*_{A2} is constant if stream temperature is constant, Eq. 1.2-5 becomes

$$\frac{d\Delta_A}{dt} = k'''_B \rho_{B2} - {}^1k'_{A2} \frac{A_{xz}}{V} \Delta_A \tag{1.2-7}$$

In the terminology used in stream pollution k'''_B is the deoxygenation coefficient with dimensions of (t^{-1}), and the group $({}^1k'_{A2})(A_{xz}/V)$ is the "reaeration coefficient" with the same dimensions. Combining Eqs. 1.2-1 and 1.2-2 we get

$$\frac{d\rho_{B2}}{dt} = -k'''_B \rho_{B2} \tag{1.2-8}$$

These final two equations represent the quantitative result of the stream reaeration analysis. A simple integration of the two equations, with $\Delta_A = \Delta_A^0$ and $\rho_{B2} = \rho_{B2}^0$ at $t = 0$ as the initial condition representing in-stream oxygen deficit and organic concentration after mixing at the waste input point, results in

$$\Delta_A = \frac{k_B''' \rho_{B2}^0}{\left({}^1k_{A2}'/h\right) - k_B'''}\left[\exp(-k_B''' t) - \exp\left(\frac{-{}^1k_{A2}' t}{h}\right)\right] + \Delta_A^0 \exp\left(\frac{-{}^1k_{A2}' t}{h}\right)$$

$$(1.2\text{-}9)$$

where V/A_{xz} has been replaced with h, the average depth of the stream. Figure 1.2-2 shows the shape of a typical dissolved oxygen sag curve. Time of flow can be related to the distance from the waste input point L and the mean stream velocity v_2 by

$$t = \frac{L}{v_2} \qquad (1.2\text{-}10)$$

The minimum point in the sag curve is usually of special interest. This is the critical location in the stream where the concentration of dissolved oxygen is a minimum. Typically this minimum should not be less than the oxygen concentration desirable to sustain some higher life forms within the stream. The minimum can be obtained in a formal fashion by setting the first derivative of the deficit in Eq. 1.2-9 to zero and solving for t_c:

$$t_c = \frac{1}{\left({}^1k_{A2}'/h\right) - k_B'''} \ln\left\{\frac{{}^1k_{A2}'}{hk_B'''}\left[1 - \Delta_A^0\left(\frac{\left({}^1k_{A2}'/h\right) - k_B'''}{k_B''' \rho_{b2}^0}\right)\right]\right\} \quad (1.2\text{-}11)$$

The downstream location of the minimum is $L_c = t_c v_2$.

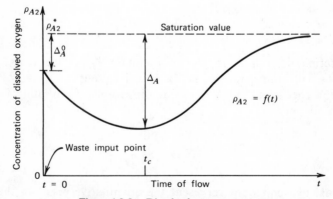

Figure 1.2-2. Dissolved oxgen sag curve.

Within its range of application this model of oxygen dynamics describes the major features of deoxygenation and reaeration. Equations 1.2-9 through 1.2-11 can be used for making calculations of the stream assimilation capacity for a particular waste input under ideal conditions. (See Problem 1.2B.) This simple stream model has several important limitations. Most streams have other oxygen sources and sinks including absorption at the mud–water interface and oxygen associated with algal respiration processes. Organic material may be released from deposits in bottom muds and backmixing usually occurs to some degree in all streams. Many streams change signficantly in form within relatively short stretches (i.e., pool and ripple streams) so that the average depth h, velocity v_2, and other quantities affected by these variables also change significantly and the model is therefore inadequate.

The Mechanism of Reaeration in Natural Streams

If there is no utilization of the dissolved oxygen within the body of water, the rate of reaeration may be obtained from Eq. 1.2-5.

$$\frac{d\rho_{A2}}{dt} = \frac{^1k'_{A2}}{h}\left(\rho_{A2}{}^* - \rho_{A2}\right) \tag{1.2-12}$$

which integrates to

$$\rho_{A2} = \rho_{A2}{}^* - \left(\rho_{A2}{}^* - \rho_{A2}{}^0\right)\exp\left(-\frac{^1k'_{A2}t}{h}\right) \tag{1.2-13}$$

These two equations show that the reaeration coefficient $^1k'_{A2}/h$ is critically important in predicting the oxygen uptake. This coefficient or $^1k'_{A2}$, the liquid phase mass transfer coefficient, and the relationship to observable stream characteristics have had the attention of many investigators. O'Connor and Dobbins[4] have successfully applied the basic theory of turbulent flow to explain many phenomena occurring in natural and artificial waterways. The following is adopted from the work of O'Connor and Dobbins.

In any stream the ability to absorb oxygen from the atmosphere, or the reaeration capacity, is a direct function of the degree of turbulent mixing. Atmospheric oxygen can be obtained only at the water surface, and the rate at which reaeration can take place is therefore directly limited by the rate of surface water replacement in a flowing stream. Thus in a relatively still pool reaeration is a very slow process, whereas the reaeration capacity of a rapids section is very great. In turn, in any stream the rate of water surface replacement at the air–water interface is controlled by the stream's physical characteristics and is related to the associated water-flow properties.

O'Connor and Dobbins considered that the controlling factor in oxygen absorption was the resistance of a liquid film at the surface, through which oxygen must be absorbed by molecular diffusion. The film was assumed to be constantly renewed by unsaturated elements from the body of the stream through the mechanism of turbulence, as originally proposed by Higbie.[5] The rate of transfer through an element of the surface depends on the length of time it has been exposed to the atmosphere. The function describing the age distribution of surface elements was taken to be that of Dankwerts[6], which is

$$\phi(t) = r \exp(-rt) \tag{1.2-14}$$

in which $\phi(t)$ is the fractional part of the exposed surface area having ages between t and $t + dt$, and r is the rate of surface renewal, dimensions of t^{-1}. These ideas along with Fick's law applied to molecular diffusion and the definition of a liquid phase mass-transfer coefficient result in an equation for that coefficient

$$^1k'_{A2} = (\mathcal{D}_{A2} r)^{1/2} \tag{1.2-15}$$

in terms of the molecular diffusivity of oxygen in water $\mathcal{D}_{A2}$ and the surface renewal rate. Some basic theory of fluid turbulence is needed to understand and interpret r in Eq. 1.2-15. An in-depth presentation of turbulent flow is presented in a later chapter.

In turbulent flow a complex secondary motion is superimposed on the primary motion of translation. Turbulence is characterized by eddies that transport parcels of fluid from one layer to another with varying velocities. The eddy motion, which is erratic and seemingly unpredictable, can only be defined in terms of probability. Thus the principles of statistics are employed to define quantitatively parameters of turbulence such as size of eddies and the velocity fluctuations.

The instantaneous velocity at any point in turbulent flow varies in magnitude and direction. Although there is no net flow in the vertical direction, because of the eddies that exist there can exist a rapidly fluctuating (i.e., up and down) velocity. The intensity of this vertical velocity fluctuation can be calculated from direct measurements and is $\sqrt{(v'_y)^2}$, where v'_y represents the instantaneous fluctuating velocity in the y direction.

Although the velocity fluctuations define the intensity of turbulence, some linear measure is required to define the scale of turbulence. In this regard the mixing length theory developed by Prandtl in 1925 is appropriate. By assuming that eddies move around in a fluid very much as molecules move about in a gas (actually a very poor analogy), Prandtl developed an expression for the fluctuating velocity in a fluid in which the mixing length plays a role roughly analogous to that of the mean free path in gas kinetic theory. This way of thinking led Prandtl to the relation

$$\sqrt{(v_y')^2} = l\frac{dv_x}{dy} \qquad (1.2\text{-}16)$$

where l is the mixing length and signifies a distance that a parcel of fluid moves from its point of departure from the mean motion until it mixes again with the main body of the fluid. It is doubtful whether it is possible to assign a more definite physical meaning to the mixing length, but it can be said that this length is a measure of the average size of the eddies responsible for the fluid mixing. The other term in Eq. 1.2-16 is the x component velocity gradient in the vertical direction. Equation 1.2-16 applies to a turbulent condition in which there exists a velocity gradient.

In the cases of comparatively deep channels it is possible that the turbulence may approach an isotropic condition near the air–water interface. An isotropic turbulent condition is one in which the intensity of the velocity fluctuations in all three directions are very nearly the same. This type of turbulence, in which there is neither a shearing stress nor a velocity gradient, is approached in the flow downstream from screens, in a hydraulic jump, and in the center of a deep and wide open channel. Nonisotropic turbulence, by contrast, is characterized by a significant correlation between the velocity fluctuations and by a velocity gradient and shearing stress. This type of turbulence is evidenced in flow in pipes and in comparatively shallow open channels.

In turbulent flow, momentum, mass, heat, or any inherent characteristic of the fluid can be transfered from one layer of fluid to another. The basic concepts of turbulence may, therefore, be used to determine the rate at which parcels of fluid at the surface layer can be replaced by parcels arising from the turbulent motion of the body of the fluid. The intensity of turbulence may be defined by some mean measure of the velocity fluctuations, such as $\sqrt{(v_y')^2}$, and the scale of turbulence by the mixing length l. The mixing length signifies a distance a parcel moves from its point of departure from the mean forward motion until it mixes again with the main body of the fluid. Therefore, only parcels within a zone defined by a mixing length from the surface will effect the renewal of this surface. Furthermore, any parcel located at a distance greater than this length from the surface will be deflected from its vertical path before reaching the surface. It may be reasoned that vertical flow exhibiting small length and high velocity characteristics will cause a greater rate of surface renewal than flow of great length and low velocity. Therefore, parcels at the surface are replaced at a rate directly proportional to the intensity of turbulence and inversely proportional to the scale of turbulence. Surface renewal may be considered to take place in a period of time defined by

$$t = \frac{l}{\sqrt{(v_y')^2}} = \frac{1}{r} \qquad (1.2\text{-}17)$$

where the values of l and $\sqrt{(v_y')^2}$ are those that prevail in the vicinity of the surface. At any point the period defined by Eq. 1.2-17 varies in time owing to the random nature of the variables involved. However, since the length and velocity are average values, t is then the average time during which such surface renewal takes place.

Although the velocity gradient may be zero at the stream surface in the case of isotropic turbulence, heat and mass can be transferred across this plane. The mixing length and the vertical velocity fluctuations likewise have finite values. The only way to obtain values of these parameters is by direct measurement. From published values of field observations of the mixing length and the vertical velocity fluctuation on the Mississippi River and estuaries it was concluded that on the average the mixing length and the vertical velocity fluctuation were approximately 10% of the average depth and the mean flow velocity, respectively. These values may be used as an approximation to determine the rate of surface renewal

$$r = \frac{\sqrt{(v_y')^2}}{l} = \frac{0.1 v_x}{0.1 h} = \frac{v_x}{h} \qquad (1.2\text{-}18)$$

Substitution of this value for r in Eq. 1.2-15 gives

$$^1 k_{A2}' = \left(\frac{\mathcal{D}_{A2} v_x}{h} \right)^{1/2} \qquad (1.2\text{-}19)$$

This equation approximates the gas interface liquid phase mass transfer coefficient for the case of isotropic turbulence in natural streams. Equation 1.2-19 was verified by comparing values of the reaeration coefficient previously reported by others (see Problem 1.2C) on five natural streams and one bay.

Table 1.2-1. Summary of Liquid Phase Mass Transfer Coefficients at Air–Water Interface of Natural Streams

Investigators[a]	Limitations, Origin	α^c	β	δ
O'Connor and Dobbins	theoretical and experimental	$\mathcal{D}_{A2}^{1/2}$	0.5	0.5
Churchill, Elmore, and	regression analysis of			
Buckingham	field data	5.026^b	0.969	0.673
Isaacs and Maag	field data	1.07^b	1.0	0.5
Langbein and Durum	laboratory and field	3.3^b	1.0	0.33
Owens, Edwards, and				
Gibbs	field data	9.47^b	0.67	0.85

[a] See related reading for literature citations.
[b] $k_{A2} = \rho_2(\alpha v_2^\beta / h^\delta) \cdot \rho_2$ in lb/ft³, v_2 in ft/s, h in ft gives $^1k_{A2}$ in lbO₂/ft²·d.
[c] α at 20°C; $k(T) = k(20) \cdot 1.016^{T-20}$, T in °C.

Most equations developed for the prediction of this mass transfer coefficient are of the form

$$^1k'_{A2} = \left(\frac{\alpha v_x^\beta}{h^\delta} \right) \tag{1.2-20}$$

Table 1.2-1 summarizes the results of several investigators. Notable alternative theories of stream aeration have been developed by Thackston and Krenkel[7] and Tsivoglou and Wallace.[8] The former is based on the average eddy viscosity in the stream and the latter is based on the energy expended when flowing water undergoes a change in elevation. The following example (Example 1.2-1) demonstrates the applicability of the material in this chapter.

Closure

This chapter has been concerned primarily with the absorption of oxygen into natural streams. The author has departed from the conventional approach and has emphasized the liquid phase mass transfer coefficient at the air–water interface, $^1k_{A2}$, rather than the "reaeration coefficient" (normally noted as k_2 in most reaeration literature). This approach was taken with the broader goals of chemodynamics in mind. There is another liquid phase mass transfer coefficient in natural streams associated with the interphase movement of chemicals at the mud–water interface. This coefficient is noted by $^3k_{A2}$ for oxygen and details are presented later in the book. Although the emphasis in this chapter has been on the absorption of a single chemical species into water the resulting coefficients presented in

Table 1.2-1 have a much wider range of utility for quantifying the exchange and movement of chemicals through the air–water interface of natural streams.

The use of a mass transfer coefficient defined by Eq. 1.2-3 using a mass concentration difference seems more appropriate. This definition maintains a parallelism with Fick's first law of diffusion and is the most unrestrictive and generally used coefficient in the field of interphase mass transfer. Mass transfer coefficients are scalars and are for a particular chemical species (hence the subscript A for O_2). Under the proper transformation the mass transfer coefficients represented in Table 1.2-1 can be used for estimating the desorption of chemical species from the stream into the overlaying air mass. Tsivoglou used this concept when he employed a krypton-85 tracer in natural stream desorption studies to obtain "reaeration coefficients."

Example 1.2-1. Recovery Time of a Stream Void of Oxygen

The time required for a stream to recover from a hypothetical state of no dissolved oxygen to some final degree of saturation gives an indication of speed of reaeration.

(a)   Calculate the time in seconds required for a stream void of oxygen to reach a state of 50% saturation. Note that this is the oxygen absorption half-life for this particular stream. Compute the distance downstream (km) and the oxygen concentration (mg/L).

(b)   Calculate the time(s) required to reach 95% saturation. This time (i.e., $t_{95\%}$) is a realistic measure of the recovery time. Compute distance and concentration. Stream data: Average velocity 0.21 m/s, average depth = 1.8 m, $\mathcal{D}_{A2} = 1.80E-5$ cm^2/s@20°C, temperature of water, 30°C.

SOLUTION

The transient oxygen concentration in a stream undergoing recovery is represented by Eq. 1.2-13. For the conditions of the problem $\rho_{A2}^0 = 0$ in oxygen deficit terminology the equation becomes

$$\frac{\Delta_A}{\Delta_A^0} = \exp\left(\frac{-^1k'_{A2}t}{h}\right) \qquad (E1.2-1)$$

where $\Delta_A \equiv \rho_{A2}^* - \rho_{A2}$ and $\Delta_A^0 \equiv \rho_{A2}^* - \rho_{A2}^0$.

(a) Half-life is $\Delta_A = \Delta_A^0/2$ and Eq. E1.2-1 becomes

$$t_{1/2} = \frac{0.693h}{^1k_{A2}'} \qquad \text{(E1.2-2)}$$

From Eq. 1.2-19 the "reaeration coefficient" becomes

$$\frac{^1k_{A2}'}{h} = \frac{(\mathscr{D}_{A2}v_x)^{1/2}}{h^{3/2}}$$

For 20°C

$$\frac{^1k_{A2}'}{h} = \left(1.80\text{E}-5\frac{\text{cm}^2}{\text{s}} \middle| \frac{0.21\text{ m}}{\text{s}} \middle| \frac{100\text{ cm}}{\text{m}}\right)^{1/2} \middle/ \left(1.8\text{ m}\middle| \frac{100\text{ cm}}{\text{m}}\right)^{3/2}$$

$$= 8.05\text{E}-6\middle/\text{s}$$

For 30°C

$$\frac{^1k_{A2}'}{h} = (8.05\text{E}-6)1.016^{30-20} = 9.44\text{E}-6\text{ s}^{-1}(0.82\text{ d}^{-1})$$

$$t_{1/2} = \frac{0.693}{9.44\text{E}-6\text{ s}^{-1}} = 7.34\text{E 4 s}(=20.4\text{ hr})$$

From Eq. 1.2-10, $L = tv_2 = (7.34\text{E 4 s})(0.21\text{E}-3\text{ km/s}) = 15.4\text{ km}$.

$$\rho_{A2}^* = 7.53\text{ mg O}_2/\text{L} \qquad \text{from Appendix C}$$

$$\rho_{A2} = \rho_{A2}^*\left(1 - \frac{\Delta_A}{\Delta_A^0}\right) = 7.53\left(1 - \tfrac{1}{2}\right) = 3.77\text{ mg O}_2/\text{L}$$

(b) Ninety-five percent recovery. $\Delta_A = 0.05\Delta_A^0$

$$t_{95\%} = \frac{3h}{^1k_{A2}'} = 3.18\text{E 5 s}(82.3\text{ hr})$$

$$L = (3.18\text{E 5 s})(0.21\text{E}-3) = 66.8\text{ km}$$

$$\rho_{A2} = 7.53(1 - 0.05) = 7.15\text{ mg O}_2/\text{L}$$

Problems

1.2A. IN-STREAM OXYGEN CONCENTRATION BELOW AN OUTFALL

A uniform natural stream 0.76 m deep flows with an average velocity of 0.12 m/s. At the point where the waste outfall enters the stream the "after-mixing" concentration of oxygen-demanding organic material is 25 mg/L (as oxygen). These oxygen-demanding substances decay and utilize oxygen according to Eq. 1.2-8. The integrated form of this equation is

$$\rho_{B2} = \rho_{B2}^0 \exp(-k_B''' t) \qquad (1.2A-1)$$

The rate constant k_B''', where $B \equiv$ organic material, is 0.3 d^{-1}(day^{-1}). Calculate the following:

1. The dissolved oxygen concentration 16.1 km downstream if the after-mixing oxygen concentration at the outfall point is 6.17 mg O$_2$/L. Assume water temperature is 20°C.
2. The concentration of the remaining organic waste material in mg/L.
3. The critical time (d) and distance downstream (km).

1.2B. STREAM ASSIMILATION CAPACITY FOR WASTE MATERIAL

The ability of a natural stream to oxidize oxygen-demanding organic (or inorganic) waste material and still maintain a dissolved oxygen content above some arbitrary minimum is termed the *stream assimilation capacity*. This capacity calculation is normally made for the time of year when the water flow rate is low and temperature is high. These conditions usually occur in late summer and an assimilation capacity calculated for this time assures a better than minimum oxygen condition throughout the remaining part of the yearly cycle.

Figure 1.2B illustrates the idea of an aqueous waste entering a stream. The notation Q_w, ρ_{A2w}, and ρ_{B2w} represents the volumetric flow rate, dissolved oxygen concentration, and O$_2$-demanding matter concentration of the waste. The terms Q_s, ρ_{A2s}, and ρ_{B2s} represent the volumetric flow rate, dissolved oxygen concentration, and O$_2$-demanding matter concentration of the stream. The in-stream conditions after mixing is represented by similar notation: Q, ρ_{A2}^0, and ρ_{B2}^0. The assimilation capacity is the quantity represented by the product $Q_w \rho_{B2w}$.

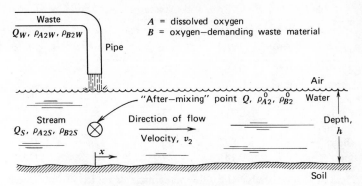

Figure 1.2B.   Stream assimilation capacity.

Material balances for the mixing point result in three useful equations. For water:

$$Q = Q_s + Q_w \qquad (1.2B\text{-}1)$$

For dissolved oxygen:

$$\rho_{A2}^0 = \frac{(Q_s\rho_{A2s} + Q_w\rho_{A2w})}{Q} \qquad (1.2B\text{-}2)$$

For oxygen-demanding material:

$$\rho_{B2}^0 = \frac{(Q_s\rho_{B2s} + Q_w\rho_{B2w})}{Q} \qquad (1.2B\text{-}3)$$

For many applications the waste flow rate is small compared to the stream flow rate, the dissolved oxygen concentration in the waste is zero, and the stream is practically void of oxygen-demanding matter prior to mixing with the waste. For this situation Eqs. 1.2B-1 and 1.2B-2 become for dissolved oxygen:

$$\rho_{A2}^0 \simeq \rho_{A2s} \qquad (1.2B\text{-}4)$$

and for oxygen-demanding material:

$$\rho_{B2}^0 = \frac{Q_w\rho_{B2w}}{Q_s} \qquad (1.2B\text{-}5)$$

Table E1.2C.   Field Data and Observed Reaeration Coefficients

River	Depth (m)	Velocity (m/s)	Temperature (°C)	Renewal Time (s)	Observed $^1k'_{A2}/h$ (d^{-1})
Elk	0.27	0.30	12	1.1^a	11.1
Clarion	0.58	0.17	13	1.7	6.00
Tennessee	1.2	0.22	23	5.3^a	3.03
Illinois	2.8	0.42	27	6.7	0.62
San Diego					
Bay	3.7	.098	20	37.5	0.111
Ohio	2.1	0.18	24.5	13	0.44

^{a}Time of renewal reduced by the percentage of rapids in the stretch.

The assimilation capacity can now be represented by the product $Q_s\rho_{B2}^0$. Do the following:

1. Verify Eqs. 1.2B-1 through 1.2B-5.
2. A chemical manufacturing company plans to build a plant on Clearwater River. Calculate the assimilation capacity of Clearwater River (kg oxygen-demanding material/d) if stream standards stipulate that the minimum dissolved oxygen in the river shall not fall below 4 mg dissolved oxygen/L under low flow conditions of 1.8E 5 m^3/d and a critical temperature of 30°C. It is assumed that the stream above the mixing point is fully saturated with oxygen and is free of oxygen-demanding organic matter. The following data apply: Average velocity 0.21 m/s, average depth 1.8 m, $k_B''' = (0.12)(1.047)^{T\text{-}20}$, d^{-1}

1.2C   VERIFICATION OF THE O'CONNOR-DOBBINS REAERATION MODEL

A portion of the field data employed by O'Connor and Dobbins[4] in the verification of their proposed model appears in Table E1.2C.

1. Calculate a $^1k_{A2}(A \equiv O_2)$ for each field data point in g/cm$^{2\cdot}$s.
2. Prepare a parity plot for comparing the calculated coefficient with the observed coefficient. [*Hint:* Plot $(\log\,^1k_{A2})_{\text{OBS}}$ versus $(\log\,^1k_{A2})_{\text{CALC}}$.]
3. Discuss this graphical demonstration of verification.

REFERENCES

1. "Scientists Probe Pesticide Dynamics," *Chem. Eng. News*, April 22, 1974, pp. 32–33.
2. J. R. Fair, B. B. Crocker, and H. R. Null, "Trace-Quantity Engineering," *Chem. Eng.*, August 7, 1972, p. 60.
3. H. W. Streeter and E. B. Phelps, "A Study of the Pollution and Natural Purification of the Ohio River," U.S. Public Health Service, Bulletin 146, 1925.
4. D. J. O'Connor and W. E. Dobbins, "The Mechanism of Reaeration in Natural Streams," *J. Sanit. Eng. Div., Proc. Am. Soc. Chem. Eng.*, **82** (SA6), 1115 (1956).
5. R. Higbie, "The Rate of Absorption of a Pure Gas into a Still Liquid during Short Periods of Exposure," *Trans. Am. Inst. Chem. Eng.*, **31**, 365 (1935).
6. P. V. Dankwerts, "Significance of Liquid-Film Coefficients in Gas Absorption," *Ind. Eng. Chem.*, **43** (6), 1469 (1951).
7. E. L. Thackston and P. A. Krenkel, "Reaeration Prediction in Natural Streams," *J. Sanit. Eng. Div., Amer. Soc. Civ. Eng.*, **95** (SA1), 65–94 (1969).
8. E. C. Tsivoglou and J. R. Wallace, "Characterization of Stream Aeration Capacity," U.S. Environmental Protection Agency Report EPA-R3-72-012, Washington, D.C., 1972.

RELATED READING

Churchill, M. A., H. L. Elmore, and R. A. Backingham. *J. Sanit. Eng. Div., Proc. Am. Soc. Chem. Eng.*, (SA4), 1 (July 1962).

Isaacs, W. P., and J. A. Maag. *Eng. Bull.* (Purdue University), **LIII** (1969).

Langbein, W. B., and W. H. Durum. U.S. Dept. of Interior, Geol. Survey, Bull. 542 (1967).

Owens, M., R. W. Edwards, and J. W. Gibbs. *Int. J. Air Water Pollut.*, **8**, 469 (1964).

CHAPTER 2

EQUILIBRIUM AT ENVIRONMENTAL INTERFACES

2.1. CHEMICAL EQUILIBRIUM AT ENVIRONMENTAL INTERFACES

Air, water, and earthen solids constitute the three major phases of the Earth's crust. The fourth phase, the biosphere, containing the living material, is not of concern in this work. On a mass basis the fourth phase is of minor importance; the bulk of the chemical residuals are usually found to be in the atmosphere or suspended in water or in surface soils, river silt, and similar earthen solids.

As we begin to address the nature of chemical transfer between the three major phases it is necessary to consider some general physical and chemical properties of each phase. From a gross point of view, the atmosphere is a thin layer of gaseous mixture surrounding the surface of the Earth and remaining attached to the Earth by the pull of gravity. The hydrosphere consists of more than 99% water, mainly in liquid form. The lithosphere is the earthen solid material underneath the air mass and the oceans. It is categorized as soil, rock, sand, mud, clay, and so on. A silt loam surface in good condition for plant growth is 20 to 30% (volume) air, 20 to 30% water, 45% minerals, and 5% organic matter. Tables 2.1-1 through 2.1-4 contain chemical data characterizing and typifying the three major phases. Table 2.1-1 gives the sea-level composition for the dry atmosphere. Table 2.1-2 is a list of approximately 45 elements, in order of decreasing concentration, present in solution in seawater. The lithosphere is comprised of rock and soil (mainly weathered rock) of chemical compositions given in Tables 2.1-3 and 2.1-4, respectively.

Table 2.1-1. Sea-Level Atmospheric Composition
for a Dry Atmosphere[a]

Constituent Gas	Mol. Fraction (%)	Molecular Weight $(O = 16,000)$
Nitrogen (N_2)	78.09	28.016
Oxygen (O_2)	20.95	32.0000
Argon (A)	0.93	39.944
Carbon dioxide (CO_2)	0.03	44.010
Neon (Ne)	1.8×10^{-3}	20.183
Helium (He)	5.24×10^{-4}	4.003
Krypton (Kr)	1.0×10^{-4}	83.7
Hydrogen (H_2)	5.0×10^{-5}	2.0160
Xenon (Xe)	8.0×10^{-6}	131.3
Ozone (O_3)	1.0×10^{-6}	48.0000
Radon (Rn)	6.0×10^{-18}	222.

Source. Reference 1. Reprinted with permission from *CRC Handbook of Chemistry and Physics*, 49th Edition, Robert C. Weast, Ed., CRC Press, Inc., Cleveland, Ohio, 1968, p. F-147. ©, The Chemical Rubber Co., CRC Press Inc.

[a]These values are taken as standard and do not necessarily indicate the exact condition of the atmosphere. Ozone and radon particularly are known to vary at sea level and above.

Table 2.1-2. Elements Present in Solution in Seawater
Excluding Dissolved Gases

Element	Concentration (g/metric ton) or Parts per Million
Cl	18,980
Na	10,561
Mg	1,272
S	884
Ca	400
K	380
Br	65
C (inorganic)	28
Sr	13
(SiO_2)	0.01–7.0
B	4.6
Si	0.02–4.0
C (organic)	1.2–3.0
Al	0.16–1.9
F	1.4
N (as nitrate)	0.001–0.7
N (as organic nitrogen)	0.03–0.2

Table 2.1-2. (*Continued*)

Element	Concentration (g/metric ton) or Parts per Million
Rb	0.2
Li	0.1
P (as phosphate)	$>0.001-0.10$
Ba	0.05
I	0.05
N (as nitrite)	0.0001–0.05
N (as ammonia)	$>0.005-0.05$
As (as arsenite)	0.003–0.024
Fe	0.002–0.02
P (as organic phosphorus)	0–0.016
Zn	0.005–0.014
Cu	0.001–0.09
Mn	0.001–0.01
Pb	0.004–0.005
Se	0.004
Sn	0.003
Cs	0.002 (approximate)
U	0.00015–0.0016
Mo	0.0003–0.002
Ga	0.0005
Ni	0.0001–0.0005
Th	<0.0005
Ce	0.0004
V	0.0003
La	0.0003
Y	0.0003
Hg	0.0003
Ag	0.00015–0.0003
Bi	0.0002
Co	0.0001
Sc	0.00004
Au	0.000004–0.000008
Fe (in true solution)	$<10^{-9}$
Ra	$2.10^{-11}-3.10^{-10}$

Source. Reference 1. Sverdrup, Johnson, Fleming, THE OCEANS, © 1942, Renewed 1970, pp. 176–177. Printed by permission of Prentice-Hall, Inc., Englewood Cliffs, New Jersey.

Table 2.1-3. Chemical Composition of Rocks

Element	Average Igneous Rock	Average Shale	Average Sandstone	Average Limestone	Average Sediment
SiO_2	59.14	58.10	78.33	5.19	57.95
TiO_2	1.05	0.65	0.25	0.06	0.57
Al_2O_3	15.34	15.40	4.77	0.81	13.39
Fe_2O_3	3.08	4.02	1.07	0.54	3.47
FeO	3.80	2.45	0.30		2.08
MgO	3.49	2.44	1.16	7.89	2.65
CaO	5.08	3.11	5.50	42.57	5.89
Na_2O	3.84	1.30	0.45	0.05	1.13
K_2O	3.13	3.24	1.31	0.33	2.86
H_2O	1.15	5.00	1.63	0.77	3.23
P_2O_5	0.30	0.17	0.08	0.04	0.13
CO_2	0.10	2.63	5.03	41.54	5.38
SO_3		0.64	0.07	0.05	0.54
BaO	0.06	0.05	0.05		
C		0.80			0.66
	99.56	100.00	100.00	99.84	99.93

Source. Reference 1. "Chemical Composition of Average Rocks" (After Clarke) in SEDIMENTARY ROCKS, 2nd Edition by F. J. Pettijohn. © 1949, 1957 by Harper and Row, Publishers, Inc. Used by permission of the publisher.

Table 2.1-4. Analysis of Representative U. S. Surface Soils

Constituents	Norfolk Fine Sand, Florida (%)	Sassafras Sandy Loam, Virginia (%)	Ontario Loam, New York (%)	Loam from Ely, Nevada (%)	Hagerstown Silt Loam, Tennessee (%)	Cascade Silt Loam, Oregon (%)	Marshall Silt Loam, Iowa (%)	Summit Clay from Kansas (%)
SiO_2	91.49	85.96	76.54	61.69	73.11	70.40	72.63	71.60
TiO_2	0.50	0.59	0.64	0.47	1.05	1.08	0.63	0.81
Fe_2O_3	1.75	1.74	3.43	3.87	6.12	3.90	3.14	3.56
Al_2O_3	4.51	6.26	9.38	13.77	8.30	13.14	12.03	11.45
MnO	0.007	0.04	0.08	0.12	0.44	0.07	0.10	0.06
CaO	0.01	0.40	0.80	5.48	0.37	1.78	0.79	0.97
MgO	0.02	0.36	0.75	2.60	0.45	0.97	0.82	0.86
K_2O	0.16	1.54	1.95	2.90	0.91	2.11	2.23	2.42
Na_2O	Trace	0.58	1.04	1.47	0.20	1.98	1.36	1.04
P_2O_5	0.05	0.02	0.10	0.18	0.16	0.16	0.12	0.09
SO_3	0.05	0.07	0.08	0.12	0.07	0.21	0.12	0.11
Nitrogen	0.02	0.02	0.16	0.10	0.27	0.08	0.17	0.09

Source. Reference 2. Copyright © 1974 by Macmillan Publishing Co., Inc.

An interface is the place at which two different systems (or subsystems) meet and interact with each other; it is also the boundary between any two phases. Among the three phases (gas, liquid, and solid) there are five types of interfaces: gas–liquid, gas–solid, liquid–liquid, liquid–solid, and solid–solid. In considering chemical movements the last type is almost unimportant; however, the remaining four interfaces constitute important environmental planes through which a host of synthetic chemicals move, eventually becoming distributed throughout the ecosystem.

A chemical introduced into the environment on one side of an interface will eventually become present, by a spontaneous process, in the other phases. As molecules of the chemical commence to accumulate on the other side a certain level is reached, after which gross accumulation ceases. At this time definite and possibly drastically different concentrations are in evidence on opposite sides of the interface. A system is in equilibrium when its state is such that it can undergo no spontaneous or unaided changes. When no further observable changes in concentration of the chemical occur on either side of the interface, equilibrium between phases has been attained. This description of phase equilibrium is strictly phenomenal.

A kinetic explanation of chemical equilibrium in an air–water system maintains that molecules of the chemical are continually crossing the interface but that the rate at which gas phase molecules condense on the liquid surface is equal to the evaporation rate of the chemical from the liquid surface into the gas. Thermodynamic considerations require that the chemical potentials of the component in both phases of a multicomponent system are identical at equilibrium for constant pressure and temperature

$$\mu_{A1} = \mu_{A2} \tag{2.1-1}$$

where $\mu \equiv$ chemical potential of constituent A. Since the fugacity of any component i is directly related to its chemical potential by the relation $(d\mu_A = RT \ln f_A)_T$, fugacity can also be used as a criterion of equilibrium between phases. This is a more useful property than chemical potential for defining equilibrium, since fugacity can be expressed as an absolute value, whereas chemical potential can be expressed only relative to some arbitrary reference state. Conceptually fugacity is closely related to pressure. For equilibrium of component A between two phases

$$f_{A1} = f_{A2} \tag{2.1-2}$$

where subscripts refer to the different phases, $1 \equiv$ air and $2 \equiv$ water. Since fugacity (f) can be expressed as a simple function of activity coefficient

(γ) mole fractions (x and y) and a reference fugacity (f^0) a more useful form of the equation results:

$$\left(y_A \gamma_A f_A^0\right)_1 = \left(x_A \gamma_A f_A^0\right)_2 \qquad (2.1\text{-}3)$$

This final equality is the starting point for calculations of equilibrium concentrations (i.e., x,y) of chemicals on both sides of an interface and can be generalized to include all four phase combinations. A thermodynamic text by Hougen, Watson, and Ragatz[3] contains details and extensions of the general criteria for equilibrium between phases. Extensions and applications of Eq. 2.1-3 for equilibriums at environmental interfaces is treated in detail and quantitative results are made available.

IDEAL SOLUTIONS

An ideal solution is defined as a solution in which the fugacity of each component i is equal to the product of its mole fraction and the fugacity of the pure component at the same temperature, pressure, and state of aggregation as those of the solution. An ideal liquid solution presupposes that when one component is mixed with another, mutual solubility results, that no chemical interaction with accompanying heat effects occur, that molecular diameters are the same, and that intermolecular forces of attraction and repulsion are the same between unlike as between like molecules. Most chemicals of concern in the environment form nonideal aqueous solutions as they exist in water. At low pressures gaseous mixtures exhibit near ideal behavior. For this reason chemicals in air as molecular dispersed species are assumed here to have ideal behavior.

Air–Water Equilibrium Occurrences

The interface between the atmosphere and hydrosphere is an important site for the transfer of chemicals because of the fluid nature of each phase. Phase contact of water with pure gases and air with pure liquid chemicals are also considered.

PURE GASES IN CONTACT WITH WATER

When equilibrium is established between a pure gas and water at constant pressure and temperature, the solubility of the gas in water is manifest. We are concerned here only with atmospheric pressure data. Equation 2.1-3 becomes

$$1 = x_A \gamma_{A2} f_{A2}^0 \qquad (2.1\text{-}4)$$

where γ_{A2} is the activity coefficient, f_{A2}^0 is the pure component fugacity of A in the water, and x_A is the mole fraction solubility of A in water. This equation is useful for calculating x_A for gases for which no solubility data are available since γ_A and f_A^0 can be obtained from further thermodynamic relations.

The system pressure can fall into one of three conditions. Some low molecular weight gases such as light hydrocarbons (butane and lighter) are unable to exert their full saturation vapor pressure at atmospheric temperature. Solubility data are given at 1 atm where Eq. 2.1-4 applies with f_{A2}^0 being the saturation vapor pressure. If the vapor pressure is atmospheric pressure, then $f_{A2}^0 = 1$ in Eq. 2.1-4. If the system temperature exceeds the critical temperature of the gas, the saturation vapor pressure (i.e., f_{A2}^0) must be estimated by extrapolating to system temperature and the result used in Eq. 2.1-4. The activity coefficient, γ_{A2}, accounts for the nonideal behavior of A in liquid water. Activity coefficients for selected compounds are available in the literature; otherwise they can be estimated by group contribution methods.[4,5] Some activity coefficients for gases in aqueous solution appear in the appendix.

Solubility data are available for common gases. Some data are given in Appendix C. Gas solubility is highly dependent on pressure. The nature of the water also affects the magnitude of gas solubility, as indicated by the difference in oxygen solubility in fresh water and seawater.

Example 2.1-1. Solubility of Ethane in Water

Estimate the solubility (g/m^3) of ethane $(C_2H_6 \equiv A)$ in water at 24°C if it is present above water as a pure gas at 1 atm pressure. Here $\gamma_{A2} = 7.02E\,2$ at 25°C.

SOLUTION

Since ethane in the gas is pure $y_A = 1$, forms an ideal gas mixture $\gamma_{A1} = 1$ and at 1 atm $f_{A1}^0 = 1$, Eq. 2.1-3 becomes

$$1 = x_A \gamma_{A2} f_{A2}^0 \qquad (2.1\text{-}4)$$

The critical pressure and temperature of ethane is 48.2 atm at 32.3°C. The problem temperature condition is below the critical temperature. The vapor pressure of ethane at 25°C is 42.9 atm = f_{A2}^0. This pressure is estimated from ethane vapor pressure data in a handbook.

$$x_A = \frac{1}{(7.02E2)(42.9)} = 3.32E-5$$

$$c_{A2} = x_A c_2 = 3.32E-5(1E\ 6\ g/m^3) = 33.2\ g/m^3$$

McAuliffe [*J. Phys. Chem.*, **70**, 1267 (1966)] reports a value of 60.4 g/m³ for these conditions.

PURE LIQUID IN CONTACT WITH AIR

When equilibrium is achieved between a pure liquid chemical and the closed air space above it at constant total pressure ($p_T = 1$ atm) and temperature the pure component vapor pressure of the chemical is manifest within the air space. Equation 2.1-3 yields the air space mole fraction of chemical A

$$y_A = \frac{p_A^0}{p_T} \tag{2.1-5}$$

where p_A^0 is the vapor pressure of liquid chemical A and p_T is the total pressure in the air space above the interface. Vapor pressure data for many pure liquids and solids at ambient temperatures can be found in *Chemical Engineer's Handbook** and other chemical data handbooks. Some of the data for common liquids and solids are available in Appendix C.

EFFECT OF TEMPERATURE ON VAPOR PRESSURE

The forces causing the vaporization of a liquid are derived from the kinetic energy of translation of its molecules. An increase in kinetic energy of molecular translation should increase the rate of vaporization and therefore the vapor pressure. The kinetic energy of translation is directly proportional to the absolute temperature. An increase in temperature should cause an increased rate of vaporization and a higher equilibrium vapor pressure. This is found to be universally the case. It must be remembered that it is the temperature of the liquid surface at the interface that is effective in determining the rate of vaporization and the vapor pressure.

Chemical Engineer's Handbook, Robert H. Perry and Cecil H. Chilton, Eds., 5th Edition, McGraw-Hill N.Y. (1973).

If the temperature does not vary over wide limits, and it is assumed that the latent heat of vaporization λ_A is constant, the volume of liquid is neglected and the ideal gas law applies; then the Clausius-Clapeyron equation is applicable

$$\ln\left(\frac{p_{A1}^0}{p_{A10}^0}\right) = \frac{\lambda_A}{R}\left(\frac{1}{T_0} - \frac{1}{T}\right) \qquad (2.1\text{-}6)$$

where R is the universal gas constant and λ_A is the molal heat of vaporization. This equation permits calculation of the vapor pressure of a substance p_{A1}^0 at a temperature T if the vapor pressure p_{A10}^0 at another temperature T_0 is known, together with the latent heat of vaporization. The latent heat of vaporization is the quantity of heat that must be added to transform a substance from the liquid to the vapor state at the same temperature. Values of the heats of vaporization at the normal boiling point of many compounds are listed in handbooks. Equation 2.1-6 gives rise to empirical formulas of the form $\ln p_A^0$ versus $1/T$ commonly used for correlating vapor pressure data. (See Problem 2.1H.)

PARTITION COEFFICIENTS FOR THE AIR–WATER SYSTEM

The need for equilibria information for the distribution of traces of chemicals between the air phase and the water phase is of primary importance, since this is the more common case, rather than the preceding cases where one or the other phase is a pure chemical.

DISTRIBUTION LAW (DEFINITION)

If a substance is added to a system of two mutually, totally, or nearly immiscible phases, in both of which it is soluble, it distributes itself between the two by dissolving in each in fixed proportions, independent of the quantity of the solute, the ratio of its concentrations in the two phases being constant at a constant temperature; this ratio (called the *distribution coefficient* or *partition coefficient*) is constant and equals the ratio of the values of the solubility of the substance in each of the two solvents. Other names are *partition law* and *Nernst theorem* or *law of distribution*. The definition for air–water is

$$\mathcal{K}_{A12}^* \equiv \frac{y_A}{x_A} \qquad (2.1\text{-}7)$$

There are many ways of expressing the partition coefficient. Equation 2.1-3 can be transformed to yield

$$\mathcal{K}^*_{A12} \equiv \frac{\gamma_{A1}p^0_{A1}}{p_T} \tag{2.1-8}$$

which gives the partition coefficient of moles of A in air to moles of A in water.

The relative volatility is another technique used to relate equilibriums of volatile liquid mixtures. The definition of the relative volatility of a species distributed between a vapor phase made of A and water vapor and a liquid water phase containing A is

$$\alpha^*_A \equiv \left(\frac{y_A}{y_B}\right)\left(\frac{x_B}{x_A}\right) \tag{2.1-9}$$

where $B \equiv H_2O$. This definition is substantially correct if air is present. If the liquid is predominantly water $x_B \simeq 1$ and $y_B = p^0_B/p_T$; using Eq. 2.1-3

$$\alpha^*_A = \frac{\gamma_{A2}p^0_{A1}}{p^0_B} \tag{2.1-10}$$

which is the relative volatility of chemical A in a dilute solution of water. The relative volatility is actually more descriptive of the absolute distribution of volatile chemical A (gas, liquid, or solid) dissolved in water than the partition coefficient. Values of $\alpha > 1$ imply that there is a preferential accumulation of A in the air and for $\alpha < 1$ chemical A prefers the water phase.

Henry's law applies for dilute solutions of chemicals, either a gas, liquid, or solid at ambient conditions, in water. Equilibrium is defined by giving the Henry's law constant and the temperature

$$p_{A1} = H_A x_A \tag{2.1-11}$$

where H_A is Henry's law constant for species A. Normally p_{A1} is the partial pressure of chemical A in the gas phase (air) measured in atmospheres and x_A is the mole fraction of chemical A in solution; therefore H_A has dimensions of atmospheres. Now by using $p_A = y_A p_T$ along with Eq. 2.1-3 the Henry's law constant is seen to be a function of the activity coefficient of A in water also;

$$H_A = \frac{\gamma_{A2} p_{A1}^0}{p_T} \qquad (2.1\text{-}12)$$

Numerical values of Henry's law constant taken from handbooks and other sources should be used with care. The dimensions of the Henry's law constant depend on the particular definition and units used. (See Problem 2.1K.)

In summary, there are three ways of expressing air–water equilibrium for dilute solutions of chemical A: the partition coefficient, relative volatility, and Henry's law. It is evident from Eqs. 2.1-8, 2.1-10, and 2.1-12 that all reflect the activity of A in water and the pure component vapor

Table 2.1-5. Vapor–Liquid Equilibriums of Selected Gases and Liquids in Water at 25°C

Components	Normal Boiling Point (°C)	Henry's Law[a] Constant H_A	Relative[b] Volatility α_A^*
Nitrogen	−195.8	86,500	2,768,000
Hydrogen sulfide (H_2S)	−59.6	54,500	1,744,000
Oxygen (O_2)	−183	43,800	1,402,000
Ethane (C_2H_6)	−88.6	30,200	966,400
Propylene (C_3H_6)	−48.	5,690	182,100
Carbon dioxide (CO_2)	−78.5	1,640	52,480
Acetylene (C_2H_2)	−84	1,330	42,560
Bromine (Br_2)	−58.8	73.7	2,358
Ammonia (NH_3)	−33.4	0.843	27.0
Acetylaldehyde	20.2	5.88	188
Acetone	56.5	1.99	63.7
Isopropanol	82.5	1.19	38.1
n-Propanol	97.8	0.471	15.1
Ethanol	78.4	0.363	11.6
Methanol	64.7	0.300	9.60
n-Butanol	117	0.182	5.82
Acetic acid	118.1	0.0627	2.01
Formic acid	100.8	0.0247	0.790
Propionic acid	141.1	0.0130	0.416
Phenol	181.4	0.0102	0.326

[a]See defining equation 2.1-11; H_A in atm/mol fraction.
[b]See defining equation 2.1-9.

pressure of A. The choice of which one to use depends in part on the application and preferences of the user. Table 2.1-5 contains Henry's law constants, and relative volatility values for common gases and liquids. Henry's law constants of other common gases can be found in Appendix C. Consult chemical handbooks as sources for additional data. Faced with no data one must obtain activity coefficients and vapor pressures. Procedures are available for obtaining estimates of activity coefficients.[4,5]

EQUILIBRIUMS IN THE NATURAL ENVIRONMENT

Unfortunately the behavior of chemicals in natural environments is rendered more complex by a number of factors. Most of the equilibrium data presented in handbooks and those reproduced in the appendix were obtained under ideal laboratory conditions. This "handbook" data may not be the most desirable but must be used in the absence of data representing the natural environmental system and its accompanying synergistic effects.

Mackay and Shiu[6] review some of the factors that render more complex the equilibrium behavior of hydrocarbons in natural aquatic environments. The presence of electrolytes generally increases hydrocarbon activity coefficients and reduces solubility (i.e., the "salting-out" effect). Appendix C contains some data on the solubility of hydrocarbons in distilled water and seawater. Generally the seawater solubility is about 70 to 80% of that in distilled water. It has been well established that hydrocarbons can exist in colloidal, micellar, or particulate form in appreciable quantities. Filtration was found to reduce the apparent solubility. The presence of surface-active organic compounds (many of which occur naturally) increases the amount of colloidal hydrocarbons. It appears that a typical oil, which may be truly soluble to the extent of 200 $\mu g/L$, may be solubilized to double this concentration by the presence of a few mg/L of dissolved organic surfactant and may be also present as particulate hydrocarbon to about 10 times the true solubility, or several mg/L.

Earthen Solid–Water Equilibrium Occurrences

Contacts of soil and water are important environmental interfaces through which many chemicals are transferred and include earth material in place under water and material suspended in water. Equilibrium contacts involving pure chemical solids with water and earthen solids with pure liquid chemicals are considered also.

PURE SOLID CHEMICALS IN CONTACT WITH WATER

The extent to which a solid substance mixes with a liquid to form a homogeneous system at a given temperature is defined as *solubility*. For the case of pure solid chemicals (also gases and liquids) water solubility is an equilibrium state. Equation 2.1-3 for the case of a solid–liquid system becomes

$$f^0_{A3} = x_{A2}\gamma_{A2}f^0_{A2} \tag{2.1-13}$$

where $2 \equiv$ water phase and $3 \equiv$ solid phase. For chemicals that are solid at the temperature in question the situation is complex since the correct reference fugacity (f^0_{A2}) is the fugacity (or vapor pressure) of the pure solid in a hypothetical liquid state. This vapor pressure can be estimated by extrapolating the liquid vapor pressure curve below the triple point. When a solid chemical is in equilibrium with an aqueous solution of concentration x_A the fugacity f^0_{A3} equals the vapor pressure of the solid; thus $\gamma_{A2}f^0_{A2}$ can be calculated as f^0_{A3}/x_A. The activity coefficient can be estimated by techniques mentioned earlier. It can be obtained from experiment. For example, solid naphthalene has a vapor pressure of 1.14E−4 atm at 25°C. Extrapolation of the liquid vapor pressure curve gives a value of $f^0_{A2} = $ 3.07E−4 atm. The solubility of solid naphthalene of 34.4 g/m^3 corresponds to a mole fraction x_A of 4.83E−6 thus, $\gamma_{A2}f^0_{A2}$ is 23.60 and γ_a is estimated to be 4.69E−4. Mackay[6] has published aqueous solubility data for solid and liquid hydrocarbons under environmental conditions. These data are reproduced in Appendix C. Other data published by Mackay for pesticides and polychlorinated biphenyls are contained in Table 4.2-1. Solubility data for common solid chemicals are given in chemical handbooks.

Although some high molecular weight organics may have low pure component vapor pressure and high boiling points relative to water, they nevertheless exhibit large relative volatilities due to large activity coefficients in water. Pierotti et al.[5] report activity coefficients of 10^3 to 10^7 for *n*-acids, *n*-primary alcohols, *sec*-alcohols, *n*-aldehydes, *n*-ketones, *n*-esters, *n*-ethers, *n*-chlorides, *n*-paraffins, and *n*-alkyl benzenes.

PURE LIQUID CHEMICALS IN CONTACT WITH EARTHEN SOLIDS

Chemical equilibrium involving pure liquids and soil surfaces is possibly not overly important from an environmental point of view. The saturation of soil surfaces with a liquid chemical spilled or otherwise placed on these surfaces occurs very frequently; however, the spatial distance for chemical movement is small. Molecules of the liquid chemical adsorb on the solid surface and diffuse very slowly into the solid material matrix. Percolation of the liquid between the solid particles (i.e., pores) can be extremely rapid and a large mass of the soil can become contaminated by this mechanism.

Although not an overly important transport mechanism in itself, the saturation of soil solids with chemicals can be the initial step in environmental movement that involves other natural mechanisms. The soil particle, with adsorbed chemical on its surface, can be dislodged and transported to a new location by moving water or wind. In its new location, in water or exposed to the air, a new interface with the environment is created and a new equilibrium approached with transfer to a more mobile phase (i.e., air and water) with eventual global movement. The contact of pure liquids with soil surfaces usually involves a finite but limited amount of chemical and soil contamination and is minor in importance when compared to other equilibrium situations considered in this work.

PARTITION COEFFICIENTS FOR THE SOIL–WATER SYSTEM

In an analogous fashion to the air–water system, chemicals in low levels or residual amounts are partitioned between soil and water to effect an equilibrium state between the phases. Equation 2.1-3 points the way for obtaining the partition between the phases

$$\left[\mathcal{K}^{*}_{A23} \equiv \right] \frac{x_{A2}}{x_{A3}} = \frac{\gamma_{A3} f^{0}_{A3}}{\gamma_{A2} f^{0}_{A2}} \tag{2.1-14}$$

where $2 \equiv$ water phase and $3 \equiv$ soil phase. The term γ_{A3} is highly dependent on the specific soil conditions and limited quantitative information is available. Specific experimental data will undoubtedly be needed in instances where precise equilibrium is desirable. Several specific applications are shown to indicate the nature of equilibrium between these two important environmental interfaces. Numerical values of the soil-water partition coefficient are expressed in various units. Weight fraction ratios (i.e. $\mathcal{K}^{*}_{A32} \equiv \omega_A / \psi_A$) are a common form of expression.

MUD

By definition, mud is a slimy, sticky mixture of solid material with water. This mixture is present on the bottom of streams, ponds, lakes, estuaries, and so on, and a fair amount of equilibrium studies have been made for certain chemicals.

STREAM BED—WATER

In many instances a specific undesirable substance such as kepone has been discharged into the water for extended periods of time rendering the stream, river, and so on, polluted. Once the health hazard is discovered and the pollutant no longer discharged, the water is proclaimed clean and healthy. However, after the pollutant faucet is turned off, the health state of the stream bed is not so quickly advanced as that of the water itself. It is important to know the equilibriums of various pollutants such as kepone, phenol, sulfates, and detergents between the water and soils comprising the muds, especially in the concentration range 5 to 10,000 ppm.

Greskovich[7] has published some water–mud equilibrium data for sodium sulfate and phenol. In the acid-mine problem areas, waters are contaminated with the sulfate ion forms such as sulfuric acid and ferric sulfate. Equilibrium studies were performed involving the sulfate system over a concentration range of 40 to 10,000 ppm by weight. Various aqueous solutions were contacted with typical central Pennsylvania clayey-silt soil until equilibrium was established. The equilibrium data shown in Fig. 2.1-1 are reported as equilibrium concentration of sulfate in the water versus the concentration of sulfate in the soil (mud) on a water-free basis. It can be seen that the soil did exhibit a slight adsorptive capacity for sulfate between 150 and 1000 ppm. A second species study

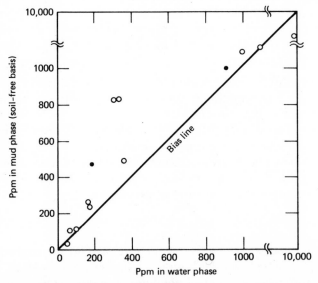

Figure 2.1-1. Equilibrium data for sodium sulfate between the soil and water at 77°F. (Reprinted by permission, **Source**. Reference 7.)

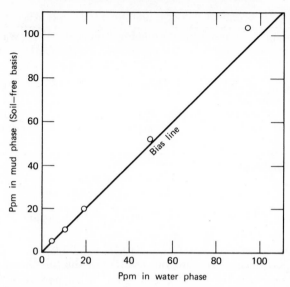

Figure 2.1-2. Equilibrium data for phenol between the soil and water at 77°F. (Reprinted by permission, **Source**. Reference 7.)

was performed with phenol. (See Fig. 2.1-2.) Little or no preferential adsorption of phenol on soil was observed for the concentration range evaluated in these studies.

It is often convenient to report values of the partition coefficient if it is indeed constant over a range of concentrations. Paris, Steen, and Baughman[8] obtained experimental values of the partition coefficients for two polychlorinated biphenyls (PCBs): Aroclors 1016 and 1242. The experimental measured values are listed in Table 2.1-6. It appears that the values of the partition coefficients are independent of Aroclor. The similarity in values for 1016 and 1242 is reasonable because the properties of these

Table 2.1-6. Partition Coefficients $(\mathcal{K}_{A32}^*)^a$ of Aroclors 1016 and 1242[b]

Sediment Type	$\mathcal{K}_{A32}^*$ of 1016	$\mathcal{K}_{A32}^*$ of 1242
Doe Run Pond	$(1.29 \pm 0.8)E\ 3$	$(1.09 \pm 0.16)E\ 3$
Hickory Hills Pond	$(1.30 \pm 0.9)E\ 3$	$(1.25 \pm 0.11)E\ 3$
USDA Pond	$(1.37 \pm 0.14)E\ 3$	$(1.21 \pm 0.12)E\ 3$

Source. Reference 8.

$^a \mathcal{K}_{A32}^* \equiv \omega_A / \phi_A$
[b]Values are the mean of six determinations.

Table 2.1-7.   Sediment–Water Equilibrium for Upper Klamath Lake

Temperature (°C)	Inorganic Phosphorus Sediment ($\mu g/g$)	Water ($\mu g/mL$)
10	413	0.23
23	406	0.64
10	793	1.66
23	723	6.08
10	504	0.19
23	499	0.58

Source.   Reference 9.

mixtures are so similar; however, the fact that all ponds display similar coefficients is surprising.

Sediment-water equilibrium was obtained for phosphorus release from moist Upper Klamath Lake, Oregon, mud.[9] For the release experiment the sediments were incubated at 10°C and 23°C for a period of 45 days and total inorganic phosphorus in the external water was measured. The equilibrium data, shown in Table 2.1-7, indicate that phosphorus is strongly adsorbed onto or within the soil solid matrix. Aerobic or anaerobic water conditions have a pronounced effect on the release and sorption of phosphate in soils and sediments. (See Problem 2.1D.)

Earthen Solid–Air Equilibrium Occurrences

Many synthetic chemicals exist in the air as dispersed molecules. As air masses containing these chemicals move over the land they become adsorbed onto earthen solids and vegetable surfaces. Chemicals placed directly onto the soil can and do vaporize into the overlying air mass. The air–earthen solid interface is large in areal extent. The size is slightly less in order of magnitude than the water–earthen solid and the air–water interfaces.

PURE GASES IN CONTACT WITH EARTHEN SURFACES

Circumstances in which chemicals in pure gaseous form come into contact with environmental soil type material are possibly minor except for laboratory adsorption studies.

PURE SOLID CHEMICALS IN CONTACT WITH AIR

In a fashion similar to pure liquids in contact with air, pure solids exert their vapor pressure in the space above. Equations 2.1-5 and 2.1-6 apply to pure solids in air. (See the section on pure liquids in contact with air for further information.)

PARTITION OF CHEMICALS BETWEEN THE AIR–SOIL PHASES

Gaseous chemical molecules in air are adsorbed onto soil as soon as they come in contact with the surface. The adsorption depends on the energy of the surfaces representing various constituents of the soil. Factors affecting the adsorption of gases (and liquids) on soils include water content, water solubility, chemical structure, nature of the surface, organic matter, and so on. Although adsorption experiments are relatively simple to perform, the results are usually difficult to interpret. The adsorption behavior of a chemical on a soil surface can be visualized with the aid of Leonard-Jones potential energy diagram (Fig. 2.1-3). As the distance of the adsorbate to the surface is decreased there is attraction; with a further decrease the potential energy increases. Here H_A is the heat of adsorption. Line I is for a physical or van der Waals type adsorption whereas II is for chemisorption. The point where both the graphs intersect is the activation energy. Usually the chemisorption process requires higher activation energy, and

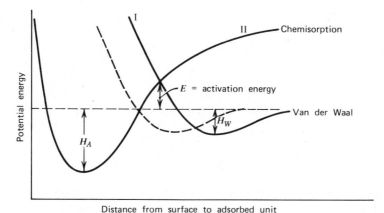

Figure 2.1-3. Adsorption process as represented by a potential energy diagram. (Reprinted by permission, **Source**. Reference 10.)

the heat of adsorption for chemisorption H_A is always greater than the heat for physisorption. For a physisorption the activation energy as well as the heat of adsorption are quite low. The pesticide–soil system generally follows a physical type of adsorption.

Adsorption data are generally represented by an equation known as an *isotherm*. An isotherm represents a relation between the amount of chemical on the soil and the chemical left in the solution (gas or liquid) after equilibrium. The three well-known isotherms are Freundlich, Langmuir, and BET (Brunauer, Emmett, and Teller).

In many cases the adsorption isotherm is satisfactorily represented by the empirical equation proposed by Freundlich:

$$\omega_A = B(\psi_A)^b \qquad (2.1\text{-}15)$$

where ω_A is the mass of chemical A adsorbed on the material or soil, ψ_A is the air phase mass fraction of chemical A in equilibrium with the surface, and B and b are empirical constants.

The simplest type of theoretical adsorption model occurs when adsorption is restricted to a single molecular layer on the solid. On this theoretical basis Langmuir developed an equation for this isotherm, assuming that at any pressure less than saturation the amount of chemical A adsorbed is proportional to the partial pressure of the gas and the fraction of the surface left uncovered. He further assumed that the adsorbed gas molecules do not dissociate or intersect on the surface. The result was

$$\omega_A = \omega_A^* \frac{b_1 \psi_A}{1 + b_1 \psi_A} \qquad (2.1\text{-}16)$$

where ω_A^* is the amount of chemical A adsorbed per unit mass of adsorbent to cover the surface with a layer one molecule thick and b_1 is an empirical constant.

The BET isotherm also has a sound theoretical foundation based on multimolecular layers of indefinite thickness. The isotherm relation is

$$\omega_A = \omega_A^* \frac{B_1(\psi_A / \psi_A^*)}{(1 - (\psi_A / \psi_A^*))[1 + (B_1 - 1)(\psi_A / \psi_A^*)]} \qquad (2.1\text{-}17)$$

where ψ_A^* is the maximum amount of concentration available for adsorption. The constant B_1 is an empirical constant but is related to the heat of adsorption of the first layer and the normal heat of condensation of the gas (or vapor) molecules.

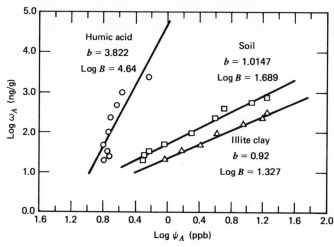

Figure 2.1-4. Freundlich plot for the adsorption of 2,4,2′,4′-tetradichlorobiphenyl on illite, humic acid, and Woodburn soil surface. (Reprinted by permission, **Source**. Reference 10.)

Hague[10] presents some results of the adsorption of pesticides in a soil environment. The majority of data representing the adsorption of pesticides on soil are represented by Freundlich-type isotherms. (See Fig. 2.1-4.) Langmuir and BET isotherms are very little used for soil–pesticide systems. In practice, the logarithm of the chemical adsorbed is plotted against the logarithm of chemical concentration at equilibrium. Such a plot usually results in a straight line and its slope and intercept give the value of the constants b and B. The constant B represents the extent of adsorption, whereas b throws much light on the nature of the adsorption as well as the role of solvent (water) in the adsorption. Table 2.1-8 gives some adsorption data typical of the air–soil system.

Temperature influences vaporization rate mainly through its effect on vapor pressure of the pesticide. Vapor pressures of pesticides follow the usual reciprocal temperature relations, $\log p_A^0 = A - B/T$, where A and B are constants. (See Eq. 2.1-6.) As temperature increases, the increase in vapor pressure of dieldrin is the same whether or not the dieldrin was in soil. The heat of vaporization of dieldrin with or without soil is 23.6 kcal/mol. Pesticides more strongly adsorbed than dieldrin may exhibit heats of vaporization in soil different from those of the pure chemical.

Every chemical has a characteristic saturation vapor density or vapor pressure that varies with temperature. Vapor pressure also is influenced by adsorption on surfaces, such as soils, and by solution in water or other

Table 2.1-8. Freundlich Isotherm Constant B and b for Adsorption of 2,4-D on Surfaces[a] (Reprinted by permission, **Source**. Reference 10.)

Surface	Temperature (°C)	b	$\log B$
Illite	0	0.685	1.11
	25	0.719	1.02
	0	0.925	−0.063
Montmorillonite	25	1.004	−0.186
	0	0.671	−0.984
Sand	25	0.827	−1.454
	0	0.97	−0.06
Alumina	25	1.01	−0.08
	0	0.90	0.58
Silica gel	25	0.95	0.11
	0	0.86	2.01
Humic acid	25	0.931	1.9

[a]Constants to be used with ω_A in ng/g and ψ_A in ppb.

liquids, such as plant oils and waxes. Vapor pressures of pesticides are greatly decreased by their interaction with soils mainly due to adsorption. How much adsorption reduces the vapor pressure of a pesticide in soil depends mainly on the nature of the pesticide, pesticide concentration in soil, soil water content, and soil properties such as organic matter and clay content.

Adsorption reduces the chemical activity, or fugacity, below that of the pure compound. This is then reflected in changes in vapor pressure of the chemical. For weakly polar or nonionic pesticides the amount of soil organic matter is the most important soil factor for increasing adsorption (see Problem 2.1I) and, consequently, for decreasing vapor pressure or potential volatility of a pesticide added to the soil. With more polar or ionic molecules, clay minerals play an increasingly important role in adsorption and volatility affects. Most of the more volatile pesticides are only weakly polar or nonionic; thus, their adsorption by soils is closely related to organic matter content. For example, dieldrin vapor pressure in five soils varied inversely with soil organic matter content. The vapor pressure of weakly polar pesticides in soil increases greatly with increases in pesticide concentration and reaches saturation vapor densities equal to that of the pesticide without soil at relatively low soil pesticide concentration. In moist Gila silt loam, saturation vapor densities for dieldrin, o,p'-DDT, trifluralin, lindane, and p,p'-DDT were reached at soil concentrations of 25, 39, 73, 55, and 15 μg/g, respectively. Typical applications of pesticides are, for example, 1 kg/ha, which is equivalent to 10

$\mu g/cm^2$, or approximately 150 ppm in the top 0.5 mm of soil. It appears that immediately after application some pesticides exert a vapor pressure equal to the vapor pressure of the pure chemical.

Pesticides are not the only chemicals that become placed on soil surfaces but have been covered in detail here because of the quantity of information available. The concentration on soil surfaces and total quantity placed into the environment is possibly extremely high compared to most chemicals. The accidental release of 2,3,7,8-Techlorodibenzo-p-dioxin (TCDD) in Italy in 1976 involved only 2 to 10 lb but severely affected an inhabited area of 150 acres.

Water–Liquid Chemical Equilibrium Occurrences

The contact of water with nonaqueous liquid chemicals, which involves two like phases (i.e., liquid–liquid), occurs with some frequency. This environmental interface is present at the underside of an oil slick on the surface of water, where the hydrocarbon phase contacts the water phase. This topic is treated in Chapters 4 and 5. The inadvertent placement of high density liquids in water bodies is another example of the creation of a liquid–liquid environmental interface.

PURE LIQUIDS IN CONTACT WITH WATER

Just as for pure gases in contact with water, Eq. 2.1-4 applies here also, and for such systems the solubility thus equals the inverse of the activity coefficient since $f_A^0 \simeq 1$. For example, the aqueous solubility of benzene is 1780 g/m^3, corresponding to a mole fraction of 4.10E−4 and an activity coefficient of 2.4E 3. A limited amount of data on solubility of liquid chemicals in water appears in Appendix C and elsewhere in the book. Data are available in handbooks.

MULTICOMPONENT LIQUIDS IN CONTACT WITH WATER

Crude oil is usually a multicomponent liquid at ambient conditions. Leinonen et al.[11,12] have studied this system in some detail. For this equilibrium situation Eq. 2.1-3 can be simplified to

$$x_A \gamma_{A2} = x_{A4} \gamma_{A4} \qquad (2.1\text{-}18)$$

where $4 \equiv$ liquid hydrocarbon phase. The reference fugacities cancel since they are both pure component fugacities of the same chemical. Equation

2.1-18 also applies to nonaqueous liquid phases other than hydrocarbons. The value of γ_{A4} is fairly close to unity; however, it is desirable to obtain its actual value particularly at a low concentration where it tends to be highest. As noted in a previous section the aqueous phase activity coefficients γ_{A2} are very high (e.g., 10^3 to 10^7), and it is difficult to predict the influence one dissolved hydrocarbon has on another. A solubility enhancement of from 1 to 25% with an average of 10.7% occurs owing to the presence of other hydrocarbon components, which apparently reduces the hydrocarbon activity coefficient in the aqueous phase. Few data are available on solubility of hydrocarbon mixtures under environmental conditions.

Phase Equilibrium, Phase Nonequilibrium, and Interface Equilibrium

Phase equilibrium has been the topic thus far in this section. Equilibrium represents a static condition. The net movement of a chemical between phases (i.e., interphase transfer) ceases at equilibrium. Viewed from another perspective, it establishes the potential for a chemical to move between phases. Besides establishing the fact, the foregoing material allows one to calculate the magnitude of chemical concentrations when equilibrium has been achieved.

Nonequilibrium conditions must exist between the phases for the net movement of a chemical species to occur. If the concentrations of chemical A existing in adjoining phases (x_A, y_A) are different from the equilibrium concentrations (x_A, y_A^*), then there will be interphase movement. (See Problems 2.1A and 2.1B.)

Even though the concentrations of the bulks of each phase may be in a nonequilibrium condition, equilibrium is assumed always to exist at the interface! The equilibrium interface (by definition) is assumed to be a hypothetical physical region two molecules thick, one monomolecular layer in each phase. This physical region is sufficiently small in scale so as to respond extremely quickly and remain at equilibrium at all times under conditions when the adjacent bulk phases are undergoing rapid fluctuations in concentration. The existence of equilibrium at the interface does not preclude interphase mass transfer when the bulk phases are not in equilibrium. (See Fig. 2.2-1.) Further developments of these concepts are presented in Chapter 3 under the topic of binary mass transfer coefficients in two phases.

Note the following on equilibrium notation: An equilibrium condition expressed in either concentration or mole fraction units should be denoted by * as a superscript. For example, the set of concentrations (c_{A1}^*, c_{A2}) signify gas phase concentration in equilibrium with a liquid phase concentration whereas (y_A, x_A^*) signifies a liquid mole fraction in equilibrium with a gas mole fraction. These superscripts were omitted in the preceding section on phase equilibrium for reasons of simplicity and clarity; however, from this point forward the designation is used. The following examples demonstrate the use of the equilibrium notation in phase equilibrium calculations.

Example 2.1-1.   Methane in Water

Obtain the solubility of methane gas in water at 20°C in molar concentration, mole fraction, mass concentration, and mass fraction.

SOLUTION

Let $CH_4 \equiv A, M_A = 16.04$ g/mol. $H_2O \equiv B, M_B = 18.01$ g/mol. According to *Chemical Engineer's Handbook*,* 0.4 cm^3 of CH_4 are soluble in 100 g of water at 20°C. Assuming a pressure of 1 atm and dilute solution,

$$c_{A2}^* = \frac{0.4 \text{ cm}^3 A}{100 \text{ g H}_2\text{O}} \left| \frac{\text{mol}}{22,400 \text{ cm}^3} \right| \frac{273°\text{K}}{293°\text{K}} \left| \frac{1000 \text{ g H}_2\text{O}}{\text{L}} \right. = 1.66\text{E}-4 \text{ mol } A/\text{L}$$

$$x_A^* = 1.66\text{E}-4 \frac{\text{mol } A}{\text{L}} \left| \frac{\text{L}}{1000 \text{ g H}_2\text{O}} \right| \frac{18.02 \text{ g}}{\text{mol}} \left| = 2.99\text{E}-6 \right.$$

The dilute solution assumption is justified for this low mole fraction. Equation (F), Table 1.1-2,

$$\rho_{A2}^* = c_{A2}^* M_A = 2.66\text{E}-3 \text{ g } A/\text{L}$$

$$\phi_A^* = \frac{x_A^* M_A}{M_B} = 2.66\text{E}-6 \qquad\qquad (1.1\text{-}7\text{A})$$

Example 2.1-2.   Choloroform Vapor in Air

Obtain the equilibrium concentration of chloroform vapor in air at 26°C and 1 atm above a pool of pure liquid chloroform at the same temperature.

SOLUTION

The handbook value of the vapor pressure of chloroform ($CHCl_3 \equiv A$) at 26°C is 200 mm Hg $M_A = 119.4$. Equation 2.1-5 must be used for the mole

*Op. cit., p. 3-38.

fraction of chloroform in air:

$$y_A^* = \frac{p_A^0}{p_T} = \frac{200 \text{ mm Hg}}{760 \text{ mm Hg}} = 0.263$$

This mole fraction according to the criteria in Fig. 1.1-4 is not a dilute solution. Assuming the air mixture is an ideal gas, the molar concentration in air is

$$c_{A1}^* = \frac{0.263 \text{ mol } A}{\text{mol mixture}} \left| \frac{\text{mol}}{22.4 \text{ L}} \right| \frac{273^\circ\text{K}}{299^\circ\text{K}} = 0.0107 \text{ mol } A/\text{L}$$

and

$$\rho_{A1}^* = 0.0107 \frac{\text{mol } A}{\text{L}} \left| \frac{119.4 \text{ g}}{\text{mol}} \right| = 1.28 \text{ g } A/\text{L}$$

The mass fraction, defined as $\psi_A \equiv$ mass A/mass mixture, necessitates the average molecular weight of the mixture. Equation (I) in Table 1.1-2 can be used for the average molecular weight of the gas phase

$$M_1 = y_A M_A + y_B M_B, \quad \text{where} \quad B \equiv \text{air}, \ M_B = 29.0 \text{ g/mol}$$
$$M_1 = (0.263)(119.4) + (1-0.263)(29.0) = 52.8$$
$$\psi_A^* = y_A^* \left(\frac{M_A}{M_1} \right) = 0.595$$

Example 2.1-3. Vinyl Chloride in Air

The Henry's law constant for vinyl chloride ($A \equiv CH_2CHCl$) at 25°C has been reported to be 50, where H_A is defined as the concentration of A in air (mg A)/(L@25°C, 760 mm Hg) divided by concentration of A in water (mg A)/(L of water). Vinyl chloride appears to have a 50 times greater distribution in air than in water under equilibrium conditions; however, the dimensions on H_A hamper this interpretation.

(a) Convert H_A to the partition coefficient given in Eq. 2.1-7 to effect a realistic partitioning of A between phases in terms of molecules and/or moles.

(b) Obtain the equilibrium concentration of A in air in mg A/L@25°C, 760 mm Hg above a wastewater treatment vessel in which the aqueous phase concentration is 2.5 g/m³.

SOLUTION

(a) From Eq. 2.1-7

$$\mathcal{K}_{A12}^* \equiv \left[\frac{y_A^*}{x_A} \right] = \frac{50 \text{ mg } A}{\text{L air}} \left| \frac{\text{L H}_2\text{O}}{\text{mg } A} \right| \frac{1000 \text{ g}}{\text{L H}_2\text{O}} \left| \frac{22.4 \text{ L air}}{\text{mol}} \right| \frac{\text{mol}}{18.01 \text{ g}} \left| \frac{298 \text{ K}}{273 \text{ K}} \right.$$

$$\mathcal{K}_{A12}^* = 67900$$

(b) From the definition of H_A

$$\rho_{A1}^* = H_A \rho_{A2} = 50(2.5 \text{ mg/L}) = 125 \text{ mg } A/\text{L} \qquad \text{in air}$$

2.2. THERMAL EQUILIBRIUM AT ENVIRONMENTAL INTERFACES

The movement of thermal energy is an important aspect of the subject of chemodynamics as an effect in itself or as it affects the movement of chemicals within environmental phases and at interfaces. Thermal energy is mechanical, potential, and kinetic energy of random motion on a molecular or microscopic scale. Heat is thermal energy in the process of being added to or removed from a given substance or moving from one portion of material substance to another by a temperature gradient. In describing thermal phenomena we find that we need an additional fundamental quantity besides mass, length, and time and that quantity is temperature.

The sensations of "hotness" and "coldness" of a given body are determined by what is called its *temperature*. Add heat to a body and its temperature ordinarily rises. Remove heat from a body (i.e., let it give thermal energy to some other body) and its temperature ordinarily goes down. Two bodies have the same temperature if when placed in contact, no heat flows from one to the other. Body A is at a higher temperature than body B if when they are placed in contact, heat flows from A to B. A group of objects, two phases of substance, and so on, placed within a well-insulated enclosure (i.e., net heat flow is zero) transfer heat in such a way that eventually they come to the same temperature. At this stage *thermal equilibrium* is said to be established, and no rearrangement of the objects, phases, and so on, can result in further transfer of heat.

The similarities and contrasts of chemical and thermal equilibrium and nonequilibrium concepts are noteworthy from the viewpoint of environmental interfaces. Figure 2.2-1 shows an environmental interface consisting of two fluids in which both heat and chemical transfer is occurring.

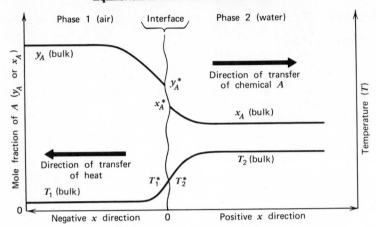

Figure 2.2-1. Heat and mass transfer through environmental interfaces.

Profiles of concentration and temperature in the region near the interface represent time-averaged values. Equilibrium, both chemical and thermal, exists at the interface. Whereas the chemical concentration profile is usually discontinuous at the interface, the thermal profile is continuous, since thermal equilibrium is satisfied when

$$T^*_{1i} = T^*_{2i} \qquad (2.2\text{-}1)$$

Temperature of phase 1 equals temperature of phase 2 at the interface, where equilibrium is always assumed to exist. The bulk phases (i.e., that region far back from the interface) are not at a thermal or chemical equilibrium condition so interphase movement of heat and mass can take place. The mass of material associated with the interface regions where temperature and concentration gradients exist is a vanishingly small fraction of the mass of the respective phases. Temperature and concentration measurements made by sampling and analyzing a particular phase yield the bulk phase temperature and concentration. Special techniques are necessary to detect the interface concentrations and temperatures and these are rarely known.

Problems

2.1A.   EQUILIBRIUM, NONEQUILIBRIUM, AND DIRECTION OF CHEMICAL MOVEMENT

Concentrations of chemicals on either side of environmental interfaces are given for the following situations. Do these concentrations represent equilibrium conditions? Give the direction of chemical movement, if any.

1. A sample of seawater (18,980 ppm chloride concentration) at 18.5°C has a dissolved oxygen concentration of 8.0 mg/L.

2. A sample of pure water in contact with pure n-butane gas at 1 atm pressure and 25°C has a concentration of 61.4 g n-butane/m^3.

3. The concentration of toluene in air above a pool of pure toluene liquid was 102 g/m^3 at 25°C and 1 atm pressure.

4. The concentration of H_2S above a septic water body was $y_A = 2E-4$ and the hydrogen sulfide in the water was $x_A = 1.5E-9$.

5. A water sample contains 60 ppm phenol and the adjoining mud contains 100 ppm. (Use Fig. 2.1-2.)

2.1B.   GRAPHICAL REPRESENTATION OF NONEQUILIBRIUM

The graphical representation in Fig. 2.1B shows the equilibrium curve and a nonequilibrium point (x,y) for chemical A. This point represents the concentrations across a certain environmental interface. In what direction is chemical A moving: G to L or L to G? Explain your answer.

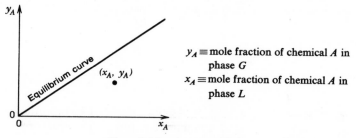

$y_A \equiv$ mole fraction of chemical A in phase G

$x_A \equiv$ mole fraction of chemical A in phase L

Figure 2.1B.   A nonequilibrium condition.

2.1C.   APPLYING CRITERIA FOR EQUILIBRIUM TO SPECIFIC CASES

State and justify the transformation of Eq. 2.1-3 to

1. Eq. 2.1-4.
2. Eq. 2.1-5.
3. Eq. 2.1-8.
4. Eq. 2.1-12.

2.1D.   DISTRIBUTION OF PHOSPHORUS BETWEEN SOIL AND WATER

Samples of Crowley and Mhoon soils were incubated under aerobic and anaerobic conditions, and the distribution of added soluble phosphorus

Table 2.1D. Soil–Water Equilibrium for Phosphate in Soils

Anaerobic		Aerobic	
Added (μg P/g soil)	Equilibrium Solution (μg P/mL)	Added (μg P/g soil)	Equilibrium Solution (μg P/mL)
Crowley Soil			
15	0.025	15	0.025
30	0.05	30	0.04
45	0.12	45	0.095
60	0.8	60	0.4
180	4.0	180	5.0
420	10.6	420	30
780	30	780	90
1080	40	1080	100.2
Mhoon Soil			
10	1.5	10	0.07
20	2.0	20	0.15
40	2.5	40	0.35
80	3.0	80	4.0
160	8.0	160	15
360	30	360	40
660	70	660	85
1020	80	1020	120

Source. Reference 13.

between the solution and solid phases was determined.[13] Each sample (300 g) was kept in suspension by use of a magnetic stirrer in 1500 mL of water in a sealed 2 L flask. Slow streams of air for the aerobic treatments and argon for the anaerobic treatments were continuously bubbled through the suspensions. The samples were incubated for 17 days at 30°C before the incremental additions of phosphorus as $Ca(H_2PO_4)_2$. Samples of the suspension were removed 24 hours after each addition of phosphorus and filtered prior to analyzing for phosphorus. The chemical analysis at equilibrium is shown in Table 2.1D. Transform this data so that plots like Fig. 2.1-1 and 2.1-2 can be constructed. Make the vertical axis μg P/g soil and the horizontal axis μg P/mL. Is phosphorus partitioned more strongly in the soil or in the water? What effect do anaerobic conditions have on the release of phosphorus from the soil?

2.1E. EQUILIBRIUM CHEMICAL CONTENT IN AIR ABOVE SOIL

1. Calculate the amount of 2,4-D (g) in a 1 m cube of air at equilibrium with illite at 25°C that contains 10 ppm.

2. Determine the amount of DDT (g) in 1 m^3 of air at equilibrium with soil at 25°C that contains 50 $\mu g/g$ (*Hint:* Use Table 4.2-1.)

2.1F.   ON CALCULATING HENRY'S LAW CONSTANTS.[14]

Henry's law constants can be calculated by

$$H \equiv \frac{\rho_{A1}^*}{\rho_{A2}^*} = 16.04 \frac{p_A^0 M_A}{T \rho_{A2}^*} \qquad (2.1F\text{-}1)$$

where ρ_{A1}^* and ρ_{A2}^* are the equilibrium concentrations in the air and water phases, respectively, p_A^0 is the vapor pressure of the pure solute in mm Hg, M_A is the molecular weight of the solute, T is temperature in K, and ρ_{A2}^* is the solubility of the solute in water in mg A/L.

1. Derive Eq. 2.1F-1 by using the ideal gas law to calculate ρ_{A1}^*. The units of ρ_{A2}^* are equal to those of ρ_{A1}^* so that H is dimensionless (i.e., mg/L ÷ mg/L)!
2. Calculate the Henry's law constant given by Eq. 2.1F-1 for the chlorohydrocarbon data in water compiled by Dilling in Table 2.1F. Compare calculated value to other values "found" in the literature.

Table 2.1F.   Equilibrium Data for Chlorohydrocarbons in Water at 25°C, 1 atm

Compound	Solubility ρ_{A2}^* (mg/L)	Vapor Pressure p_A^0 (mm Hg)	Henry's Law Constant H (found)
CH_3Cl	6270	760	0.30
CH_2Cl_2	19400	438	0.11
$CHCl_3$	7840	192	0.13
CCl_4	800	113	0.87
$CH_2Cl\,CH_2Cl$	8700	82	0.040
CH_3CCl_3	730	99	1.4
$CH_2=CHCl$	60	760	50
$CH_2=CCl_2$	400	497	6.3
$CHCl=CCl_2$	1100	59	0.365
$CCl_2=CCl_2$	140	18.6	0.50
$CCl_2=CCl_2$	120	14.2	0.82

Source. Reference 14.

2.1G.   MULTICOMPONENT GAS–LIQUID EQUILIBRIUM

The sea-level composition of dry air is shown in Table 2.1-1. Compute the equilibrium concentration of nitrogen, oxygen, and carbon dioxide in water at 20°C. Give the numerical results in mole fraction (x_A) and concentration (ρ_{A2} in mg A/L).

2.1H.　EFFECT OF TEMPERATURE ON VAPOR PRESSURE

The tendency for a pure chemical to vaporize is strongly dependent on temperature. Temperature at a particular location on the surface of the Earth can be quite variable. Daytime temperatures can reach 100°F and nighttime temperatures often go down to 32°F and lower. The vapor pressure dependence with temperature is usually highly nonlinear.

Naphthalene in air: This chemical is a solid at normal environmental temperatures and melts at 80.2°C. Its vapor pressure in mm Hg is given by

$$\log_{10}p_A^0 = 11.45 - \frac{3729.3}{T}, \qquad 0°C \leqslant t \leqslant 80°C \qquad (2.1H\text{-}1)$$

where $T = 273.1 + t$, t in °C.
Compute the mole fraction (y_A^*) of naphthalene in air at 32 and 100°F.

2.1I. SOIL–WATER PARTITION COEFFICIENT FOR AN ORGANIC PESTICIDE

The organic content of the soil strongly affects the partitioning of organic pesticides between soil and water phases. Table 2.1I contains experimental results of an equilibrium study with soils of organic matter $\geqslant 1\%$.

1. Compute the following partition coefficients for each soil.
   Total soil partition coefficient:

$$\mathcal{K}_{TS}^* \equiv \frac{\text{concentration in solution}}{\text{concentration in total soil}}$$

Total soil = organic plus inorganic solids

Table 2.1I.　Sorption of 4-Amino-3,5,6-Trichloropicolinic Acid[a] by Soils at pH 2

Soil	Organic Matter Content of Soil (Wt. %)	Acid in Solution at Equilibrium (Wt. %)	Acid Sorbed on Soil at Equilibrium (Wt. %)
D_1	1.0	51	49
N_1	2.7	23	77
B_2	4.1	11	89
B_1	10.7	5.8	94
Q_1	32.2	2	98

Source: Reference 15.
[a] 4 ml of a 1.0 ppm solution + 1g soil adjusted to pH 2 with HNO_3 and incubated 1 hr.

Organic matter partition coefficient:

$$\mathcal{K}^*_{OM} \equiv \frac{\text{concentration in solution}}{\text{concentration in organic phase}}$$

Compute liquid concentration in g A/cm^3 H_2O and soil concentration in g A/g soil or g A/g soil organic.

2. Compare the results of the two partition coefficients. Which one is almost constant for all soil types? What does the constant coefficient imply about the adsorption?

2.1J. RUNOFF CONTAMINATION FROM AN ORGANIC PESTICIDE

A certain large farming operation plans to use 4-amino-3, 5, 6-trichloropicolinic (ATP) acid for the control of a certain pest. It is expected that the concentration of ATP at the soil surface will be 15 $\mu g/g$. The soil contains 5% organic matter. The state environmental protection department is concerned about the water runoff contamination. Calculate the maximum water concentration of ATP likely to occur in the runoff in mg ATP/L. Additional data: $\mathcal{K}^*_{A23} = 0.00183$. Definition of the partition coefficient is $\mathcal{K}^*_{A23} = \rho_{A2}/\omega_A$, where ρ_{A2} is g A/cm^3 water, and ω_A is g A/g organic matter in soil.

2.1K. THE MANY FORMS OF HENRY'S LAW

Henry's law for expressing the equilibrium distribution of a chemical between dilute gas and liquid phases enjoys several formulations. The mole fraction form is dimensionless and can be expressed as

$$y_A = H_{Ax} x_A \qquad (2.1K\text{-a})$$

A common form is given by Eq. 2.1-11

$$p_{A1} = H_A x_A \qquad (2.1\text{-}11)$$

where H_A is in atm/mol fraction. Other forms are

$$p_A = H\rho_{A2} \qquad (2.1K\text{-b})$$

where H is in atm·cm^3/g, and

$$\rho_{A1} = H_\rho \rho_{A2} \qquad (2.1K\text{-c})$$

where H_ρ is in cm^3 water/cm^3 air STP. In the preceding expressions y_A and x_A are in mole fraction, p_{A1} is in atm, ρ_{A1} and ρ_{A2} are in g/cm^3 air STP and g/cm^3 water, respectively. Find the multipliers a, b, and c for the equality

$$H_{Ax} = aH_A = bH = cH_\rho \qquad (2.1K\text{-}d)$$

2.1L. CUMENE PARTITIONING BETWEEN WATER AND OIL SLICK PHASES

Commencing with Eq. 2.1-18, show that the concentration of cumene in water is given by

$$c_{A2} = c_{A2}^* x_{A4} \gamma_{A4} \qquad (2.1L\text{-}1)$$

where c_{A2}^* is the solubility of cumene in water, x_{A4} is the mole fraction cumene within the oil slick, and γ_{A4} is the activity coefficient of cumene within the oil slick. (*Hint:* In your development, apply Eq. 2.1-18 twice. Apply it to a two-phase system consisting of pure cumene and water. Apply it to a two-phase system consisting of the oil slick and water.)

REFERENCES

1. R. C. Weast, Ed., *Handbook of Chemistry and Physics*, Vol. 49, The Chemical Rubber Co., Cleveland, Ohio, 1968.

2. N. C. Brady, *The Nature and Properties of Soils*, Macmillan, New York, 1974, p. 24.

3. O. A. Hougen, K. M. Watson, and R. A. Ragatz, *Chemical Process Principles, II—Thermodynamics*, Wiley, New York, 1959, p. 849–856, 862–879.

4. G. M. Wilson and C. H. Deal, "Activity Coefficients and Molecular Structure," *Ind. Eng. Chem. Fund.*, 1 (1), 20 (1962).

5. G. J. Pierotti, C. H. Deal, and E. L. Derr, "Activity Coefficients and Molecture Structure," *Ind. Eng. Chem.*, 51 (1), 95–102 (1959).

6. D. Mackay and W.-Y. Shiu, "The Aqueous Solubility and Air–Water Exchange Characteristics of Hydrocarbons under Environmental Conditions," *Chem. Phys. Aqueous Gas. Sol.*, ASTM, Philadelphia, Pa. 1974, p. 93–110.

7. E. J. Greskovich, "Equilibrium Data for Various Compounds between Water and Mud," *Am. Inst. Chem. Eng. J.*, 20 (5) 1024 (1974).

8. D. F. Paris, W. C. Steen, and G. L. Baughman, "Role of Physico-Chemical Properties of Aroclors 1016 and 1242 in Determining their Fate and Transport in Aquatic Environments," 172nd American Chemical Society National Meeting, Paper PEST 110, San Francisco, 1976.

9. R. E. Wilding and R. L. Schmidt, "Phosphorus Release from Lake Sediments," U.S. Environmental Protection Agency, EPA-R3-73-024, Washington, D.C. 1973.

10. R. Hague, "Role of Adsorption in Studying the Dynamics of Pesticides in a Soil Environment," in R. Hague and V. H. Freed, Eds., *Environmental Dynamics of Pesticides*, Plenum, New York, 1975, p. 97–114.

11. P. J. Leinonen, D. Mackay, and C. R. Phillips, "A Correlation for the Solubility of Hydrocarbons in Water," *Can. J. Chem. Eng.*, **49**, 288–290 (April 1971).

12. P. J. Leinonen and D. Mackay, "The Multicomponent Solubility of Hydrocarbons in Water," *Can. J. Chem. Eng.*, **51** 230–233 (April 1973).

13. W. H. Patrick, Jr., and R. A. Khalid, "Phosphate Release and Sorption by Soils and Sediments: Effect of Aerobic and Anaerobic Conditions," *Science*, **186**, 53–55 (October 1974).

14. W. L. Dilling, "Interphase Transfer Processes II Evaporation Rates of Chloro Methanes, Ethanes, Ethylenes, Propanes, and Propylenes from Dilute Aqueous Solutions: Comparison with Theoretical Predictions," *Environ. Sci. Tech.*, **11** (4), 405–409 (1977).

15. J. W. Hamaker, C. A. I. Goring, and C. R. Youngson," Sorption and Leaching of 4-Amino-3,5,6-Trichloropicolinic Acid in Soils," in R. F. Gould, Ed., "Organic Pesticides in the Environment" Advances in Chemistry Series 60, American Chemical Society, Washington, D.C., 1966, p. 23–37.

CHAPTER 3

TRANSPORT
FUNDAMENTALS

On the topic of fate of and tracking of elusive pollutants in an aqueous medium, it has been flippantly written that the "physical mechanisms are all expressions of the pressure to escape from the aqueous to a gaseous or solid phase." Presumably the author was referring to the expression

$$\text{rate of movement of chemical } A = K(\Delta p_A)$$

If one interprets *pressure* to mean the equilibrium driving force, some appropriate rate constant K is still needed to complete the equality. This chapter is concerned with presenting mechanisms of chemical transport in the regions near environmental interfaces. The task encompasses finding the correct rate constant K but, in general, entails formulating and quantifying the correct rate of movement expression.

The word phrases *chemical movement*, *fate of chemical A*, and *chemical transport* have been used in generic sense. More specific meanings are desirable at this point. On performing a material balance for component A on an infinitesimally small-volume element $\Delta x \, \Delta y \, \Delta z$ located at a fixed position within any one of the three geospheres, the following form of the continuity equation results:

$$\underbrace{\frac{\partial c_{Aj}}{\partial t}}_{\text{accumulation}} + \underbrace{\overline{\nabla} \cdot (\bar{v} c_{Aj})}_{\substack{\text{convection} \\ \text{(or advection)}}} = \underbrace{\overline{\nabla} \cdot (\mathcal{D}_{Aj} \overline{\nabla} c_{Aj})}_{\text{diffusion}} + \underbrace{R_{Aj}}_{\text{reaction}}, \quad j = 1, 2, 3 \quad (3\text{-}1)$$

This expression reflects the entire fate of chemical A within any phase of the environment. It is written in vector notation to depict the three-dimensional nature of the movement terms. Rarely is the equation employed in the general form shown. When applied to solids $\bar{v} = 0$.

Convection (also called *advection*) is the term commonly employed to account for movement due to bulk flow of the phase. In the field of meteorology advection denotes the process of transport of an atmospheric property solely by the mass motion of the atmosphere. In the field of oceanography it denotes the process of transport of water, or of any aqueous property, solely by the mass motion of the oceans, most typically via horizontal currents. The velocity concentration products (e.g., $c_{Aj}\bar{v}$) give rise to the preceding interpretations. In the case of laminar fluid flow and solids the diffusion term accounts for the movement of chemical A within the phase because of a concentration gradient. If turbulence is present the diffusion term accounts for the randomlike exchange of fluid parcels between regions of space. When applied to an interfacial region the diffusion term accounts for the interphase transfer of chemical A. The reaction term accounts for the generation or depletion of A by a homogeneous chemical (or biochemical) reaction within the phase. As the reaction term appears in Eq. 3-1 it is positive for the net production of A within the phase.

Depending on the magnitude and polarity (i.e., positive, zero, or negative) the total effect of advection, diffusion, and reaction on chemical A may result in it increasing, remaining constant, or decreasing within the phase. Therefore, the magnitude and polarity of the accumulation term reflects the fate of chemical A within the respective phase of the environment.

Beginning with basic ideas in molecular diffusion, topics of mass transfer and turbulence in the environment are developed here and related to the natural physical forces that are the prime movers behind the transport of chemicals across the interfaces of the geospheres.

Figure 3-1 illustrates the three dominant environmental interfaces and the interface exchanges sites that are of concern. Each interface has at least one fluid phase associated with it. The specific type of movement of the fluid near the interface regulates to some degree the rate of interphase transfer through the particular boundary. For example, the fluid movements near the air–water interface of the oceans are only remotely related to the fluid movements near the same interface of a fast-moving land-locked stream. The physical flow characteristics on the bottom of an estuary are unlike those at the bottom of a stream, and air flow above the ocean does not compare directly to that above earthen solids and vegetation. As to the extent to which information is available on the particular aspects of the interphase transport and fluid movement, then that body of

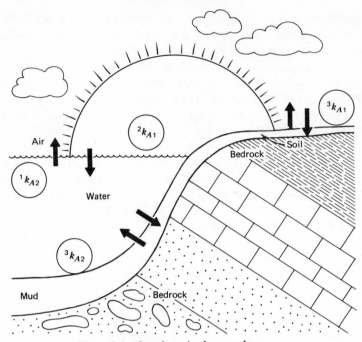

Figure 3-1. Interfaces in the geospheres.

knowledge, with its special adaptations should be applied. The special adaptations are presented for the most part in Chapter 4 and following chapters. This chapter attempts to unify the transport concepts into a somewhat general framework so that each of the example applications to be presented will not be interpreted as an isolated case.

From the undergraduate point of view of practical application, this chapter is theoretical. Those students who wish to get on to the meat of the subject of chemodynamics may skip this chapter now and return to it at various times when insight into particular mechanisms of transport is needed or desirable.

3.1. DIFFUSION AND MASS TRANSFER

The advective-type movement of a chemical, either within the atmosphere or within the hydrosphere, is a process of transport affected solely by the mass motion of the atmosphere or the hydrosphere. Nonequilibrium conditions of a chemical species, between phases or within a single phase, must exist for the diffusion-type movement of the chemical to occur. Non-equilibrium conditions between phases has been presented in Chapter 2. Within a single phase, nonequilibrium conditions are usually characterized by a concentration gradient (i.e., $\partial c_A / \partial x$).

An all-encompassing theory capable of explaining and quantifying diffusion-type movement is incomplete. The incompleteness is due in large part to the complexity of the fluid turbulence associated with the mass motion of fluids. Turbulence is almost always present in both the atmosphere and hydrosphere. A diverse body of knowledge has been developed to explain and quantify diffusion-type chemical movements in both the human-influenced and the natural environment. The following is a review of fundamental diffusion-type processes with restrictions to dilute solutions and environmental temperatures and pressures.

General Description of Molecular Diffusion

At any temperature above absolute zero, the individual molecules of a substance move incessantly and at random, apparently independently of each other. Frequent collisions occur between particles, so that the path of a single particle is a zigzag one. However, an aggregation of diffusing particles has an observable drift, from places of higher to places of lower concentration. For this reason diffusion is known as a transport phenomena.

Molecular diffusion in liquids can be observed by placing a layer of iodine solution, avoiding any convection or stirring, underneath pure water. At first only the lower part, containing the iodine, is colored, but then the color is observed to spread slowly upward. Eventually the upper part becomes colored too, the intensity of the color decreasing from bottom to top. After a long time the entire mass is uniformly colored.

Diffusion in gases can be similarly observed. In gases the process is much more rapid than in liquids. Diffusion in solids is generally the slowest type.

The average path traveled by a molecule in the interval between collisions is known as the *mean free path*. It decreases with increasing concentration. Another quantity characterizing the diffusing substance is the displacement. By displacement is meant the distance between the original position of a particle and its position after a certain period of time t. The mean displacement is zero, since in the absence of a difference in concentration, positive and negative displacement are equally probable. For this reason the mean square displacement $\overline{x^2}$ is introduced.

Consider a hypothetical horizontal plane passing through the iodine water system described earlier. Iodine molecules move in both directions through the plane. If one considers all the particles in a layer of a certain thickness on either side of a plane it will be observed that there are more iodine molecules in the same thickness below than above the plane. Since the same percentage of molecules crosses the hypothetical boundary per second this results in more molecules moving upward than downward until there is no longer a concentration difference across the plane. There results an overall flow from positions of higher to positions of lower concentrations. In this way the concentration is equalized. This equalization is macroscopically observable and is called *diffusion*. It is obvious that the mean square displacement and the diffusion time are a measure for the rate of diffusion. The ratio of the mean square displacement and the diffusing time is the diffusion coefficient as follows:

$$\mathcal{D} = \frac{\overline{x^2}}{2t} \tag{3-1.1}$$

The mathematical description of the process of diffusion is accomplished with this quantity $\mathcal{D}$, as is shown below.

FICK'S FIRST LAW

Qualitative observations of diffusion preceded quantitative descriptions. Robert Brown provided in 1827 the closest thing to direct, visual evidence for the motion of molecules. He observed that very minute particles suspended in a gas or liquid and viewed under a microscope were seen to be in a state of continual, random motion. Random molecular motion is sufficient to bring about diffusion, as noted earlier. Thomas Graham (1805–1869) quantified that the relative rates at which gases diffuse is inversely proportional to the square roots of their respective densities or molecular weights.

In a paper in 1855 Fick finally put Graham's experiments on a qualitative basis. Fick's introduction of his basic idea is "the diffusion of the dissolved material... is left completely to the influence of the molecular forces basic to the same law... for the spreading of warmth in a conductor and which already been applied with great success to the spreading of electricity." In other words, diffusion can be described on the same mathematical basis as Fourier's law of heat conductance or Ohm's law for electrical conduction. He defined a one-dimensional flux

$$J_{Az} = -c_2 \mathcal{D}_{A2} \frac{dx_A}{dz} \tag{3.1-2}$$

of component A through an area A across which diffusion is occurring, J_{Az}

Table 3.1-1. Experimental Diffusivities

System	Temperature (°K)	$\mathcal{D}_{ij}$ (cm²/s)
Gas pair, CO_2–N_2O	273.2	0.096
Liquid pair, 5% CH_3CH_2OH–H_2O	295	1.13E–5
Solid state, H_2 in S_iO_2	773	0.6-2.1E–8

is the flux per area c_2 is the molar density of the fluid, x_A is the mole fraction of A in the fluid, and z is the distance. The quantity $\mathcal{D}_{A2}$, which Fick called the "constant depending on the nature of the substances," is the diffusion coefficient or binary diffusivity of A in 2. The units of the mass diffusivity $\mathcal{D}_{Aj}, j = 1$, 2, and 3, are cm²/s.

Table 3.1-1 contains typical diffusivities. A more complete table of diffusivities appears in Appendix C. Relationships, based in part on theory and in part on experimental observations, are available for estimating diffusivities.[1] Gas diffusivities are dependent on the chemical species diffusing, temperature, and pressure. Liquid diffusivities are dependent on the species, temperature, and viscosity of the mixture. More on this later.

DIFFUSION OF TRACE CHEMICALS IN STAGNANT MEDIA

Molecular diffusion is a slow mechanism for the movement of chemicals through the atmosphere and the hydrosphere. Consider the simplistic problem of the diffusion of carbon dioxide into a stagnant layer of air lying above a water body containing a sufficient quantity of the dissolved gas so that the air interface is always saturated at 20°C. (See Problem 4.1P on CO_2 present in geothermal effluents for related information.)

This particular diffusion problem is called semi-infinite solid diffusion and requires Fick's second law for adequate description

$$\frac{\partial c_{A1}}{\partial t} = \mathcal{D}_{A1} \frac{\partial^2 c_{A1}}{\partial z^2} \tag{3.1-3}$$

A region of air enriched with CO_2 grows upward from the interface as the air mass remains in contact with the water body. The solution of Eq. 3.1-3 yields an equation for the CO_2 distribution in the stagnant air. The term y_A is the mole fraction concentration of $CO_2 (\equiv A)$ for a given distance of penetration z and time t

$$\frac{y_A - y_A^*}{y_A^0 - y_A^*} = \frac{2}{\sqrt{\pi}} \int_0^{z/\sqrt{4\mathcal{D}_{A1}t}} e^{-n^2} dn \tag{3.1-4}$$

and y_A^* and y_A^0 are the saturation mole fraction at the air–water interface and the initial mole fraction, respectively. $\mathcal{D}_{A1}$ is the diffusivity of CO_2 in air at 20°C, 1 atm. The right-hand side is commonly known as the Gauss error integral or probability function. Tabularized values of this function appear in Appendix B. Equation 3.1-4 is often written as

$$\frac{z_A - y_A^*}{y_A^0 - y_A^*} = \mathrm{erf}\left(\frac{z}{\sqrt{4\mathcal{D}_{A1}t}}\right) \tag{3.1-5}$$

Table 3.1-2 contains calculated CO_2 concentrations within the stagnant air mass. The table entries indicate the slowness of the molecular diffusion process in air.

Table 3.1-2. Carbon Dioxide Enrichment of a Stagnant Air Mass (CO_2 Concentration in Mole Fraction, y_{A1})[a]

Time,	Penetration Distance, z (cm)				
t (s)	0.001	0.01	0.10	1.00	10.0
1	0.0657	0.0654	0.0606	0.0326	0.0300
10	0.0658	0.0656	0.0642	0.0503	0.0300
60 (1 min)	0.0658	0.0657	0.0651	0.0592	0.0307
300 (5 min)	0.0658	0.0658	0.0655	0.0628	0.0406
3600 (1 hr)	0.0658	0.0658	0.0657	0.0649	0.0574

[a]$y_A^* = 0.0658$, $y_A^0 = 0.030$, $\mathcal{D}_{A1} = 0.153$ cm^2/s at 20°C, 1 atm.

Now consider a stagnant layer of oxygen-rich water suddenly placed above an anaerobic mud layer capable of consuming oxygen and depleting the water of its oxygen content. Just as with the air mass, the water layers adjacent to the mud becomes void of oxygen and this voided region gradually increases with time. Table 3.1-3 contains oxygen concentrations

Table 3.1-3. Oxygen Depletion from a Stagnant Water Body (O_2 Concentration in mg/L)[a]

Time,	Penetration Distance, z (cm)				
t (s)	0.001	0.01	0.1	1.0	10.0
300 (5 min)	0.069	0.70	6.1	9.17	9.17
3600 (1 hr)	0.021	0.21	2.1	9.12	9.17
36,000 (10 hr)	[b]	0.064	0.64	5.69	9.17
86,400 (1 day)	[b]	0.041	0.41	3.92	9.17
172,800 (2 days)	[b]	0.028	0.29	2.87	9.17

[a]$\mathcal{D}_{O_2, H_2O} = 1.80E{-5}$ cm^2/s at 20°C.
[b]Denotes less than 0.001 mg/L.

within the overlaying water. The entries in this table indicate the extreme slowness of the molecular diffusion process in water.

The gas transfer process in nature, for both the air and the water, is much faster than reflected in Tables 3.1-2 and 3.1-3. Molecular diffusion, although present, is dominated by a more rapid transport mechanism. Fluid mixing to some degree is always occurring in nature. Mixing near environmental interfaces speeds up the interphase mass transfer process. It is due in large part to this flow-induced mixing or turbulence that regions of the environment, such as rivers, can assimilate large quantities of some waste organic chemicals and retain some degree of diversity of biological life. (See Section 1.2.)

Eddy Diffusion

As a fluid moves parallel to a fixed surface, as, for example, air across the surface of the soil, it can be in laminar flow or turbulent flow. Only very slow-moving air is laminar, the connotation being that adjacent layers of fluid remain distinct and identifiable and do not intermix. As the wind speed increases, fluid layers become irregular and eddies develop. An eddy is thought of as an irregular but somehow identifiable material and wind structure, perhaps similar to a "puff of wind" or to a "cat's paw" over open water, having the ability to transfer air properties across the flow in a way that can conveniently be thought of as analogous to transfer by the air molecules on a much smaller scale.

THE EDDY DIFFUSION COEFFICIENT

Mass transfer in a turbulent stream is essentially a mixing process, whereby mass is transported by the mixing and blending of the eddies. For brevity, it is referred to as "eddy diffusion," though its similarity to molecular diffusion lies only in the fact that the molar flux in a binary mixture is proportional to the concentration gradient under many but not all conditions. It is usually very rapid, though near a phase boundary where the eddy motion is damped it may be unimportant as compared with the parallel process of diffusion by molecular motion.

The eddy diffusion coefficient $\mathcal{D}_{A2}^{(t)}$ is defined to follow the pattern of molecular diffusion as relating the molar flux of a species to the concentration gradient of the same species. Including the small contributions of molecular diffusion the total flux relationship is

$$J_{Ay} = -c_2\left(\mathcal{D}_{A2}^{(t)} + \mathcal{D}_{A2}^{(t)}\right)\frac{dx_A}{dy} \tag{3.1-6}$$

The difficulty in this approach to practical environmental problems lies

Table 3.1-4. Diffusion Coefficients in Various Aqueous Environments

Environment region	$\mathcal{D}_{A2}^{(l)}$ or $\mathcal{D}_{A2}^{(t)}$ (cm^2/s)
Dispersion: horizontal surface waters	100–1,000,000
Eddy diffusion: in pipes and flat ducts, normal to flow	0.1–100
Turbulent diffusion: vertical, thermocline, and deeper regions in lakes and oceans	0.01–1.0
Molecular diffusion: salts and gases in water	10^{-4}–10^{-5}
Molecular diffusion: ionic solutes in sediments and soils	10^{-6}–10^{-8}

Source. Reference 2.

not only in the complex dependence of $\mathcal{D}_{A2}^{(t)}$ on the properties of the turbulent flow field but also in the fact that the flux is not always proportional to the concentration gradient.

The magnitude of the eddy diffusion coefficient in the natural environment is usually many times larger than the molecular diffusivity. Table 3.1-4 contains typical ranges for eddy diffusivities observed in nature. A similarly defined eddy diffusion coefficient exists in the air (i.e., $\mathcal{D}_{A1}^{(t)}$).

Eddy diffusion is initiated and propagated by fluid turbulence. In a similar fashion to Eq. 3.1-6 it is possible to define an eddy viscosity. In the interest of understanding turbulent shear in flow near fixed boundaries Prandtl developed eddy movements into a quantitative model by employing *mixing lengths*. A mixing length is roughly related to eddy diameters. One outcome of the Prandtl mixing length model is that eddy viscosity increases linearly with distance from the fixed boundary. Details of the Prandtl mixing length model are presented later in this chapter.

Eddy diffusion coefficients are impossible to use in many cases where diffusion-type chemical movements are of interest. This is particularly true near phase boundaries. $\mathcal{D}_{Aj}^{(t)}$, $j=1,2$, are not constant and concentration gradients are difficult to obtain and to estimate. This approach is abandoned for many complex situations near environmental interfaces. Faced with these difficulties a simpler flux equation has been created similar to Newton's law of heat conduction.

BINARY MASS TRANSFER COEFFICIENTS IN ONE PHASE

A more general form of Fick's first law in terms of N_A, the molar flux relative to stationary coordinates, is

$$N_{Az} = x_A(N_{Az} + N_{Bz}) - c_2\big(\mathcal{D}_{A2}^{(l)} + \mathcal{D}_{A2}^{(t)}\big)\frac{\partial x_A}{\partial z} \qquad (3.1-7)$$

where A is the chemical of concern and B is water. This equation shows that the diffusion flux N_{Az} relative to stationary coordinates is the resultant of two vector quantities: the vector $x_A(N_{Az} + N_{Bz})$, which is the molar flux of A resulting from the bulk motion of the fluid, and the vector

$$J_{Az} \equiv -c_2\left(\mathcal{D}_{A2}^{(l)} + \mathcal{D}_{A2}^{(t)}\right)\frac{\partial x_A}{\partial z}$$

which is the molar flux of A resulting from the diffusion superimposed on the bulk flow, both in the z direction.

Based on a previous argument concerning the difficulty of obtaining the relative contributions of $\mathcal{D}_{A2}^{(l)}$ and $\mathcal{D}_{A2}^{(t)}$ near the interface, it has proved to be convenient to replace the diffusion term with the product of a mass transfer coefficient $k_{A2}^{\bullet}$ and some characteristic composition difference Δx_A between the fluid at the interface and the fluid far removed from the interface:

$$\dot{N}_{A0} = x_{A0}(N_{A0} + N_{B0}) + k_{A2}^{\bullet}\Delta x_A \tag{3.1-8}$$

Note that N_{A0} and N_{B0} are the components of the fluxes at the interface (i.e., $z = 0$) measured into the phase of interest. The black dot ($\bullet$) on $k_{A2}^{\bullet}$ serves as a reminder that the mass transfer coefficient itself depends on the mass transfer rate.[3] Equation 3.1-8 constitutes a definition of any liquid phase mass transfer for component A in a binary system. The analogous expression for any gas phase is

$$\dot{N}_{A0} = y_{A0}(N_{A0} + N_{B0}) + k_{A1}^{\bullet}\Delta y_A \tag{3.1-9}$$

The dimensions of the molar mass transfer coefficients are $\text{mol}/L^2 \cdot t$. Typical units are $\text{mol}/\text{cm}^2 \cdot \text{s}$.

For most environmental applications Eqs. 3.1-8 and 3.1-9 may be simplified without any loss of definiteness. When applied to dilute solutions (e.g., y_{A0} and $x_{A0} \leqslant 0.05$) and low mass transfer rates, which is usually the case in environmental chemistry, the bulk flow term is usually small compared to the diffusion term and the coefficient is independent of the mass transfer rate. The molar flux expression commonly employed under the preceding assumption takes the form

$$\dot{N}_{A0} = \dot{k}_{A1}\Delta y_A \tag{3.1-10a}$$

for the gas phase and

$$\dot{N}_{A0} = \dot{k}_{A2}\Delta x_A \tag{3.1-10b}$$

for the liquid phase. The dot ($\cdot$) shown in $\dot{N}_A$ and $\dot{k}_A$ denotes a local value of the coefficients. For surfaces of finite area A, one can parallel Eq. 3.1-10 and define the molar flux over the entire surface:

$$W_{A0} = k_{A1} A \, \Delta y_A \qquad (3.1\text{-}11a)$$

and

$$W_{A0} = k_{A2} A \, \Delta x_A \qquad (3.1\text{-}11b)$$

where W_{A0} has dimensions of mol/t and A is the surface area (L^2) through which chemical A is moving. The mass transfer coefficient without the overhead dot denotes an average value for the entire surface area A. This completes the general definition of individual phase mass transfer coefficients.

With respect to the movement of chemicals across the major interfaces of the geospheres there are four individual mass transfer coefficients. These coefficients and their approximate interface locations are shown in Fig. 3-1. Note that coefficients are only associated with the fluid sides. There are two gas phase and two liquid phase coefficients as noted by the subscripts 1 and 2. The superscript denotes the coexisting phase of the interface.

Equations like 3.1-11 have proved to be extremely utilitarian for computing flux rates across phase boundaries. Conventional techniques of qualitative and quantitative chemistry along with phase equilibrium principles (Chapter 2) allow x_A's, y_A's, and hence Δx_A and Δy_A to be determined. Many experimental observations, both in the laboratory and in the field, along with much theoretical work have resulted in a host of semiempirical equations from which fairly accurate values of the individual mass transfer coefficients can be obtained. Specific correlations and applications are presented in later chapters. The interface or surface area A is usually the plane area separating the phases. For cases in which the area is difficult to define, such as a swarm of gas bubbles or an irregular geometric form, a modification of Eq. 3.1-11 is used:

$$W_{A0} = k_{A1} a_v V \, \Delta y_A \qquad (3.1\text{-}12a)$$

and

$$W_{A0} = k_{A2} a_v V \, \Delta x_A \qquad (3.1\text{-}12b)$$

Here the coefficient and the area per unit volume product, $k_{A1} a_v$, is retained as a single term. The term a_v has dimensions of (L^2/L^3) or L^{-1} and is the interfacial area per unit volume of the mass transfer unit.

Equations 3.1-11 and 3.1-12 do not contain a concentration gradient so a mathematical analysis cannot yield spatial concentrations of chemicals in the various fluid phases. The mass transfer coefficients are usually associated with the region of the phase near the interface where concentration gradients are presumed to be greatest.

Although the mass transfer coefficients defined by Eq.3.1-8 and 3.1-9 using a mole fraction driving force are the most correct for all applications, other coefficients have been defined and are used extensively. In most environmental applications of interest in this book, temperature and pressure variations are small so that a concentration driving force may be used. Equation 3.1-10a can be written as

$$\dot{N}_{A0} = \dot{k}'_{A1} \Delta c_{A1} \tag{3.1-10a}$$

where $\dot{k}'_{A1}$ has dimensions of L/t and is related to $\dot{k}_{A1}$ by $\dot{k}_{A1} = \dot{k}'_{A1} c_1$, since $\Delta c_{A1} = c_1 \Delta y_A$. In a similar fashion, Eq. 3.1-10b can be written

$$\dot{N}_{A0} = \dot{k}'_{A2} \Delta c_{A2} \tag{3.1-10b}$$

where $\dot{k}'_{A2}$ has dimensions of L/t and is related to $\dot{k}_{A2}$ by $\dot{k}_{A2} = \dot{k}'_{A2} c_2$, since $\Delta c_{A2} = c_2 \Delta x_A$.

As noted earlier, the coefficients $\dot{k}'_{A1}$ and $\dot{k}'_{A2}$ have dimensions of velocity. This has led to these coefficients being called *piston velocities*. Apparently the interpretation is that a piston of this velocity moves through the fluid and sweeps all the molecules ahead of it.

There can be confusion as to which coefficient to use in a particular application. Obviously the driving force, whether it is mole fraction, molar concentration, mass fraction, or mass concentration, dictates which form of the mass transfer coefficient to use. Confusion is alleviated if one checks the dimensions of the mass transfer coefficient and driving force product to assure that the units of mass or molar flux result. Molar flux has dimensions of $mol/L^2 \cdot t$, and mass flux has dimensions of $m/L^2 \cdot t$.

Mass Transfer Theories

The mass transfer coefficients introduced in Eqs. 3.1-8 through 3.1-12 are nothing more than a proportionality constant at this point. The coefficients are termed *conductances* and the reciprocals, $1/k_{A2}$, are conventionally termed *resistances*. These coefficients are important quantities for calculating the rate of chemical transfers at environmental interfaces and require a degree of qualitative and quantitative interpretation.

It is important to understand what natural forces and occurrences within the environment influence the magnitude of these coefficients to the same degree that chemical potential and the displacement from equilibrium aided the understanding of the driving force Δx_A.

Flow happenings near environmental interfaces are of paramount importance but they are difficult to describe mathematically because of the complexities of turbulent flow; however, several conceptual models have been developed to aid our understanding. Although simple in concept these models have been highly successful in helping us gain insight into natural mechanisms. They are all speculations and are continually being revised as more observational data become available. The most relevant facts pertaining to mass transfer into or away from a turbulent stream are that the resistance is confined largely to a thin region adjacent to the interface and that the concentration gradients are steep near the interface.

THE STAGNANT-FILM MODEL

The film model is over 70 years old, having been proposed by Nernst[4] in 1904 and applied to gas absorption by Whitman[5] in 1923. When a fluid flows in a turbulent manner past a solid surface, the velocity is zero at the surface; there must be a viscous layer or film in the fluid very near the surface. Beyond this film the flow is turbulent so that the rate of mass transfer is limited by the rate of molecular diffusion through the viscous layer.

The film idea can also be applied to both fluid surfaces near a gas–liquid or a liquid–liquid interface. There is assumed to be a stagnant film in each phase. The rate equation for the film model at low mass transfer rates across a liquid film is

$$N_{A0} = \left(\frac{\mathcal{D}_{A2}^{(l)} c_2}{\delta_{A2}} \right)(x_{A0} - x_A) \tag{3.1-13}$$

where c_2 is the average molar density of the liquid mixture (mol/cm^3) and δ_{A2} is the effective film thickness (cm). The mass transfer coefficient k_{A2} is equal to the group $\mathcal{D}_{A2}^{(l)} c_2 / \delta_{A2}$ (mol/s · cm^2).

A prediction of this theory is that mass transfer coefficients for different solutes being transfered under the same fluid flow conditions are directly proportional to the molecular diffusivities $\mathcal{D}_{A2}^{(l)}$ of the solutes. Various experimental observations have shown that the mass transfer rate is proportional to $\mathcal{D}_{A2}^n$, $0.1 \lesssim n \lesssim 0.9$. It was recognized very early that the film concept was a gross oversimplification of the actual conditions near a phase boundary. Although it has been useful in several applications, the film theory has been largely discredited, except as a limiting case.

One useful result of the film model has been the development of the Sherwood number Sh. A dimensionless ratio can be created by forming a quotient of the Eq. 3.1-10b and 3.1-13 to yield the defined Sherwood number

$$\text{Sh}_{A2} \equiv \frac{\dot{k}_{A2}L}{\mathcal{D}^{(l)}_{A2}c_2} \qquad (3.1\text{-}14)$$

where L is an appropriate length that reflects the geometry of the system. The term L replaces the fictitious film thickness δ_{A2}. The Sherwood number has been found to be a convenient dimensionless group useful in correlating experimental data. The Sherwood number may be expected to depend on the Reynolds number $\text{Re} \equiv Lv/v_2$ and the Schmidt number $\text{Sc} \equiv v_2/\mathcal{D}^{(l)}_{A2}$ as

$$\text{Sh} = f(\text{Re}, \text{Sc}, \text{and geometry}) \qquad (3.1\text{-}15)$$

v is average fluid velocity (cm/s) and v_2 is the kinematic viscosity of the fluid (cm^2/s). Numerous correlations for different geometric shapes and flow conditions have employed this form.

THE PENETRATION THEORY

Higbie[6] developed a basic idea about the mechanism of mass transfer near fluid–fluid interfaces in 1935 that established what is known as the *penetration theory*. Turbulent flow produces eddies that may be visualized as parcels of fluid consisting of an enormous number of molecules that exists together for a while as a well-defined entity and move about continuously, then eventually "dissolve" into the surrounding fluid. The parcels are continuously coming into being and dissolving throughout the entire fluid body. Only molecular movement occurs within the parcels. As a parcel of uniform concentration x_A contacts an interface, one side is in direct contact with the other phase for a short period of time $\bar{t}$ (i.e., the average exposure time). The equilibrium concentration at the interface is x_{A0}. During this exposure time molecular diffusion results in the penetration of chemical A into (or out of) the parcel. The depth of penetration is small compared to the dimensions of the parcel.

Fick's second law (i.e., Eq. 3.1-3) describes the mathematics of the process. With the semi-infinite solid boundary conditions, Eq. 3.1-4 is the solution that gives the concentration of A as a function of penetration depth z and exposure time $\bar{t}$. The average flux rate over the exposure time is

$$N_{A0} = 2\sqrt{\frac{\mathcal{D}^{(l)}_{A2}}{\pi \bar{t}}}\ (c_{A20} - c_{A2}) \qquad (3.1\text{-}16a)$$

or

$$N_{A0} = 2c_2 \sqrt{\frac{\mathcal{D}_{A2}^{(l)}}{\pi \bar{t}}} \ (x_{A0} - x_A) \qquad (3.1\text{-}16\text{b})$$

This theory predicts that the mass transfer coefficient k_{A2} is equal to $2c_2\sqrt{\mathcal{D}_{A2}^{(l)}/\pi \bar{t}}$. The k_{A2} should vary as the square root of the molecular diffusivity, whereas the film model indicates the first power. The square root is nearer the truth in many instances. In practice the average exposure time $\bar{t}$ is seldom known, so that the model cannot be used to predict mass transfer rates except in special cases. The same difficulty is encountered in the film theory, which involves the unknown film thickness δ_{A2}.

THE SURFACE-RENEWAL THEORY

The assumption that all parcels had the same exposure time was revised by Danckwerts[7] in 1951. He assumed that parcels can remain in contact with the surface for variable times that may be anything from zero to infinity. He created the fractional renewal rate, s (s^{-1}), and used it in the surface-age distribution function $\phi = s\exp(-st)$. Here ϕ represents the probability that a parcel will be exposed the time t before being replaced by fresh mixed fluid from the bulk. The mean steady-state flux normal to the phase boundary was shown to be

$$N_{A0} = \sqrt{\mathcal{D}_{A2}^{(l)} s} \ (x_{A0} - x_A) \qquad (3.1\text{-}17)$$

Since values of s are not generally available, its appearance in the model presents the same problems as δ_{A2} and $\bar{t}$ of the film and penetration models.

Extensions and unifications of the preceding three models have resulted in qualitative understanding and quantitative expressions capable of explaining many experimental observations. Reviews of these further developments can be found in textbooks by Treybal[8] and Sherwood et al.[9]

The preceding models help explain some of the occurrences at natural fluid–fluid and fluid–solid interfaces that affect the magnitude of the individual phase coefficients. Although the expressions are given for the liquid side of the interface similar expressions apply for cases in which the fluid is a gas.

On the Generality of Mass Transfer Coefficients

Once an experimental value of a mass transfer coefficient has been obtained for a certain chemical in a particular phase it may be used for other chemical species in the same phase by employing the proper transformation techniques based on the best available mass transfer theories. Individual phase mass transfer coefficients are scalars and are therefore independent of the direction of the concentration driving force. This means that a coefficient measured under absorption conditions can be transformed and used in a desorption application. (See, for example, Problem 4.1C.) The equation most commonly used to transform coefficients given for one chemical to another chemical is

$$k_{B2} = k_{A2} \left(\frac{\mathcal{D}_{B2}^{(l)}}{\mathcal{D}_{A2}^{(l)}} \right)^n \qquad (3.1\text{-}18)$$

where $n = 1$ for the film theory and $n = \frac{1}{2}$ for the penetration and surface renewal theory. Boundary layer theory suggests $n = \frac{2}{3}$ and this is possibly the best estimate to use in most cases. Care must be taken that the geometry and boundary conditions are identical when using published coefficients and correlations. In some cases the mass transfer coefficient is reported in the form of dimensionless number correlations of the form of Eq. 3.1-15. In this case the diffusivity transformation is inherent in the correlating equation and Eq. 3.1-18 is not needed.

It should also be noted that semi-empirical correlations of the general form of Eq. 3.1-15 are phase independent. For example, if experimental measurements are made on a liquid phase coefficient above a flat plate and correlated correctly with appropriate dimensionless groups, then it is logical to assume that one can obtain reasonable estimates of gas phase coefficients from this same correlation. Specifically the correlation for tangential flow along a sharp-edged semi-infinite flat plate with mass transfer into the stream is the well-known equation[10]

$$\text{Sh} = 0.664 \, \text{Re}^{1/2} \text{Sc}^{1/3}, \qquad \text{Re} \leqslant \text{E5} \qquad (3.1\text{-}19)$$

for laminar flow. (See Figure 3.1-1.) The Sherwood number in this case is defined as

$$\text{Sh} \equiv \frac{{}^{3}k_{A1}}{v_{1\infty}c_1} \equiv \frac{{}^{3}k_{A2}}{v_{2\infty}c_2} \qquad (3.1\text{-}20)$$

where v_{∞} is the undisturbed velocity well removed from the plate. The Reynolds number is defined as $\text{Re} \equiv v_1 L / \nu \equiv v_2 L / \nu$, where L is the length of the plate. Using the appropriate velocity, molar density, and other fluid properties, as specified for the phase, we obtain the appropriate individual

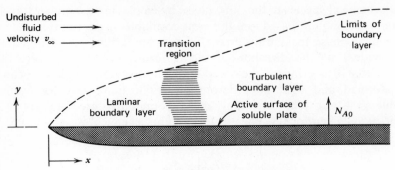

Figure 3.1-1. Boundary layer on a flat plate.

phase coefficient. If turbulent flow[11] exists, the correlation is

$$\text{Sh} = 0.036 \, \text{Re}^{0.8} \text{Sc}^{1/3}, \qquad \text{Re} > 5\text{E}\,5 \tag{3.1-21}$$

and the Sherwood number in this case is defined as

$$\text{Sh} \equiv \frac{k_{A1}L}{\mathcal{D}^{(l)}_{A1}c_1} \equiv \frac{k_{A2}L}{\mathcal{D}^{(l)}_{A2}c_2} \tag{3.1-22}$$

The Reynolds and Schmidt numbers are defined as earlier. These correlations are some of the results of boundary layer theory.

MOLECULAR WEIGHT AND TEMPERATURE CORRECTIONS

According to Thomas Graham (British: 1805–1869), at constant temperature the relative rates at which gases diffuse is inversely proportional to the square roots of their respective densities or molecular weights. Mathematically stated, Graham's law in terms of diffusivities is

$$\left(\frac{\mathcal{D}_{A1}}{\mathcal{D}_{B1}} \right) = \sqrt{\frac{M_B}{M_A}} \tag{3.1-23}$$

where M_A and M_B are the molecular weights of species A and B, respectively.

For accurate calculations of the diffusivity in gases at low density, the Chapman-Enskog kinetic theory should be used. The Chapman-Enskog formula for the gaseous state at low density employing the ideal gas law becomes

$$\mathcal{D}_{A1} = 0.0018583 \frac{\sqrt{T^3((1/M_A)+(1/M_1))}}{p\sigma_{A1}{}^2 \Omega_{D,A_1}} \qquad (3.1\text{-}24)$$

in which $\mathcal{D}_{A1}[=] \text{ cm}^2/s$, $T[=]K$, $p[=]$ atm, $\sigma_{A1}[=]$ angstrom units, and $\Omega_{D,A1}$ is a dimensionless function of the temperature and of the intermolecular potential field for one molecule of A and one of B. Consult mass transfer textbooks[3,9,10] for techniques of estimating σ_{A1} and $\Omega_{D,A1}$. Equation 3.1-24 is useful for calculating gas diffusivities and has been written for the case of air as the solvent (i.e., air noted by subscript 1). Consult Appendix C for chemical diffusivities in air. In light of Eq. 3.1-24 Graham's law is a reasonable approximation.

For liquids the Stokes-Einstein equation has been shown to be fairly good for describing the diffusion. The simple hydrodynamic approach gives expressions for the diffusion coefficient for spherical molecules in dilute solutions and also for the coefficient of self-diffusion. Wilke and Chang[12] have developed an approximate analytical relation based on the Stokes-Einstein equation that gives the diffusion coefficient in cm^2/s for small concentrations of A in water (H_2O liquid denoted by subscript 2):

$$\mathcal{D}_{A2} = 7.4\text{E} - 8 \frac{(\Psi_2 M_2)^{1/2} T}{\mu_2 \tilde{V}_A{}^{0.6}} \qquad (3.1\text{-}25)$$

Here $\tilde{V}_A$ is the molar volume of the solute A in cm^3/mol as liquid at its normal boiling point, μ_2 is the viscosity of the solution in centipoises, Ψ_2 is the "association parameter" for water, and T is the absolute temperature in K. The recommended value of Ψ_2 for water is 2.6. This equation is good only for dilute solutions of nondissociating solutes; for solutions it is usually good within $\pm 10\%$.

Temperature corrections of diffusivities for gases and liquids can be inferred from Eqs. 3.1-24 and 3.1-25, respectively. The $\frac{3}{2}$ power of the absolute temperature is used for gases. For liquids, since viscosity is strongly temperature dependent, the following proportionality should be used:

$$\mathcal{D}_{A2}@T_2 = \mathcal{D}_{A2}@T_1 \left(\frac{T_2 \mu_1}{T_1 \mu_2} \right) \qquad (3.1\text{-}26)$$

where the subscripts 1 and 2 on the viscosity refer to temperatures T_1 and

T_2. The molecular weight correction for liquid diffusivities can be approximated by

$$\left(\frac{\mathcal{D}_{A2}}{\mathcal{D}_{B2}}\right) = \left(\frac{M_B}{M_A}\right)^{1/2} \tag{3.1-27}$$

according to Eq. 3.1-25. All the preceding relations apply to molecular diffusivities only and not turbulent diffusion coefficients or individual mass transfer coefficients.

Binary Mass Transfer Coefficients in Two Phases and the Two-Resistance Theory of Interphase Mass Transfer

In many environmentally important mass transfer systems there is an interface with concentration gradients on both sides. Consider, as an example, the contacting of liquid water, which contains species A, with air into which A is desorbing at constant temperature. The general shape of the concentration gradients in the two phases are sketched in Fig. 3.1-2. Concentrations appear as mole fractions and, as is typical of most environmental mass transfer applications dilute solutions of A, exist in both phases. In this development any resistance to mass transfer at the interface is assumed to be negligible so that the major resistances reside in the gas phase and in the liquid phase. If the interface resistance is negligible and if equilibrium is assumed at the interface, then $y_{Ai}^* = f(x_{Ai})$, where the functional form of the relationship is derived solely on the thermodynamics of

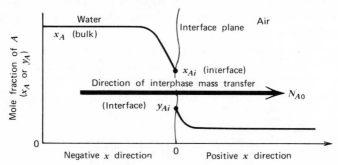

Figure 3.1-2. Interphase mass transfer.

gas–liquid equilibriums as presented in Chapter 2. An analysis similar to that given earlier holds for other fluid–fluid interfaces.

The molar flux of species A across each phase can be expressed in terms of the individual mass transfer coefficients and the concentration difference in each phase. For slow mass transfer rates in a dilute solution Eqs. 3.1-11a and 3.1-11b apply:

$$\left.\frac{dW_{A0}}{dA}\right|_1 = {}^2k_{A1}(y_{Ai} - y_A) \tag{3.1-28}$$

and

$$\left.\frac{dW_{A0}}{dA}\right|_2 = {}^1k_{A2}(x_{Ai} - x_A) \tag{3.1-29}$$

The local average flux rates may be defined as $(dW_{A0}/dA)_1 \equiv N_{A01}$ so that average individual coefficients may be used. Note that since the mass transfer process is at steady state $N_{A01} = -N_{A02}$. The rate at which A enters the gas phase is equal in magnitude but opposite in sign to the rate at which A enters the liquid phase.

For convenience one may define overall mass transfer coefficients based on an overall concentration difference:

$$N_{A01} \equiv {}^2K_{A1}(y_A^* - y_A) \tag{3.1-30}$$

and

$$N_{A02} \equiv {}^1K_{A2}(x_A^* - x_A) \tag{3.1-31}$$

where ${}^2K_{A1}$ is the overall gas phase mass transfer coefficient for species A located at the liquid interface, ${}^1K_{A2}$ is the overall liquid phase mass transfer coefficient for species A located at the gas interface, y_A^* is a nonexistent gas phase mole fraction concentration in equilibrium with x_A, and x_A^* is a nonexistent liquid phase mole fraction concentration in equilibrium with y_A. In application either Eq. 3.1-30 or 3.1-31 may be used; the choice is left to the student.

The relationship between the overall coefficients and the individual phase coefficients are easily derived. This derivation is left as an exercise for the student. (See Problem 3.1B.) The relationships are

$$\frac{1}{{}^2K_{A1}} = \frac{H_{XA}}{{}^1k_{A2}} + \frac{1}{{}^2k_{A1}} \tag{3.1-32}$$

and

$$\frac{1}{{}^1K_{A2}} = \frac{1}{{}^1k_{A2}} + \frac{1}{{}^2k_{A1}H_{XA}} \tag{3.1-33}$$

where H_{AX} is the slope of the equilibrium line in the neighborhood of (x_{Ai}, y_{Ai}). For dilute solution H_{XA} is the Henry's law constant in mole fraction form (i.e., $y_A^* = H_{XA}x_A$). If concentration driving forces such as those of Eqs. 3.1-10a and b are used instead of mole fraction driving

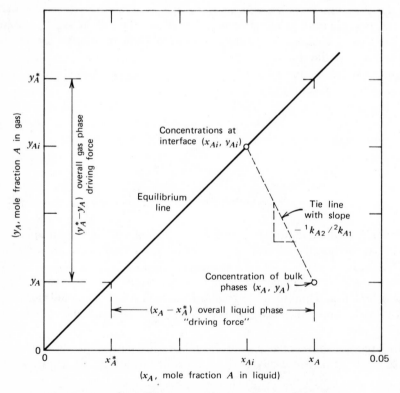

Figure 3.1-3. Graphical interpretation of the two-resistance theory.

forces, then Eqs. 3.1-32 and 3.1-33 should not be used. (See Problem 3.1I for the correct relationship in this case.)

The preceding is the essence of the two-resistance theory for interphase mass transfer. The overall resistance for mass transfer across a fluid–fluid interface is the sum of the resistance in each phase as given by Eqs. 3.1-32 or 3.1-33. Figure 3.1-3 contains a summary of the theory in the form of a graphical interpretation. The equilibrium line is shown and the point set representing the bulk phase concentrations (x_A, y_A) indicates that a potential for mass transfer exists. The interface concentration point set (x_{Ai}, y_{Ai}) is obtained from a tieline, the equation for which can be obtained from the ratio of Eqs. 3.1-28 and 3.1-29.

CONTROLLING RESISTANCE FOR INTERPHASE MASS TRANSFER

The reciprocal of the mass transfer coefficient is conventionally termed a resistance. Equation 3.1-32 can be interpreted as the overall resistance $(1/^2K_{A1})$, and it is equal to the sum of the resistance in the liquid phase $(H_{XA}/^1k_{A2})$ and the gas phase $(1/^2k_{A1})$. It frequently occurs that one phase or the other dominates the overall resistance and therefore controls the rate of interphase mass transfer. A variation in individual coefficients $^2k_{A1}$ and $^1k_{A2}$ can cause this; however, the numerical values of H_{XA}, the Henry's law constant, cover a wide range, and it is this term that usually determines which phase resistance controls the rate of mass transfer. For example, if H_{XA} is large (i.e., $H_{XA} \gg 1$) the liquid phase resistance dominates. If H_{XA} is small $(H_{XA} \ll 1)$ the gas phase resistance dominates. Normally insoluble gases such as nitrogen, hydrogen sulfide, and oxygen are liquid phase controlling, whereas soluble chemicals such as phenol and propionic acid are gas phase controlling. Some chemicals such as methanol encounter significant resistances in each phase. (See Table 2.1-5 for the range of values H_{XA} can have.)

Problems

3.1A. DERIVATION OF THE FILM THEORY RESULT

Show that Eq. 3.1-7 may be transformed to Eq. 3.1-13 for the region near and interface between phases where a stagnant film of thickness δ_{A2} is assumed to exist. (*Hint*: Assume N_A in Eq. 3.1-7 is constant through the film and see Fig. 3.1A. Note there is a similar development for the gas film.)

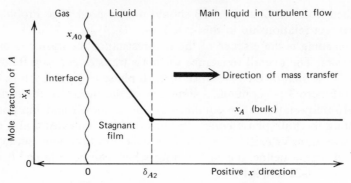

Figure 3.1A. Diffusion through a stagnant film.

3.1B. COMPLETION OF THE TWO-RESISTANCE THEORY DERIVATION

Show that Eqs. 3.1-28, 3.1-29, and 3.1-31 can be combined to yield Eq. 3.1-33. *Hint*: Use $N_{A01} = -N_{A02}$ and start with the tautology

$$x_A - x_A^* = (x_A - x_{Ai}) + (x_{Ai} - x_A^*) \tag{3.1B}$$

3.1C. EFFECTIVE FILM THICKNESS, PENETRATION TIME, AND SURFACE RENEWAL RATE

The mass transfer coefficient for the dissolution of furfural into water was found to be 0.44 lb mol/hr·ft². Calculate the following:

1. The effective thickness of the stagnant film (cm),
2. The average penetration time (s), and
3. The surface renewal rate (s⁻¹). Use 1.31E−5 cm²/s for the diffusivity of furfural in water.

3.1D. ESTIMATING LIQUID PHASE DIFFUSIVITY

Calculate the diffusivity of furfural in water at 25°C (cm²/s).

3.1E. ESTIMATING THE INDIVIDUAL LIQUID-PHASE MASS TRANSFER COEFFICIENT

Calculate the furfural liquid phase mass transfer coefficient for a flat plate geometry. Assume Re = 1E 12 and $L = 100$ cm. Report the answer in mol/cm²s at 25°C.

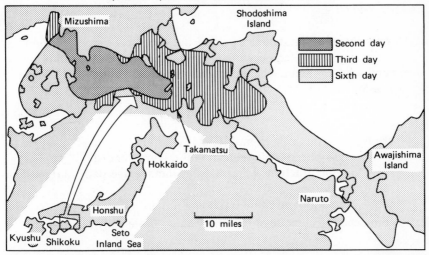

Spilled oil spread some 70 miles within a week

Figure 3.1F. Spread of oil slick. (Reprinted with permission of copyright owner, The American Chemical Society, from the June 2, 1975 issue of *Chemical and Engineering News*.)

3.1F. DISPERSION COEFFICIENT FOR HEAVY FUEL OIL ON SEA SURFACE

Failure of a storage tank at a harborside refinery resulted in the release of 270,000 bbl of heavy fuel oil into Mizushima harbor and a substantial part of Japan's Seto Inland Sea. Figure 3.1F shows the progressive spread of the oil slick. Compute the dispersion coefficient for the oil on horizontal surface waters (cm^2/s) and compare to the similar dispersion coefficient in Table 3.1-4. (*Hint*: Use Eq. 3.1-1.)

3.1G. MOLECULAR DIFFUSION OF BENZENE IN WATER FROM SEA-SURFACE SLICK

Benzene was spilled on water to form a sea-surface slick. Assume that both the slick and the water underneath are perfectly still and that molecular diffusion is the only operative mass transfer mechanism. Calculate the benzene concentration in the water at each centimeter down to 10 cm underneath the slick after 24 hours diffusion time. Assume the water contained a background concentration of 10 g/m^3 benzene prior to the spill and is at 25°C.

$$\mathcal{D}_{Bz, H_2O} = 1.02E + 5 \quad cm^2/s \text{ at } 293°K$$

3.1H. APPROPRIATENESS OF USING AVERAGE TEMPERATURE*

1. Discuss the appropriateness of using the vapor pressure corresponding to the average daily temperature for computing the volatilization rate of a pure solid or liquid chemical into air. (See Figure 3.1H) The rate expression is

$$N_A = {}^3k_{A1}(y_A^* - y_A) \qquad (3.1H)$$

where N_A = the flux rate in mol A/cm²·d
  ${}^3k_{A1}$ = gas phase coefficient in mol mixture/cm²·d
  y_A^* = mole fraction A in air from the average daily surface temperature
  y_A = background mole fraction A in air

2. Recommend a technique of obtaining an accurate daily flux rate.

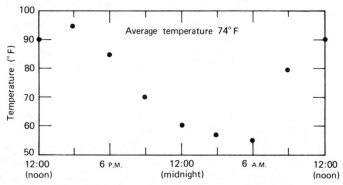

Figure 3.1H. Typical summer day temperature cycle.

3.1I. TWO-RESISTANCE LAW WITH CONCENTRATION DRIVING FORCES

If the phase concentrations are in mole fraction form, then Eqs. 3.1-32 and 3.1-33 are valid for computing the overall mass transfer coefficient. If phase concentrations are in units of moles per volume (i.e., c_{A1} and c_{A2}), show that the correct forms of the two-resistance law are

$$\frac{1}{{}^2K_{A1}'} = \frac{1}{{}^2k_{A1}'} + \frac{H_{Ax}c_1}{c_2\,{}^1k_{A2}'} \qquad (3.1I\text{-a})$$

*Refer to Problem 2.1H before attempting to work this problem.

and

$$\frac{1}{{}^1K'_{A2}} = \frac{1}{{}^1k'_{A2}} + \frac{c_2}{H_{Ax}c_1{}^2k'_{A1}} \qquad (3.1\text{I-b})$$

3.2. TURBULENCE IN THE ENVIRONMENT

In the preceding section the transport processes of molecular diffusion were introduced. The slowness of this process of chemical movement was demonstrated by examples of the movement of CO_2 into stagnant air above the air–water interface and the depletion of oxygen from a stagnant water layer above the mud–water interface. Interphase chemodynamics are much more rapid than the molecular diffusion process. The turbulent (or eddy) diffusion process was briefly mentioned in Section 3.1. The idea that mass is transferred in turbulent flow by a mixing or blending of eddies was presented. The body of knowledge on turbulence and on turbulent flow is incomplete, and this severely hampers the use and procurement of eddy diffusion coefficients. The eddy diffusion approach was abandoned, and mass transfer coefficients were created. By this creation the unknown eddy diffusivity and unknown concentration gradient were replaced by a single constant and known concentration difference. Although highly utilitarian, the mass transfer coefficient lumps together and hides much of our ignorance of the turbulent transport processes. Section 3.1 ended with a presentation of models that relate the mass transfer coefficient to molecular diffusivity.

Section 3.2 readdresses turbulence and turbulent processes in the environment. Fundamental turbulent flow concepts are presented and simple models are developed. The models are then extended to the derivation of useful analytical results. These results include reasonably accurate mathematical relationships for describing flow and turbulent transport of chemicals in the regions of environmental interfaces. Only mechanical turbulence is discussed in this section. Thermal turbulence is discussed in Section 6.1.

The topic of turbulence and turbulent flow in the environment cannot begin without introducing the idea of laminar flow, if for no other reason than to create a contrast. This section is concerned with the flow of fluids in general, and it is concerned specifically with the flow of air and water in the regions near solid surfaces. General equations embodying the laws of conservation of mass (the continuity equations) and Newton's second law of motion (the Navier-Stokes equation) are presented to provide a unified introduction to laminar flow and later turbulent flow. The equation of energy and the multicomponent continuity equation for dilute solutions

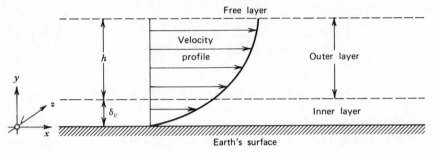

Figure 3.2-1. The atmospheric boundary layer.

are introduced also, to complete the presentation of the equations of change. Turbulent flow ideas and simple models are then extended to the development of useful results. The results include reasonably accurate mathematical relationships plus a sound qualitative understanding of some fundamentals of turbulence.

In general, flow of air or water in the region near the earth's surface can be divided into two layers often referred to as *boundary layers*. Looking at Fig. 3.2-1, the atmospheric boundary layer is generally divided into two distinct layers: an inner layer of height δ_v of the order of a few centimeters in which the velocities relative to the earth's surface are close to zero, and the viscous stresses dominate. The flow in this layer is laminar. In the outer layer of height h, of the order of a kilometer, Reynolds stresses (a turbulent flow stress to be defined later) dominate. Because of the relative importances of the various forces, the behavior of these two layers is different.

The movement of water near a solid surface results in a profile similar to that shown in Fig. 3.2-1. In the case of the benthic boundary layer, the three bottommost hydraulic layers are termed from bottom upward: *viscous layer*, *logarithmic layer*, and *outer layer*.

Presentation of the Equations of Change

In this section the *equations of change* are presented and not derived. Procedures of derivation from first principles may be found in any one of several textbooks on transport phenomena. These equations provide a starting point for both qualitative discussions and quantitative formulations concerning interface regions.

THE EQUATION OF CONTINUITY

The equation of continuity is developed by writing a total mass (i.e., all chemical species) balance over a stationary volume element $\Delta x \, \Delta y \, \Delta z$

through which a fluid is flowing (see Fig. 1.1-5), and is

$$\frac{\partial \rho}{\partial t} = -\left[\frac{\partial}{\partial x}(\rho v_x) + \frac{\partial}{\partial y}(\rho v_y) + \frac{\partial}{\partial z}(\rho v_z) \right] \tag{3.2-1}$$

This is the *equation of continuity*, which describes the rate of change of density at a fixed point resulting from the changes in the mass fluxes at the point. It is frequently desirable to modify Eq. 3.2-1 by performing the indicated operations and collecting all derivatives of ρ on the left side:

$$\frac{\partial \rho}{\partial t} + v_x \frac{\partial \rho}{\partial x} + v_y \frac{\partial \rho}{\partial y} + v_z \frac{\partial \rho}{\partial z} = -\rho \left(\frac{\partial v_x}{\partial x} + \frac{\partial v_y}{\partial y} + \frac{\partial v_z}{\partial z} \right) \tag{3.2-2}$$

Remember that this equation is simply a statement of the conservation of total mass.

THE EQUATION OF MOTION

The *equation of motion* is the result of a momentum balance for the volume element $\Delta x \, \Delta y \, \Delta z$, such as that used in the previous section. Momentum and forces acting on the volume element are summed for each space dimension yielding three equations. The x component of the equation of motion is

$$\frac{\partial}{\partial t}(\rho v_x) = -\left(\frac{\partial}{\partial x}(\rho v_x v_x) + \frac{\partial}{\partial y}(\rho v_y v_x) + \frac{\partial}{\partial z}(\rho v_z v_x) \right)$$

$$-\left(\frac{\partial}{\partial x}\tau_{xx} + \frac{\partial}{\partial y}\tau_{yx} + \frac{\partial}{\partial z}\tau_{zx} \right) - \frac{\partial p}{\partial x} + \rho g_x \tag{3.2-3a}$$

The y and z components are similar:

$$\frac{\partial}{\partial t}(\rho v_y) = -\left(\frac{\partial}{\partial x}(\rho v_x v_y) + \frac{\partial}{\partial y}(\rho v_y v_y) + \frac{\partial}{\partial z}(\rho v_z v_y) \right)$$

$$-\left(\frac{\partial}{\partial x}\tau_{xy} + \frac{\partial}{\partial y}\tau_{yy} + \frac{\partial}{\partial z}\tau_{zy} \right) - \frac{\partial p}{\partial y} + \rho g_y \tag{3.2-3b}$$

$$\frac{\partial}{\partial t}(\rho v_z) = -\left(\frac{\partial}{\partial x}(\rho v_x v_z) + \frac{\partial}{\partial y}(\rho v_y v_z) + \frac{\partial}{\partial z}(\rho v_z v_z) \right)$$

$$-\left(\frac{\partial}{\partial x}\tau_{xz} + \frac{\partial}{\partial y}\tau_{yz} + \frac{\partial}{\partial z}\tau_{zz} \right) - \frac{\partial p}{\partial z} + \rho g_z \tag{3.2-3c}$$

Collectively the equations state

$$
\left\{\begin{array}{l}\text{rate of increase}\\\text{of momentum per}\\\text{unit volume}\end{array}\right\} = -\left\{\begin{array}{l}\text{rate of momentum}\\\text{gain by convection}\\\text{per unit volume}\end{array}\right\} - \left\{\begin{array}{l}\text{rate of momentum}\\\text{gain by viscous}\\\text{forces per volume}\end{array}\right\}
$$

$$
-\left\{\begin{array}{l}\text{pressure forces}\\\text{on element per}\\\text{unit volume}\end{array}\right\} + \left\{\begin{array}{l}\text{gravitational force}\\\text{on element per}\\\text{unit volume}\end{array}\right\}
$$

The terms τ_{xx}, τ_{xy}, and so on, are the nine momentum fluxes known as stresses. Thus τ_{xx} is the normal stress on the x face (see Fig. 1.1-5), and τ_{yx} the x-directed tangential (or shear) stress on the y face resulting from viscous forces. These equations of motion state that a small volume of element moving with the fluid is accelerated because of forces acting on it. In other words, this is a statement of Newton's second law in the form: Mass $\times$ acceleration = sum of forces.

VISCOSITY

For Newtonian fluids the stresses may be expressed in terms of Newton's law of viscosity of the form

$$
\tau_{yx} = -\mu\frac{dv_x}{dy} \tag{3.2-4}
$$

Atmospheric air and liquid water are both Newtonian fluids since the viscous shear stress can be described by Eq. 3.2-4, which is also a definition of viscosity. Table 3.2-1 contains some experimental viscosity

Table 3.2-1. Viscosity of Water and Air at 1 Atm Pressure[a]

Temperature T (°C)	Water (liquid)		Air (gas)	
	Viscosity $\mu \times E\,3$ (N·s/m²)	Kinematic Viscosity $\nu \times E\,6$ (m²/s)	Viscosity $\mu \times E\,3$ (N·s/m²)	Kinematic $\nu \times E\,6$ (m²/s)
0	1.787	1.787	0.01716	13.27
20	1.0019	1.0037	0.01813	15.05
40	0.6530	0.6581	0.01908	16.92
60	0.4665	0.4744	0.01999	18.86
80	0.3548	0.3651	0.02087	20.88
100	0.2821	0.2944	0.02173	22.98

Source. Reference 4.

[a]Viscosity is a physical property of a simple fluid that characterizes the flow resistance.

and kinematic viscosity, $\nu \equiv \mu/\rho$, data for water and air. The SI unit for viscosity is the newton-second per square meter.

The preceding equations (i.e., Eq. 3.2-3) in their complete form are seldom used to set up flow problems. Usually restricted forms of the equations of motion are used for convenience. For constant density and viscosity and Newton's law, these equations may be simplified to give

$$\rho\left(\frac{\partial v_x}{\partial t} + v_x\frac{\partial v_x}{\partial x} + v_y\frac{\partial v_x}{\partial y} + v_z\frac{\partial v_x}{\partial z}\right) = \frac{-\partial p}{\partial x} + \mu\left(\frac{\partial^2 v_x}{\partial x^2} + \frac{\partial^2 v_x}{\partial y^2} + \frac{\partial^2 v_x}{\partial z^2}\right) + \rho g_x$$

$$(3.2\text{-}5a)$$

$$\rho\left(\frac{\partial v_y}{\partial t} + v_x\frac{\partial v_y}{\partial x} + v_y\frac{\partial v_y}{\partial y} + v_z\frac{\partial v_y}{\partial z}\right) = \frac{-\partial p}{\partial y} + \mu\left(\frac{\partial^2 v_y}{\partial x^2} + \frac{\partial^2 v_y}{\partial y^2} + \frac{\partial^2 v_y}{\partial z^2}\right) + \rho g_y$$

$$(3.2\text{-}5b)$$

$$\rho\left(\frac{\partial v_z}{\partial t} + v_x\frac{\partial v_z}{\partial x} + v_y\frac{\partial v_z}{\partial y} + v_z\frac{\partial v_z}{\partial z}\right) = \frac{-\partial p}{\partial z} + \mu\left(\frac{\partial^2 v_z}{\partial x^2} + \frac{\partial^2 v_z}{\partial y^2} + \frac{\partial^2 v_z}{\partial z^2}\right) + \rho g_z$$

$$(3.2\text{-}5c)$$

for the x, y, and z components, respectively. Here p is pressure, and g_x is the x component of gravitational acceleration. Equation 3.2-5 is the celebrated Navier-Stokes equation, and it has been widely used for describing flow systems in which viscous effects are dominant.

THE EQUATION OF ENERGY

The equation of energy is the result of the conservation of energy or the first law of thermodynamics. The equation is obtained by writing the law of conservation of energy for the fluid contained within the stationary volume element $\Delta x \Delta y \Delta z$. The equation of energy in terms of the fluid temperature T is

$$\rho\hat{c}_v\left\{\frac{\partial T}{\partial t} + v_x\frac{\partial T}{\partial x} + v_y\frac{\partial T}{\partial y} + v_z\frac{\partial T}{\partial z}\right\} = -\left\{\frac{\partial q}{\partial x} + \frac{\partial q}{\partial y} + \frac{\partial q}{\partial z}\right\}$$
$$- T\left(\frac{\partial p}{\partial T}\right)_{\hat{v}}\left\{\frac{\partial v_x}{\partial x} + \frac{\partial v_y}{\partial y} + \frac{\partial v_z}{\partial z}\right\}$$

$$(3.2\text{-}6)$$

A term accounting for the irreversible rate of internal energy increase by viscous dissipation has been omitted from the right-hand side of Eq. 3.2-6. This term is needed only in special situations such as extremely high shear rates.

THERMAL CONDUCTIVITY

The local heat flow per unit area (heat flux) in the positive y direction is designated q_y. Fourier's law of heat conduction states that the heat flux by conduction is proportional to the temperature gradient, or mathematically,

$$q_y = -k\frac{dT}{dy} \qquad (3.2\text{-}7)$$

This is also a definition of thermal conductivity k. In addition to the thermal conductivity, a quantity known as the thermal diffusivity α is widely used in the heat transfer literature; it is defined as

$$\alpha \equiv \frac{k}{\rho\hat{c}_p} \qquad (3.2\text{-}8)$$

Thermal conductivity data for air, water, and some solid materials are available in Appendix D. Density and heat capacity data are also available there. The SI units of thermal conductivity are the joule per meter-second-kelvin.

For most uses in chemodynamics, the fluids may be considered incompressible ($\rho=$ constant), $\hat{c}_p = \hat{c}_v$, and ($\overline{\nabla}\cdot v$) is zero. After substituting the heat flux terms for each direction, Eq. 3.2-6 becomes

$$\rho\hat{c}_p\left\{\frac{\partial T}{\partial t} + v_x\frac{\partial T}{\partial x} + v_y\frac{\partial T}{\partial y} + v_z\frac{\partial T}{\partial z}\right\} = k\left\{\frac{\partial^2 T}{\partial x^2} + \frac{\partial^2 T}{\partial y^2} + \frac{\partial^2 T}{\partial z^2}\right\} \qquad (3.2\text{-}9)$$

For solids, velocities are zero, and Eq. 3.2-9 becomes

$$\rho\hat{c}_p\frac{\partial T}{\partial t} = k\left\{\frac{\partial^2 T}{\partial x^2} + \frac{\partial^2 T}{\partial y^2} + \frac{\partial^2 T}{\partial z^2}\right\} \qquad (3.2\text{-}10)$$

Equations 3.2-9 and 3.2-10 are the starting points for most of our subsequent uses of heat transfer near environmental interfaces.

THE MULTICOMPONENT CONTINUITY EQUATION FOR DILUTE SOLUTIONS

The multicomponent continuity equation for dilute solutions was presented early in Chapter 3 and appears in vector form. (See Eq. 3-1.) This early presentation was desirable to define the concept of the fate of chemical A in the environment and to develop heuristic ideas concerning chemical movements by diffusion and mass transfer. Equation 3-1 in expanded form for the case of chemical A in air becomes

$$\frac{\partial c_{A1}}{\partial t} + v_x \frac{\partial c_{A1}}{\partial x} + v_y \frac{\partial c_{A1}}{\partial y} + v_z \frac{\partial c_{A1}}{\partial z} = \mathcal{D}_{A1} \left\{ \frac{\partial^2 c_{A1}}{\partial x^2} + \frac{\partial^2 c_{A1}}{\partial y^2} + \frac{\partial^2 c_{A1}}{\partial z^2} \right\} + R_{A1}$$

$$(3.2\text{-}11)$$

Similar equations for chemicals in water and in a solid phase can be written. For solids, velocities are zero, and Eq. 3.2-11 becomes

$$\frac{\partial c_{A3}}{\partial t} = \mathcal{D}_{A3} \left\{ \frac{\partial^2 c_{A3}}{\partial x^2} + \frac{\partial^2 c_{A3}}{\partial y^2} + \frac{\partial^2 c_{A3}}{\partial z^2} \right\} \qquad (3.2\text{-}12)$$

This completes the presentation of the equations of change. As was pointed out earlier, these equations are usually simplified considerably when the analysis of a particular problem is desired. This simplification process will be demonstrated throughout the textbook as these general equations are transformed and applied to specific chemodynamic problems. (See Problem 3.2A for an application.) The preceding set of general equations therefore serves as a unified starting point for the many specific chemodynamic applications to follow. This procedure often avoids the necessity of resorting to first principles and rederivation to affect a new problem analysis. Besides specific applications, the equations of change taken together as a group convey a certain amount of information.

An inspection and comparison of the equations of motion, energy, and multicomponent continuity reveal many similarities. This aspect is one of the important lessons conveyed by the transport phenomena concepts. Transport phenomena are concerned with the transfer of momentum, energy, and mass from one point in the universe to another. Specifically it should be noted that the units of molecular diffusivity $\mathcal{D}_{Aj}$ are cm^2/s. Kinematic viscosity ν and the thermal diffusivity α also have the same units. The ways in which these three quantities are analogous can be seen from the following equations for the fluxes of mass, momentum, and energy in one-dimensional systems:

$$j_A = -\mathcal{D}_{A1} \frac{d}{dy}(\rho_{A1}) \qquad \text{(Fick's law)} \qquad (3.2\text{-}13)$$

$$\tau_{yx} = -\nu_1 \frac{d}{dy}(\rho_1 v_x) \qquad \text{(Newton's law)} \qquad (3.2\text{-}14)$$

$$q_y = -\alpha_1 \frac{d}{dy}(\rho_1 \hat{c}_{p_1} T) \qquad \text{(Fourier's law)} \qquad (3.2\text{-}15)$$

It can now be appreciated that the similarity of the equations of change is due in large part to the mathematics of expressing the basic phenomenological processes. These equations state, respectively, that mass, momentum, and energy transports occur because of gradients in mass concentration, momentum concentration, and energy concentration. The three variables denoted by $\mathcal{D}_{A1}$, ν_1, and α_1 are fundamental transport properties dependent only on molecular processes. (See Hirschfelder, Curtiss, and Bird.[14]) This fact places the major limitation on the equations of change in that they do not apply if turbulence is present. They apply only to solids and fluids in laminar (i.e., viscous) flow.

The Viscous Sublayer (or Inner Layer)

Except for a thin layer of air close to surfaces, the atmosphere is essentially always turbulent. There is also a viscous sublayer on the sea floor that is distinguished from the zone above in that turbulent transfer of momentum is less important than the molecular transfer of momentum. Figure 3.2-2 illustrates the flow characteristics occurring in the viscous sublayer next to a solid when the mass of fluid is moving in the positive x direction.

A simplification of the equation of motion for the case of one-dimensional, steady-state, constant pressure flow of a constant density Newtonian fluid results in

$$\mu_1 \frac{d^2 v_x}{dy^2} = 0 \tag{3.2-16}$$

Since a real fluid is compelled by molecular attraction to adhere to a solid boundary, $v_x = 0$ at $y = 0$. The viscous sublayer is assumed to have a "film thickness" of δ_v at which $v_x = v_x(\delta_v)$. These two boundary conditions

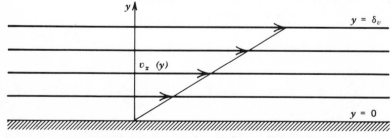

Figure 3.2-2. Flow in the viscous sublayer.

combined with Eq. 3.2-16 yield that the velocity profile is linear as depicted in Fig. 3.2-2:

$$v_x(y) = \frac{v_x(\delta_v)y}{\delta_v} \tag{3.2-17}$$

The momentum transfer rate to the solid wall (τ_0) is obtained by use of Eq. 3.2-14 and is

$$\tau_0 = \frac{-\mu_1 v_x(\delta_v)}{\delta_v} \tag{3.2-18}$$

This τ_0 is also the shear stress exerted by the fluid onto the surface and has SI units of newton per square meter.

Because of increasing agitation of the molecules there is a continuous transfer of momentum from regions of high bulk velocity to regions of low bulk velocity, and the rate of transfer of momentum across unit area of a plane surface in the fluid is expressed by the product of the viscosity and the gradient (Eq. 3.2-18). From a molecular point of view, it should be observed that although there is no net transfer of molecules across a plane parallel to the direction of flow, the existence of a gradient of bulk velocity ensures that the random motion brings about a continuous cross-stream transfer of momentum.

Some Characteristics of Turbulent Flow

Turbulence is the property, easy to recognize but difficult to define, of irregular, chaotic motion possessed by almost all natural fluid flows. Nearly all natural motion, whether of air or water, is turbulent. Turbulent flow has been defined: motion of fluids in which local velocities and pressures fluctuate irregularly, in a random manner. Figure 3.2-3 shows a portion of an anemometer record of the natural wind velocity at an elevation of 35 m. In contrast, the laminar flow pictured in Fig. 3.2-2 is highly ordered. For practical purposes we can best define a turbulent fluid flow as one that has the ability to disperse particles and/or molecules embedded within itself quite rapidly, at a rate orders of magnitude greater than can be accounted for by molecular diffusion.

The air velocity increases with distance from the air-soil interface; the laminar flow pattern, with its steady advance in separate layers, is not maintained: the flow becomes unsteady, with chaotic movements of parts of the air in different directions superimposed on the main flow of the air,

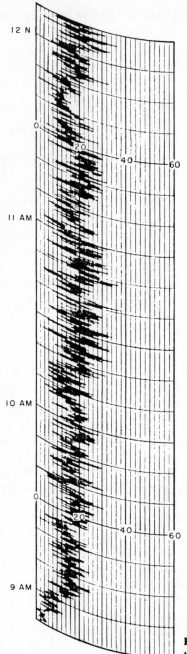

Figure 3.2-3 Wind velocity and velocity fluctuations.

104

as in Fig. 3.2-3. Movement of any particular element of fluid is now very complicated, and it can be described in terms of averages. This is called turbulent flow.

In turbulent flow, transfers of momentum between neighboring pulses of the fluid are of primary importance, as is described in detail later. These inertial (momentum) effects in turbulent flow (as contrasted with purely viscous effects in laminar flow) cause the velocity and density of the flowing fluid to assume great importance. Turbulent flow replaces laminar flow when these inertial effects are great compared with viscous effects. For the flow of fluid in a pipe of diameter d, Reynolds (1883) used the ratio of inertial force to viscous force to characterize the change of flow from laminar to turbulent and thus created the dimensionless ratio known as the Reynolds number:

$$Re \equiv \frac{v_x d}{\nu} \tag{3.2-19}$$

Here ν denotes the ratio μ/ρ, which occurs frequently in hydrodynamic theory and has been introduced previously.

The effect of random disturbances (and there are many in the natural environment) on a fluid in laminar flow is shown in Fig. 3.2-4. In frame (a) the flow is pure laminar and in a streamline flow pattern in the viscous sublayer and well into regions beyond. In frame (b) a random disturbance is shown, the density and velocity gradient effects (momentum effect) being shown by the arrow. In frame (c) more momentum is being transferred into the disturbance than is being damped out, and an eddy is formed. Laminar motion is stable as long as there is no net transfer of

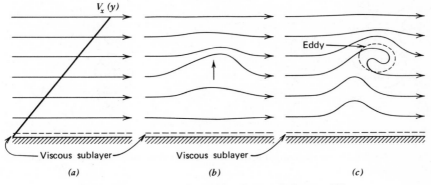

Figure 3.2-4. Origin of an eddy. (**Source.** Reference 15.)

energy from the primary flow into any super imposed random disturbances. All disturbances are damped by the viscosity μ, and the closeness of the solid wall. Figure 3.2-5 shows the turbulent eddies above the air–soil interface as viewed by an observer moving from left to right with the mean velocity. This is the case for fully developed turbulent flow. Near the wall, the strong velocity gradients within the fluid tear the fluid into small eddies. Some of these migrate upward, where larger eddies are also to be found.

Experiments with many fluids in smooth, circular pipes of different diameters have confirmed that Re does indeed characterize the velocity of flow at which laminar breaks down to turbulent flow. At Reynolds numbers up to about 2000, the flow in a smooth pipe is always laminar. Between 2000 and 4000 (the so-called transition region), there is usually a gradual change to turbulent flow. Generally turbulence is fully established when Re > 4000. For geometries other than that of a uniform circular pipe, the "characteristic length" d in Eq. 3.2-19 can be suitably assigned.

For the case of fully developed turbulent air flow above the air–soil interface, Slade[16] employs a semiquantitative development based on the five dimensional quantities (i.e., $dv_x/dy, v, y, \rho,$ and τ_0) and the well-known Π theorem of dimensional analysis to obtain the Reynolds number

$$\text{Re} \equiv \frac{v_* \delta_v}{\nu} \tag{3.2-20}$$

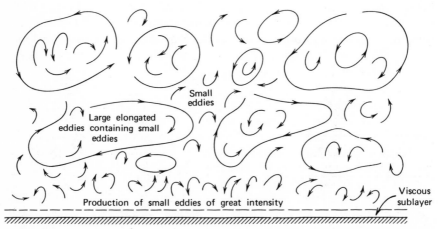

Figure 3.2-5. Eddy production patterns of turbulent flow above the air–soil interface. (Source. Reference 15.)

where δ_v is some fixed elevation above which turbulence is present and v_* is the friction velocity equal to $\sqrt{\tau_0/\rho}$. It has been found experimentally in numerous specific cases, pipe flow being one, that if the "natural reference length:' is used in the Reynolds numbers, values of the number above about 1E2 or 1E3 are always turbulent.

What length scale applies in the atmosphere? Because flows for which Re$\gg$1E2 are ordinarily turbulent, it follows that an upper limit to δ_v, the depth of the laminar sublayer for the atmosphere, will be the order of a millimeter, since v_* is known from observations to be of the order of 100 cm/s and ν for air equals about 1E$-$1 cm^2/s. This means that blades of grass, grains of dirt, sticks, twigs, people, and so forth, all protrude through the laminar sublayer. Anyone who has lost a hat in a sudden gust of wind has inadvertently experienced that δ_v is small and that one does indeed protrude through the viscous sublayer. People live at the air–soil interface and therefore have experienced many aspects of the fluid flow phenomena, including laminar motion, eddies, and convection. This experience together with the qualitative treatment presented here should provide a degree of understanding of the turbulent processes at work at all interfaces over which a fluid is moving, including ocean, lake, and river bottoms. These turbulent processes are important to the subject of chemodynamics.

The Turbulent Equations of Change

THE STATISTICAL NATURE OF TURBULENT FLOW

We must now focus our attention on the fluid behavior at one point above the air–soil interface where turbulent flow exists. Consider the anemometer record shown in Fig. 3.2-3, the average velocity in the x direction $\bar{v}_x$, and the velocity fluctuation v_x'. We can imagine that while we are watching this one spot above the surface, the average wind velocity increases slowly from a value of 3 km/hr at 8:50 A.M. to a value of 20 km/hr at 10:30 A.M. Beyond this point the average velocity fluctuates about 20 km/hr with a period much greater than the rapidly fluctuating instantaneous velocity v_x'. We define the time-smoothed (i.e., average) velocity $\bar{v}_x$ by taking a time average of v_x over a time interval t_0 large with respect to the time of rapid turbulent oscillation but small with respect to the larger period oscillations:

$$\bar{v}_x = \frac{1}{t_0} \int_t^{t+t_0} v_x \, dt \qquad (3.2\text{-}21)$$

The instantaneous velocity may be written as the sum of the time-smoothed velocity $\bar{v}_x$ and a velocity fluctuation v_x':

$$v_x = \bar{v}_x + v_x' \tag{3.2-22}$$

A similar expression can be written for the pressure, which is also fluctuating; clearly $\bar{v}_x' = 0$ by the foregoing definitions. But $\overline{v_z'^2}$ will not be zero, and in fact $\sqrt{\overline{v_x'^2}} / \langle \bar{v}_x \rangle$ is a measure of the magnitude of the turbulent disturbance and is known as the *intensity of turbulence*. Notice that the fluctuating, or turbulent, component of the wind in Fig. 3.2-3 is of the same order of magnitude as the mean (i.e., intensity of about unity). This is a characteristic of atmospheric turbulence and distinguishes it sharply from wind tunnel turbulence, where the intensity is likely to be 0.01 to 0.001.

A convenient concept of the turbulent wind velocity at a fixed point is a velocity v_x in the x direction, representing the mean wind and three mutually perpendicular vectors in the x, y, and z directions with time varying components v_x', v_y', and v_z' to give

$$v_y = \bar{v}_y + v_y' \tag{3.2-23}$$

and

$$v_z = \bar{v}_z + v_z' \tag{3.2-24}$$

along with Eq. 3.2-22. The v_x' vector represents the fluctuation or "eddy velocity" in the direction of the mean wind, v_y' the vertical fluctuation, and v_z' the fluctuation across the mean wind. Using these definitions, measurements of turbulence in the atmosphere reveal: (1) intensity of turbulence is essentially independent of wind speed, but does depend on atmospheric stability; (2) at low levels turbulence is nonisotropic (i.e., $\sqrt{\overline{v_x'^2}} / \bar{v}_x \neq \sqrt{\overline{v_y'^2}} / \bar{v}_y \neq \sqrt{\overline{v_z'^2}} / \bar{v}_z$); and (3) some authorities have concluded that turbulence is isotropic at heights in excess of about 25 m.

TIME SMOOTHING OF THE EQUATIONS OF CHANGE FOR AN INCOMPRESSIBLE FLUID

In this section we present the equations that describe the time-smoothed velocity and pressure for an incompressible fluid. A brief outline of the definition is presented, and details are left as an exercise for the student. (See problem 3.2C.) Starting with the x component of the equation of motion in the following form:

$$\frac{\partial}{\partial t}(\rho v_x) + \frac{\partial}{\partial x}(\rho v_x v_x) + \frac{\partial}{\partial y}(\rho v_y v_x) + \frac{\partial}{\partial z}(\rho v_z v_x)$$

$$= \frac{-\partial p}{\partial x} + \mu \left[\frac{\partial^2 v_x}{\partial x^2} + \frac{\partial^2 v_x}{\partial y^2} + \frac{\partial^2 v_x}{\partial z^2} \right] + \rho g_x$$

$$(3.2\text{-}25)$$

replace v_x by Eq. 3.2-22 and p by $\bar{p} + p'$ everywhere they occur. Now take the time average of Eq. 3.2-25 according to Eq. 3.2-21 for v_x and a similar integral average for p. The following equation results:

$$\frac{\partial}{\partial t}(\rho \bar{v}_x) + \frac{\partial}{\partial x}(\rho \bar{v}_x \bar{v}_x) + \frac{\partial}{\partial y}(\rho \bar{v}_y \bar{v}_x) + \frac{\partial}{\partial z}(\rho \bar{v}_z \bar{v}_x)$$

$$+ \frac{\partial}{\partial x}(\rho \overline{v_x' v_x'}) + \frac{\partial}{\partial y}(\rho \overline{v_y' v_x'}) + \frac{\partial}{\partial z}(\rho \overline{v_z' v_x'})$$

$$= \frac{-\partial p}{\partial x} + \mu \left[\frac{\partial^2 \bar{v}_x}{\partial x^2} + \frac{\partial^2 \bar{v}_x}{\partial y^2} + \frac{\partial^2 \bar{v}_x}{\partial z^2} \right] + \rho g_x \qquad (3.2\text{-}26)$$

This is the time-smoothed x component of the equation of motion. All terms appearing in Eq. 3.2-25 appear in this time-smoothed equation replaced with average values; but, in addition, new terms arise, that are associated with the turbulent velocity fluctuations. These new terms are the components of the turbulent momentum flux and are usually referred to as the *Reynolds stresses*:

$$\bar{\tau}_{xx}^{(t)} = \rho \overline{v_x' v_x'}; \qquad \bar{\tau}_{xy}^{(t)} = \rho \overline{v_y' v_x'}; \qquad \bar{\tau}_{xz}^{(t)} = \rho \overline{v_z' v_x'} \quad (3.2\text{-}27a, b, c)$$

With the preceding definitions and $\tau = \bar{\tau}^{(l)} + \bar{\tau}^{(t)}$, Eq. 3.2-26 appears like Eq. 3.2-3a. The equation of motion can therefore be used for turbulent flow problems provided one changes all v_i to $\bar{v}_i$, ρ to $\bar{p}$, and τ_{ij} to $\bar{\tau}_{ij}^{(l)} + \bar{\tau}_{ij}^{(t)}$. Similar time-smoothing results can be obtained for the y and z components of the equation of motion.

When considering the energy balance it is necessary to use a point temperature made up of an average temperature and a fluctuating component: $T = \bar{T} + T'$. Time smoothing the equation of energy in the same manner as was done for the equation of motion results in additional fluxes also. The development is restricted to a fluid of constant ρ, $\hat{c}_p$, μ, and k.

(See problem 3.2C, part 3, for directions of derivations.) Additional terms created have led to the definition of the turbulent energy fluxes with components:

$$\bar{q}_x^{(t)} = \rho\hat{c}_p\overline{v_x'T'}; \qquad \bar{q}_y^{(t)} = \rho\hat{c}_p\overline{v_y'T'}; \qquad \bar{q}_z^{(t)} = \rho\hat{c}_p\overline{v_z'T'}$$

$$(3.2\text{-}28\text{a, b, c})$$

The similarity between the components of $\bar{q}^{(t)}$ and those of $\bar{\tau}^{(t)}$ should be noted.

Time smoothing the multicomponent continuity equation for dilute solutions also results in new terms. (See problem 3.2C, part 4.)

$$\bar{J}_x^{(t)} = \overline{v_x'c_{A1}'}; \qquad \bar{J}_y^{(t)} = \overline{v_y'c_{A1}'}; \qquad \bar{J}_z^{(t)} = \overline{v_z'c_{A1}'} \quad (3.2\text{-}29\text{a, b, c})$$

where c_{A1}' is the fluctuating component of concentration around the mean c_{A1}. The new terms describe the turbulent mass transport. All fluxes are defined with respect to the mass average velocity.

Interpretation of the Reynolds Stresses

The meaning of the Reynolds stresses can be made clear by considering the case of air flow in the x direction near a soil surface. The mean velocity is distributed as shown in Fig. 3.2-6 with $\bar{v}_y$ and $\bar{v}_z$ zero. The quantity $\bar{\tau}_{yx}^{(t)} = -\rho\overline{v_x'v_y'}$ represents the Reynolds stress in the x direction acting on a plane perpendicular to the y direction. We would like to show that this term is different from zero and of the correct sign.

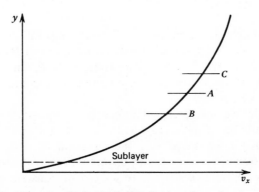

Figure 3.2-6. Time-smoothed velocity distribution above the air–soil surface.

Suppose a small pocket of fluid having a mean velocity represented by A in Fig. 3.2-6 is transported, because of a negative v_y', to a region B, where v_x is smaller. Since the pocket will retain approximately its original velocity $\bar{v}_{xA}$, at B there is created a positive v_x', and $v_x'v_y'$ is negative. On the other hand, if v_y' happens to be positive, the fluid is transported to a region C, where the mean velocity is greater than $\bar{v}_{xA}$; a negative v_x' is created, and $v_x'v_y'$ is again negative. It is thus easy to see that $\overline{v_x'v_y'}$ will be a negative quantity so that the mean turbulent stress $\bar{\tau}_{yx}^{(t)}$ is negative as shown in Eq. 3.2-26.

THE PRANDTL MIXING LENGTH THEORY

By analogy with the kinetic theory of gases, Prandtl (1926) postulated that as the masses of fluid migrated laterally they carried with them the mean velocity (and hence the momentum concentration) of their point of origin. Still considering a flow above a soil surface and referring to Fig. 3.2-6, we define the mixing length in the following way. Consider again a small pocket of fluid that is displaced from A to B in the y direction with a velocity v_y'. In reality, the lump of fluid will gradually lose its identity, but in the definition of the mixing length it is assumed to regain its identity until it has traveled a distance l defined as the Prandtl mixing length. For the small distance involved, we can write

$$\frac{d\bar{v}_x}{dy} = \frac{\bar{v}_{xB} - \bar{v}_{xA}}{l} \tag{3.2-30}$$

As was mentioned earlier, the pocket of fluid can be assumed to retain its original velocity, so that $\bar{v}_{xB} - \bar{v}_{xA}$ is approximately $-v_x'$. Therefore, we have

$$v_x' = -l\frac{d\bar{v}_x}{dy}$$

and, in general,

$$|\overline{v_x'}| = l\left|\frac{d\bar{v}_x}{dy}\right| \tag{3.2-31}$$

Prandtl also assumed that $|\overline{v_y'}|$ was about the same absolute magnitude as $|\overline{v_x'}|$, so that

$$\overline{v_x'v_y'} = l^2\left|\frac{d\bar{v}_x}{dy}\right|^2$$

Since the sign of $\overline{v_x' v_y'}$ is negative and the opposite of $d\overline{v}_x/dy$ (i.e., the slope of $\overline{v}_x$ versus y in Fig. 3-2.6), we write the preceding equation for the turbulent shear stress as

$$\tau_{yx}^{(t)} = -\rho \overline{v_x' v_y'} = l^2 \left| \frac{dv_x}{dy} \right| \frac{dv_x}{dy} \tag{3.2-32}$$

Equation 3.2-32 is a main result of the Prandtl mixing length theory. Here l is a quantity having the dimension of length. According to Prandtl, l is the average distance traveled by fluid lumps involved in the turbulent mixing process. According to more recent views, however, l is to be interpreted as a quantity proportional to the average size of macroturbulent eddies and/or to the average length of their displacement during turbulent mixing. The average size of macroturbulent eddies and the average length of their displacements vary along the thickness of the flow, that is, l is a function of the distance from the wall. At the present state of knowledge, the exact form of the variation of l with y is not known.

By analogy with the molecular viscosity, an *eddy kinematic viscosity* or *turbulent coefficient of viscosity* can be defined for parallel flow as

$$\overline{\tau}_{yx}^{(t)} = -\rho \nu^{(t)} \frac{d\overline{v}_x}{dy} \tag{3.2-33}$$

The quantity $\nu^{(t)}$, introduced by Boussinesq (1877), unlike molecular viscosity, is not a function of state but depends strongly on position. We have no way of calculating $\nu^{(t)}$ a priori, although it can be determined experimentally from a given distribution of $\overline{v}_x$ versus y.

Comparing Eqs. 3.2-32 and 3.2-33, it may appear that the only result of the Prandtl mixing length has been to replace one empirical, nondeterminable quantity with another, but the mixing length is easier to estimate than $\nu^{(t)}$. In the next section it is shown that some valuable results for the velocity distribution for turbulent flow near environmental interfaces can be obtained with the simple relation

$$l = \kappa_1 y \tag{3.2-34}$$

where y is the distance from the surface and κ_1 is a universal constant, 0.4.

VELOCITY DISTRIBUTION IN A TWO-DIMENSIONAL TURBULENT FLOW WITH A FREE SURFACE

In this section the result of the Prandtl mixing length theory is employed to obtain the velocity profile for water flowing in a stream. Figure 3.2-7

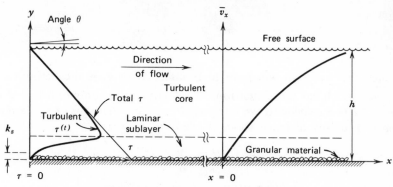

Figure 3.2-7.   Flow of a natural stream.

shows a section of a natural stream with a free surface at the air–water interface and a bottom made of a granular medium.

The left side of Fig. 3.2-7 shows schematically the distribution of the shear stress $\tau^{(t)}$ and $\tau^{(l)} = \tau - \tau^{(t)}$ along the depth of the flow. Here δ_v is the thickness of the laminar sublayer and k_s is the roughness height of the granular material. It can be shown that above the laminar sublayer the shear stress is almost equal to the turbulent shear stress $\tau^{(t)}$, the component $\tau^{(l)}$ being negligible. Within the sublayer the opposite is true. The total shear stress within the fluid may be obtained from an x-momentum balance, and the result is

$$\tau_{yx} = \rho_2 hg \tan\theta\left(1 - \frac{y}{h}\right) \tag{3.2-35}$$

where $\tan\theta$ is the slope of the free surface. (See Problem 3.2D.) Note that the shear stress at the bottom, τ_0, is $\rho_2 gh \tan\theta$.

Now in the turbulent core, that section of the stream from the laminar sublayer to the free surface, the total stress is equal (almost) to the turbulent stress. Using Eq. 3.2-34 for the mixing length and equating Eqs. 3.2-32 and 3.2-35, we obtain

$$\rho_2 \kappa_1^2 y^2 \left(\frac{dv_x}{dy}\right)^2 = \tau_0\left(1 - \frac{y}{h}\right) \tag{3.2-36}$$

Prandtl made a mathematical simplification (physically indefensible) at this point by setting the right side equal to τ_0. This simplifies the mathematics somewhat, and it has been shown that the results differ very little from that obtained by integrating Eq. 3.2-36. After taking the square root we get.

$$\frac{dv_x}{dy} = \pm \frac{v_*}{\kappa_1} \frac{1}{y} \qquad (3.2\text{-}37)$$

in which the plus sign will have to be used. Here v_* is the definition of $\sqrt{\tau_0/\rho_2}$, which has the dimensions of velocity and is commonly called the *shear* or *friction velocity*. If we integrate from the outer edge of the laminar sublayer δ_v to any distance above the bottom, we get

$$\bar{v}_x = \bar{v}_x|_{\delta_v} + \frac{v_*}{\kappa_1} \ln \frac{y}{\delta_v} \qquad y \geqslant \delta_v \qquad (3.2\text{-}38)$$

This is the relation between velocity and distance from the bottom. At the surface $y = h$ and $\bar{v}_x$ is the maximum, $\bar{v}_x|_h$. For this case

$$\bar{v}_x|_h = \bar{v}_x|_{\delta_v} + \frac{v_*}{\kappa_1} \ln \frac{h}{\delta_v} \qquad (3.2\text{-}39)$$

and subtracting from Eq. 3.2-38, we obtain

$$\frac{\bar{v}_x|_h - \bar{v}_x}{v_*} = -\frac{1}{\kappa_1} \ln \frac{y}{h} \qquad (3.2\text{-}40a)$$

which is a form of the *universal velocity distribution law*.

UNIVERSAL VELOCITY DISTRIBUTION

Equation 3.2-40 and the more general form

$$\frac{\bar{v}_x}{v_*} = \frac{1}{\kappa_1} \ln y + \text{constant} \qquad (3.2\text{-}40b)$$

have been found to be a very good approximation for quantifying the velocity profiles in the turbulent regions above almost all environmental interfaces. This relation mimics the profiles in streams, in the air layers above both the soil surface and ocean surface, and in the benthic boundary layer. (See Problem 3.2F.)

The integration constant in Eq. 3.2-40 is usually defined so as to introduce the effect of surface roughness by requiring that $\bar{v}_x = 0$ when $y = y_0$; y_0 is called the *roughness length* because it expresses the effect of varying interface roughness on the velocity profile:

$$\bar{v}_x = \frac{v_*}{\kappa_1} \ln \frac{y}{y_0} \qquad (3.2\text{-}41)$$

Table 3.2-2. Universal Velocity Profile Parameters

Type of Surface	y_0 (cm)	v_* (m/s)
(a) For wind near the Earth's surface		
Smooth mud flats; ice	0.001	0.16
Smooth snow	0.005	0.17
Smooth sea	0.02	0.21
Level desert	0.03	0.22
Snow surface, lawn to		
1 cm high	0.01	0.27
Lawn grass to 5 cm	1–2	0.43
Lawn grass to 60 cm	4–9	0.60
Fully grown root crops	14	1.75
(b) For the benthic boundary layer		
Deep sea	2.0	0.001
Shelf	–	0.01

Sources. Part (a): Reference 16. Part (b) Reference 17.

Figure 3.2-8. Three-dimensional sand ripples (Euphrates River). (**Source.** Reference 18.)

Figure 3.2-9. A selection of aerial photographs showing the effects of increasing wind speeds (in knots) near the surface on waves. Note the increasing size of the whitecaps and the spray as the wind speed increases. (Courtesy of the U.S. Naval Oceanographic Office, Washington, D.C.)

25 kts.

30 kts.

Figure 3.2-9. (continued)

117

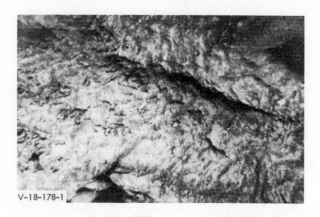

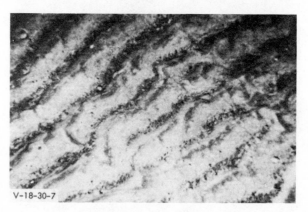

Figure 3.2-10. Four photographs taken on the eighteenth voyage of Columbia University's research vessel VEMA. The identification, for example V-18-30-7, means the seventh exposure of the thirtieth camera station on the eighteenth voyage of VEMA. (Photographs courtesy of Maurice Ewing, Lamont Geological Observatory.)

This equation is valid only for $y \geqslant y_0$ and applies only above the laminar sublayer.

Sometimes y_0 is chosen so that $\bar{v}_x = 0$ when $y = 0$. If this is done, the velocity profile equation takes the form

$$\bar{v}_x = \frac{v_*}{\kappa_1} \ln\left(\frac{y + y_0}{y_0}\right) \tag{3.2-42}$$

Since, as a practical matter, interest is ordinarily centered on velocity at

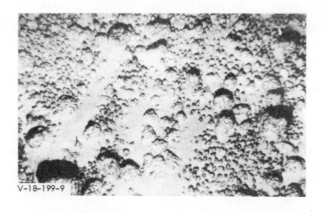

Figure 3.2-10. (continued)

heights where $y \gg y_0$, the two forms immediately above are substantially equivalent. Values of y_0 and v_* found from field experiments appear in Table 3.2-2.

ROUGHNESS OF ENVIRONMENTAL INTERFACES

Since we live at the air–soil interface we are familiar with the makeup and variations of the surface structure, and it therefore needs no elaboration here. The other environmental interfaces are not so familiar. Figures 3.2-8, 3.2-9, and 3.2-10 show, respectively, photographs of stream bottoms, the surface of the sea, and the sea bottom.

Problems

3.2A.   TRANSFORMATION AND/OR SIMPLIFICATION OF THE EQUATIONS OF CHANGE

When transformations and/or simplifications are made on a general (i.e., all-encompassing) set of equations, certain restrictions, assumptions, and specifications are implied.

1. Starting with Eq. 3-1, obtain Eq. 3.2-11. List all restrictions, assumptions, and other specifications necessary for this transformation.
2. Starting with Eq. 3.2-11, obtain Eq. 3.1-3. List all restrictions, assumptions, and other specifications necessary for this transformation.
3. Starting with Eq. 3.2-5, obtain Eq. 3.2-16. List all restrictions, assumptions, and other specifications necessary for this simplification.
4. Starting with Eq. 3.2-16, obtain Eqs. 3.2-17 and 3.2-18.

3.2B.   APPROXIMATE HEIGHT OF THE VISCOUS SUBLAYER

1. Verify that the height of the sublayer at the air–soil interface is of the order of a millimeter.
2. Compute the height of the sublayer to be found at the water–soil interface.

3.2C.   TIME SMOOTHING THE EQUATIONS OF CHANGE

1. Following the procedure outlined in the text, show details of the transformation from Eqs. 3.2-25 to 3.2-26 that were omitted.
2. Using a similar procedure as in part (1), time-smooth the continuity equation, Eq. 3.2-1.
3. Starting with the following form of the equation of energy

$$\rho \hat{c}_p \frac{\partial T}{\partial t} = - \overline{\nabla} \cdot \left(\rho \hat{c}_p T \bar{v} \right) + k \overline{\nabla}^2 T \qquad (3.2C\text{-}1)$$

and $T = \bar{T} + T'$, time-smooth and show the origin of Eqs. 3.2-28a, b, and c.
4. Starting with the following form of the multicomponent equation of continuity

$$\frac{\partial c_{A1}}{\partial t} = - \overline{\nabla} \cdot \bar{v} c_{A1} + \mathcal{D}_{A1} \overline{\nabla}^2 c_{A1} - k_A''' \, c_{A1} \qquad (3.2C\text{-}2)$$

and $c_{A1} = \bar{c}_{A1} + c'_A$, time-smooth and show the origin of Eqs. 3.2-29a, b, and c.

3.2D. SHEAR STRESS RELATION IN A NATURAL FLOWING STREAM

Show that Eq. 3.2-35 is valid for natural stream flow. Start by performing an x-component momentum (or force) balance on an element of fluid of thickness Δx, bounded by the free surface and bottom and extending a distance w in the z direction.

3.2E. THE UNIVERSAL VELOCITY DISTRIBUTION FOR TURBULENT FLOW NEAR ENVIRONMENTAL INTERFACES

Refer to text material starting with Eq. 3.2-32 and show all steps in the development of the universal velocity distribution (Eq. 3.2-40).

3.2F. VELOCITY PROFILE NEAR NATURAL INTERFACES

1. A stream. Using the following data, verify graphically the form of the universal velocity profile. Determine v_* (cm/s) and y_0 (cm) from the graphical construction. Assume $h = 10$ m.

y/h	$\bar{v}_x$ (cm/s)	y/h	$\bar{v}_x$ (cm/s)
0.034	51.8	0.50	79.3
0.069	59.4	0.68	83.8
0.17	68.6	0.85	88.4
0.25	73.2	0.92	88.4
0.36	76.2		

2. The sea. The following measurements were made in the southern entrance to the Dardanelles, which is sufficiently wide for the current to be unaffected by the lateral boundaries. Determine v_* (cm/s) and y_0 (cm).

$y(m)$	$\bar{v}_x$ (cm/s)	$y(m)$	$\bar{v}_x$ (cm/s)
2	0.3	17	5.5
7	2.8	22	6.5
12	4.6	27	7.2

3.2G. DIMENSIONS OF TRANSPORT COEFFICIENTS

1. Since Eq. 3.2-4 is a definition of viscosity, determine its dimensions. What are the dimensions of kinematic viscosity?
2. Use Eq. 3.2-7 and determine the dimensions of thermal conductivity. What are the dimensions of thermal diffusivity?
3. Use Eq. 3.1-2 and determine the dimensions of molecular diffusivity.

3.2H. MIXING LENGTH THEORY EXPRESSIONS FOR HEAT AND MASS TRANSFER

Develop the mixing length theory expressions for heat and mass transfer by the following procedure.

1. Heat transfer. Commencing with Eq. 3.2-28b, parallel the Prandtl mixing length theory for momentum transfer and show that the turbulent energy flux expression is

$$\bar{q}_y^{(t)} = -\rho\hat{c}_p l^2 \left| \frac{d\bar{v}_x}{dy} \right| \frac{d\bar{T}}{dy} \tag{3.2H-1}$$

in which l is the Prandtl mixing length. (*Hint:* Draw a time-smoothed temperature profile to accompany the time-smoothed velocity shown in Fig. 3.2-6.)

Prandtl recommends setting $l = \kappa_1 y$, in which y is the distance from the surface. Using $l = \kappa_1(y + y_0)$, where y_0 is a small roughness length, and Eq. 3.2-42, show that the turbulent thermal diffusivity is

$$\alpha^{(t)} = \kappa_1(y + y_0)v_* \tag{3.2H-2}$$

where $\alpha^{(t)}$, the turbulent thermal diffusivity defined by

$$\bar{q}_y^{(t)} \equiv -\alpha^{(t)} \frac{d}{dy}\left(\rho\hat{c}_p \bar{T}\right) \tag{3.2H-3}$$

2. Mass transfer. Commencing with Eq. 3.2-29b, parallel the Prandtl mixing length theory for momentum transfer, and show that the turbulent mass flux expression is

$$\bar{J}_{Ay}^{(t)} = -l^2 \left| \frac{d\bar{V}_x}{dy} \right| \frac{d\bar{c}_{A1}}{dy} \tag{3.2H-4}$$

in which l is the Prandtl mixing length. (*Hint:* Draw a time-smoothed concentration profile to accompany the time-smoothed velocity shown in Fig. 3.2-6.)

Prandtl recommends setting $l = \kappa_1 y$, in which y is the distance from the surface. Using $l = \kappa_1(y + y_0)$, where y_0 is a small roughness length, and Eq. 3.2-42, show that the turbulent diffusivity is

$$\mathcal{D}_{A1}^{(t)} = \kappa_1(y + y_0)v_*. \tag{3.2H-5}$$

where $\mathcal{D}_{A1}^{(t)}$ is defined in Eq. 3.1-6.

The turbulent diffusivities given by Eqs. 3.2H-2 and 3.2H-5, unlike the molecular counterparts α and $\mathcal{D}_{A1}^{(t)}$, which are properties of the fluids, are functions of the external quantities y and v_*.

3.3. OTHER TRANSPORT TOPICS

In this section the topic of heat transfer is extended and the analogy theories are presented.

Some Fundamentals of Heat Transfer

Heat transfer and temperature play a major role in the movement of chemicals near environmental interfaces, and the basic topics presented in the previous section, including the equation of energy and Fourier's law, need to be extended. A quantitative description of heat transfer processes in the environment requires several types of rate equations and several corresponding transfer coefficients. The developments given here for heat transfer parallel those given for mass transfer. The fundamentals of heat transfer are reviewed here. In later sections this information is followed by information on the specifics of the heat exchange at the air–water interface and the air–soil interface.

FOURIER'S LAW AND THERMAL CONDUCTIVITY

Fourier observed that the heat flux by conduction is proportional to the temperature gradient, or, to put it somewhat pictorially, heat "slides downhill on the temperature versus distance graph." Fourier's law is given in Eq. 3.2-7. The thermal conductivity k, defined by this equation, applies to fluids at rest, solids or fluids in laminar flow situations where the heat flux is by molecular movements and rearrangements. Thermal conductivity is a fluid property reflecting only molecular aspects, and for this reason

Eq. 3.2-7 is of limited utility in fluids undergoing turbulent flow. However, heat transfer in soil is one application where the equation can be used.

The thermal diffusivity α, defined by Eq. 3.2-8, is a more useful quantity than the thermal conductivity for some purposes. Since density ρ and heat capacity $\hat{c}_p$ of the material are properties, thermal diffusivity is therefore a property of the medium through which heat is moving. The thermal diffusivity of water at 20°C is 1.43E−3 cm^2/s.

In the fluid regions near environmental interfaces molecular movements play a minor role. The dominant mechanism by which heat is transferred in a flowing system (i.e., stream, lake, air, etc.) is by random bulk movements, known as *turbulent energy flux*. By analogy with Fourier's law of heat conduction one may write

$$q_y = -\left(k^{(l)} + k^{(t)}\right)\frac{dT}{dy} \qquad (3.3\text{-}1)$$

the quantity, $k^{(t)}$ being called the *turbulent coefficient of thermal conductivity* or *eddy conductivity*. It is not a physical property of the fluid like $k^{(l)}$, but depends on position, direction, and the nature of the turbulent flow. An *eddy thermal diffusivity* can be defined

$$\alpha^{(t)} \equiv \frac{k^{(t)}}{\rho \hat{c}_p} \qquad (3.3\text{-}2)$$

The eddy thermal diffusivity has the same dimensions as eddy kinematic viscosity $\nu^{(t)}$ (Eq. 3.2-33) and the eddy diffusion coefficient $\mathcal{D}_A^{(t)}$ (Eq. 3.1-6). These quantities may be compared for turbulent momentum and energy transport. The ratio $\nu^{(t)}/\alpha^{(t)}$ is of the order of unity. Values in the turbulence literature vary from 0.5 to 1.0.

Observations of heat conduction in water have yielded much information on the magnitude and character of this eddy thermal diffusivity. There is a wide spread in measured values, from as low as 4E−2 to as high as 200 cm^2/s. Table 3.3-1 summarizes some turbulent diffusivity observations from inland water bodies and the seas. Thermal gradients are very pronounced in the vertical direction in water bodies. Eddy thermal diffusivities reported for the horizontal direction are typically an order of magnitude greater than those in the vertical direction. Koh and Fan[20] have performed a literature review and present a summary of $\alpha^{(t)}$ values in equation and graphical form. The review closes with the following paragraph:

In general $\alpha_y^{(t)}$ has its maximum value in the surface layer: In the ocean $\alpha_y^{(t)}$ at the surface varies between 10–200 cm^2/s; in coastal areas 10–50

Table 3.3-1. Effective Vertical Thermal Diffusion Coefficients $\alpha_2^{(t)}$ in Water Bodies

Water Bodies	Surface	Thermocline	Depth
(a) Lakes and reservoirs	$\alpha_2^{(t)}$ (cm²/s)		
Fontana Reservoir	0.81	0.14	1.15 @ 29 m
Hungry Horse Reservoir	3.7	0.12	1.7 @ 54 m
Lake Tahoe	9.2	0.1	1.2 @ 90 m
Castle Lake	2.6	0.02	1.5 @ 22 m
	$\alpha_2^{(t)}$ (cm²/s)	Depth of Layer (m)	
(b) Oceans and seas			
Philippine Trench	2.0–3.2	5000–9788	
Mediterranean	42	0–28	
California Current	30–40	0–200	
Caribbean Sea	2.8	500–700	
(c) Water in laminar flow $\alpha_2^{(t)} = 0.00143$ cm²/s			

Sources. Part (a) Reference 21. Part (b) Reference 22.

cm²/s; in lakes $\sim$10 cm²/s. Below the surface mixed layer (or epilimnion) $\alpha_y^{(t)}$ drops to its minimum in the thermocline (of the order of 1 cm²/s in open ocean); in lakes $\alpha_y^{(t)}$ may drop as low as 0.05 cm²/s. Below the thermocline, $\alpha_y^{(t)}$ may increase again.

Equation 3.3-1 is difficult to use. In general, the reasons for the difficulty are identical to those given for turbulent mass diffusivities. The thermal diffusivity is highly variable near environmental interfaces, and the thermal gradient is almost never known. For these reasons and others Eq. 3.3-1 is not generally used for quantifying heat flux rates near or at environmental interfaces.

SENSIBLE, LATENT, AND RADIANT ENERGY TRANSFER AT INTERFACES

The three fundamental mechanisms of energy transfer at interfaces and their rate expressions are presented here. Topics concerned with specific interfaces are presented in those chapters concerned specifically with those interfaces.

SENSIBLE HEAT TRANSFER

Sensible heat is heat given up or absorbed by a body on being cooled or heated, as the result of the body's ability to hold heat; this excludes latent heats of fusion or vaporization. An alternate form of Eq. 3.2-7 for expressing the sensible heat flux rate across an interface is by use of

Newton's law of cooling:

$$q_{s0} = \dot{h}(T_0 - T_b) \tag{3.3-3}$$

Here q_{s0} is the sensible heat transfer rate per unit area across the interface, $\dot{h}$ is the local value of the heat transfer coefficient, T_0 is the surface or interface temperature, and T_b is the bulk or "cup mixing" fluid temperature. Newton's law of cooling is not really a law but rather a defining equation for h, which is called the *heat transfer coefficient*. Just as with mass transfer at phase boundaries, Eq. 3.1-8, the equivalent heat transfer relation finds major utility at these boundaries where the turbulent thermal conductivity and the thermal gradient are rarely known with certainty.

For surfaces of finite area A, one can parallel Eq. 3.3-3 and define an average heat transfer coefficient

$$Q_{s0} = hA(T_0 - T_b) \tag{3.3-4}$$

Here Q_{s0} is the sensible heat transfer rate and h is the average coefficient. Typical SI units for Q are joule per second (J/s) and for q_{s0} are joule per second square meter (J/s·m^2). Just as for mass transfer coefficients, $\dot{h}$ and h are scalars. Q and q are vectors (i.e., they have magnitude and direction) and take the direction from the sign of $T_0 - T_b$. Heat transfer into a body is a positive number whereas heat transfer from a body to its surroundings is a negative number.

See Sections 4.3, and 6.4 for specific heat transfer applications at the air–water, and air–soil interfaces, respectively.

LATENT HEAT TRANSFER

Latent heat is defined as the amount of heat absorbed or evolved by 1 mole, or a unit mass, of a substance during a change of state (such as fusion, sublimation, or vaporization) at constant temperature and pressure. The latent heat transfer rate is important at air–water and air–soil interfaces. At these planes water is present and changes phases, either from a liquid to a gas (vaporization) or from a gas to a liquid (condensation). As one may expect, this relationship of energy movement rate is related to the water movement rate

$$q_{l0} = n_{A0}\lambda_A \tag{3.3-5}$$

Here q_{l0} is the rate of latent heat transfer at the interface in units of J/m^2·s, n_{A0} is the mass flux rate of water ($A \equiv H_2O$) moving across the interface in units of g/m^2·s, and λ_A is the latent heat of vaporization of water in units of J/g. The latent heat of vaporization of water at 24°C is 2.44E+6 J/kg. Relations for computing the mass flux rate n_{A0} have been presented in Eqs. 3.1-8 and 3.1-9.

Even though other chemicals may be moving across the interface with accompanying heat of vaporization or condensation effects, their contribution to the latent heat transfer is negligible because of their dilute nature. It is therefore necessary to account for the water only when quantifying latent heat effects across the air–water or the air–soil interface.

RADIANT ENERGY TRANSFER

Radiant energy is defined as energy transmitted by electromagnetic waves through space or some medium. This energy, or electromagnetic radiation, becomes heat only after interaction with matter upon which it is absorbed. As a solid body exists at a temperature above absolute zero, some constituent molecules and atoms at the surface are raised to "excited states." There is a tendency for the atoms or molecules to return spontaneously to lower energy states. When this occurs, energy is emitted in the form of electromagnetic radiation. Because the emitted radiation may result from changes in the electronic, vibrational, and rotational states of the atoms and molecules, the radiation is distributed over a range of wavelengths. Thermal radiation involves wavelengths primarily in the range 0.1E−6 to 10E−6 m. The preceding is a brief description of the *emission* of radiant energy. The reverse process, which is known as *absorption*, occurs when the addition of radiant energy to a molecule or atomic system causes the system to go from a low to a high energy state.

Radiation impinging on the surface of an opaque solid is either absorbed or reflected. The fraction of the incident radiation that is absorbed is called the *absorptivity* and is given the symbol a. Kirchhoff's law states that at a given temperature, the emissivity e and the absorptivity a of any solid surface are the same when the radiation is in equilibrium with the solid surface. This allows us to conclude that

$$e = a \qquad (3.3\text{-}6)$$

Emissivities are usually measured and reported. Typical values are: rough, unglazed silica brick, 0.80; candle soot, 0.952; planed oak, 0.895; and water 0.95 to 0.963.

It has been shown experimentally that the total emitted energy from a real surface is

$$\mathbf{q_{e0}} = e\sigma T^4 \qquad (3.3\text{-}7)$$

in which T is the absolute temperature. The Stefan-Boltzmann constant σ

has been found to have a value $5.67E-12$ $J/s \cdot cm^{2} \cdot {}^{\circ}K^{4}$. The emissivity e must be evaluated at temperature T (absolute).

A reasonably accurate quantitative treatment of radiation between non-black surfaces can be treated as follows. Consider the air–water interface of area A in which the air is "surface 1" and the water is "surface 2." Since over a limited region A is flat, it intercepts none of its own rays. The rate of energy emission from surface 1 to surface 2 is given by

$$\overrightarrow{Q_{12}} = e_1 A\sigma T_1^{4} \qquad (3.3\text{-}8)$$

and the rate of energy absorption from surface 2 by surface 1 is

$$\overrightarrow{Q_{21}} = a_1 A\sigma T_2^{4} \qquad (3.3\text{-}9)$$

The net radiation rate from 1 to 2 is therefore

$$Q_{12} = \sigma A\left(e_1 T_1^{4} - a_1 T_2^{4}\right) \qquad (3.3\text{-}10)$$

where e_1 is the value of emissivity of surface 1 at T_1. The absorptivity a_1 is usually estimated as the value of e of surface 1 at T_2.

Analogy Theories of Momentum, Heat, and Mass Transfer

The analogy theories extend the experimental data collected on one environmental transport for use in others. For example, data on heat transfer coefficients between air and a soil surface can be extended by the use of the analogy theories to calculate mass transfer coefficients of chemical A between air and the soil surface. The first hint of analogy between the three transport phenomena was pointed out in Section 3.2 following the presentation of the equations of change. (See specifically Eqs. 3.2-13, 3.2-14, and 3.2-15.) In this section the analogy theories are presented and their utility in interphase chemodynamic problems is illustrated.

First the Prandtl mixing length theory is extended to heat and mass transfer to develop turbulent temperature and concentration profiles in that region above the laminar sublayer. Next, the Reynolds analogy is developed based on a proportionality. The von Kármán analogy theory, which considers not only the turbulent zone but the buffer layer and the laminar sublayer, is presented and discussed. Finally, utility and limitations of the analogy theories are considered.

SIMILAR TURBULENT PROFILES

According to Prandtl's theory, momentum and energy are transferred in turbulent flow by the same mechanism. By employing an approach similar to that developed for momentum transfer, the turbulent energy flux expression (Eq. 3.2-28b) can be related to the time-smoothed velocity and temperature gradients in the fluid above a solid interface

$$\bar{q}_y^{(t)} = +\rho\hat{c}_p l^2 \left| \frac{d\bar{v}_x}{dy} \right| \frac{d\bar{T}}{dy} \tag{3.3-11}$$

in which l is the Prandtl mixing length. Implicit in this development is $\nu^{(t)}/\alpha^{(t)} = 1$. Prandtl recommends setting $l = \kappa_1 y$, in which y is the distance from the interface. Equation 3.3-11 is similar to Eq. 3.2-32 at this point.

In the surface boundary layer, at steady state, it is assumed that the energy flux is independent of height, which then gives

$$\bar{q}_y^{(t)} = q_0 \tag{3.3-12}$$

Recall a similar simplification (i.e., $\tau_{yx}^{(t)} = \tau_0$) was made in the development of velocity profiles. If now Eqs. 3.3-11 and 3.2-36, with the right side equal to τ_0, are divided one by the other, we get

$$\frac{q_0}{\tau_0} = \frac{\hat{c}_p d\bar{T}}{d\bar{v}_x} \tag{3.3-13}$$

which may be integrated from a point beyond the edge of the sublayers where $\bar{T} = \bar{T}|_\delta$ and $\bar{v}_x = \bar{v}|_\delta$ to yield.

$$\frac{\hat{c}_p(\bar{T} - \bar{T}|_\delta)}{q_0} = \frac{\hat{v} - v|_\delta}{\tau_0} \tag{3.3-14}$$

This result simply states that within the turbulent zone above the laminar sublayers the velocity and temperature profiles are similar; this agrees roughly with field observations.

Now Eq. 3.2-38 can be arranged to the form

$$\frac{\bar{v}_x - \bar{v}_x|_\delta}{\tau_0} = \frac{1}{\kappa_1 \rho v_*} \ln\left(\frac{y}{\delta}\right) \tag{3.3-15}$$

with the aid of the friction velocity. Clearly, from Eq. 3.3-14,

$$\overline{T} - \overline{T}|_\delta = \frac{q_0}{\kappa_1 \rho \hat{c}_p v_*} \ln\left(\frac{y}{\delta}\right) \tag{3.3-16}$$

Hence one deduces that the temperature profile in the turbulent portion of the surface boundary layer will also be a logarithmic function.

Without the chemical reaction term, Eq. 3.2-11 is of exactly the same form as Eq. 3.2-5a in a constant pressure flow field with $g_x = 0$. The result obtained above will then be valid for mass transfer. With the appropriate notational changes we can immediately write down the turbulent concentration profiles by analogy with Eq. 3.3-16:

$$\bar{c}_A - \bar{c}_A|_{\delta_A} = \frac{N_{A_0}}{\kappa_1 v^*} \ln\left(\frac{y}{\delta}\right) \tag{3.3-17}$$

The sublayer thickness δ appearing in Eqs. 3.3-15, 3.3-16 and 3.3-17 is replaced by (y/y_0) or $[(y+y_0)/y_0]$, as in Eq. 3.2-42; the y_0 values observed in the field are not necessarily identical for the three transport phenomena. In general, y_0 values are chosen so that the profiles fit the field data. (See Problem 3.2F.)

THE REYNOLDS ANALOGY

The following development of the Reynolds analogy is adopted from Bennett and Myers.[23] The proportionality of heat and momentum transfer can be stated in terms of four quantities, which we define with reference to a fluid at a bulk temperature T_b flowing past and losing heat to an interface, which is at temperature T_0. The four quantities are

(a) The heat flux from the fluid to the interface, $h(T_b - T_0)$ J/m²·s.
(b) The momentum flux at the wall, τ_0, kg·m/m²·s².
(c) The rate at which energy available for transfer as heat is transported parallel to the interface, $n\hat{c}_p(T_b - T_0)$, J/m²·s, where n is the mass flow rate in kg/s·m².
(d) The rate at which momentum is transported parallel to the interface, nv_b, kg·m/m²·s².

It is then postulated that these four quantities can be described by the following proportionality:

$$\frac{(a)}{(c)} = \frac{(b)}{(d)}$$

Substituting the respective quantities yields

$$\frac{h}{\hat{c}_p} = \frac{\tau_0}{v_b} \qquad (3.3\text{-}18)$$

If the variables D, a characteristic length, k, thermal conductivity, and μ, fluid viscosity, are introduced to preserve the equality and create dimensionless groups, Eq. 3.3-18 becomes

$$\frac{Nu}{Pr} = \frac{Re\,\tau_0}{\rho v_b^2} \qquad (3.3\text{-}19)$$

where Nu is the Nusselt number for heat transfer,

$$Nu \equiv \frac{hD}{k} \qquad (3.3\text{-}20)$$

Pr is the Prandtl number,

$$Pr \equiv \frac{\hat{c}_p \mu}{k} \qquad (3.3\text{-}21)$$

and Re is the Reynolds number

$$Re \equiv \frac{D v_b \rho}{\mu} \qquad (3.3\text{-}22)$$

Equation 3.3-19 is a statement of the Reynolds analogy between heat and momentum transfer, which is known to be very good for $Pr = 1$.

The Reynolds analogy gives best results when applied to gases, for which Pr is approximately 1. As an application, consider the problem of determining the heat flux from warm air to a cool soil surface. If temperature T_b and v_b are measured at a fixed height and τ_0 estimated from a velocity profile data (Eq. 3.2-42), the Reynolds analogy yields a value of the heat transfer coefficient. The heat flux can be estimated by use of Eq. 3.3-4 if the soil surface temperature T_0 is known.

Although Reynolds was concerned only with the analogy between heat and momentum transfer, the equations referred to as the Reynolds analogy can be readily extended to cover mass transfer. A proportionality development similar to the one above but of mass and momentum transfer at an air–soil environmental interface results in

$$^3k_{A1}M_1 = \frac{\tau_0}{v_b} \qquad (3.3\text{-}23)$$

where M_1 is the molecular weight of the air–chemical mixture. Introducing cc_1, fluid molar density; $\mathcal{D}_{A1}$, molecular diffusivity; D, a characteristic length; μ_1, viscosity; and creating dimensionless groups yields

$$\frac{\text{Nu}_{A1}}{\text{Sc}_{A1}} = \frac{\text{Re}\,\tau_0}{\rho v_b^2} \tag{3.3-24}$$

where Nu_{A1} is the Nusselt number for mass transfer,

$$\text{Nu}_{A1} \equiv \frac{{}^3k_{A1}D}{c_1\mathcal{D}_{A1}} \tag{3.3-25}$$

Sc_{A1} is the Schmidt number

$$\text{Sc}_{A1} \equiv \frac{\mu_1}{\mathcal{D}_{A1}\rho_1} \tag{3.3-26}$$

As with the heat transfer analogy result, Eqs. 3.3-23 and 3.3-24 are limited to the pure turbulent regions of the surface boundary layer except for cases where the Schmidt number is unity (Sc = 1.0). For fluid with Pr = Sc = 1, the Reynolds analogy applies not only to the turbulent region, but also to the laminar sublayer.

OTHER ANALOGIES

The restriction Pr = Sc = 1 limits the general utility of the Reynolds analogy. This restriction is critical in the case of environmental chemodynamic applications since the Prandtl number for water is 7.7 at 60°F and the Schmidt number for most chemicals in water ranges from 300 to 2700! It is obvious that considering only the transport processes in the turbulent region and neglecting the laminar sublayer has its faults. Major improvements on the Reynolds analogy have been made, resulting in the Prandtl-Taylor equation, which was further extended by von Kármán. Von Kármán considered the resistance of heat transfer to be composed of three parts: the laminar sublayer, the buffer zone, and the turbulent region. The von Kármán equation for heat transfer is

$$\frac{\text{Nu}}{\text{Pr}} = \left(\frac{\tau_0}{\rho v_b^2}\right)\frac{\text{Re}}{1 + 5\sqrt{\tau_0/\rho v_b^2}\ \left\{\text{Pr} - 1 + \ln\left[(1 + 5\,\text{Pr})/6\right]\right\}} \tag{3.3-27}$$

The numerator on the right-hand side is seen to be a correction to Eq. 3.3-19, the Reynolds analogy. The von Kármán equation can also be expressed in terms of mass and momentum transfer

$$\frac{Nu_{A1}}{Sc_{A1}} = \left(\frac{\tau_0}{\rho v_b{}^2}\right) \frac{Re}{1 + 5\sqrt{\tau_0/\rho v_b{}^2}\,\left\{Sc_{A1} - 1 + \ln\left[(1 + 5\,Sc_{A1})/6\right]\right\}}$$

$$(3.3\text{-}28)$$

Both forms of the von Kàrmàn equation reduce to the Reynolds analogy when $Pr = Sc = 1$. Equations 3.3-27 and 3.3-28 should be used to correct the deficiencies of the Reynolds analogy.

One final analogy worth discussing is the empirical j-factor relation proposed by Chilton and Colburn.[24] This analogy may be obtained by replacing the entire denominator in Eq. 3.3-27 by the expression $Pr^{2/3}$. The equation is then rearranged to present it in its customary form:

$$\frac{h}{v_b \rho \hat{c}_p} Pr^{2/3} = \frac{\tau_0}{\rho v_b{}^2} \equiv j_H \qquad (3.3\text{-}29a)$$

The term j_H is called the j-factor for heat transfer. An alternative form of expression is

$$\frac{Nu}{Re\,Pr^{1/3}} = \frac{\tau_0}{\rho v_b{}^2} \equiv j_H \qquad (3.3\text{-}29b)$$

The analogy can be written for mass transfer

$$\frac{Nu_{A1}}{Re\,Sc^{1/3}} = \frac{\tau_0}{\rho v_b{}^2} \equiv j_D \qquad (3.3\text{-}30)$$

The term j_D is known as the j-factor for mass transfer. All forms of this analogy reduce to the Reynolds analogy for $Pr = Sc = 1$.

The j-factor relation has been shown experimentally to have considerable merit in correlating heat and mass transfer data. It agrees closely with the predictions of the boundary layer theory for the flat plate when Pr and Sc exceed 0.5 and appears to be fairly good for turbulent flow. Both j_H and j_D are frequently correlated as a function of Re, geometry, and boundary conditions.

LIMITATIONS OF USING ANALOGIES

The importance of analogies is that they permit one to obtain mass transfer correlations from heat transfer correlations for equivalent boundary conditions by merely substituting Nu_{A1} for Nu and Sc_{A1} for Pr. The same can be done for any flow geometry and for laminar or turbulent flow. Note that to obtain a valid analogy, one has to assume (1) constant physical properties, (2) a small rate of mass transfer, (3) no chemical reactions in the fluid, (4) no viscous dissipation, (5) no emission or absorption of radiant energy, and (6) no pressure diffusion, thermal diffusion, or forced diffusion. All of these assumptions apply for many chemodynamic applications. A summary of analogous quantities for heat and mass transfer is given in Table 3.3-2.

Table 3.3-2. Heat and Mass Transfer Analogies at Low Mass Transfer Rates

Group or Parameter	Heat Transfer Quantities	Mass Transfer Quantities
Profiles	T	x_A
Diffusivity	$\alpha \equiv \dfrac{k}{\rho \hat{c}_p}$	$\mathcal{D}_{Aj}$
Effect of profiles on density	$\beta \equiv -\dfrac{1}{\rho}(\dfrac{\partial \rho}{\partial T})_{\rho, x_A}$	$\zeta = \dfrac{-1}{\rho}(\dfrac{\partial \rho}{\partial x_A})_{\rho, T}$
Transfer coefficient	$h = \dfrac{Q}{A \Delta T}$	$^j k_{Ai} = \dfrac{W_A}{A \Delta x_A}$
Groups that are the same	$Re \equiv \dfrac{Dv\rho}{\mu}$	$Re \equiv \dfrac{Dv\rho}{\mu}$
	$\dfrac{L}{D}$	$\dfrac{L}{D}$
Groups that are different	$Nu \equiv \dfrac{hD}{k}$	$Nu_{Aj} \equiv \dfrac{^j k_{Ai} D}{c_j \mathcal{D}_{Aj}}$
	$Pr \equiv \dfrac{\hat{c}_p \mu}{k} = \dfrac{\nu}{\alpha}$	$Sc_{Aj} \equiv \dfrac{\mu}{\rho \mathcal{D}_{Aj}} = \dfrac{\nu}{\mathcal{D}_{Aj}}$
	$Gr \equiv \dfrac{D^3 \rho^2 g \beta \Delta T}{\mu^2}$	$Gr_{Aj} \equiv \dfrac{D^3 \rho^2 g \zeta \Delta x_A}{\mu^2}$

Source. Reference 10, p. 646.

Example 3.3-1. Converting a Mass Transfer Coefficient to a Heat Transfer Coefficient

It has been estimated the liquid phase mass transfer coefficient for phosphorus ($A \equiv P$) at the bottom of Lake Washington to be $^3 k_{A2}/c_2 = 36$ m/yr. Estimate the heat transfer coefficient at the lake bottom in $J/m^2 \cdot s \cdot K$. Assume the diffusivity for phosphorus is 1.0×10^{-5} cm^2/s and the water is at 20°C.

SOLUTION

In obtaining a heat transfer coefficient from a mass transfer coefficient, it is important to realize that a major message of the analogy theory is

$$\text{Nu} = \text{a function of } \left(\text{Re}, \text{Pr}, \frac{L}{D} \right) \tag{E3.3-1}$$

$$\text{Nu}_{A2} = \text{the same function of } \left(\text{Re}, \text{Sc}_{A2}, \frac{L}{D} \right) \tag{E3.3-2}$$

The functional form to use in this case is the j factor:

$$j_H = j_D \tag{E3.3-3}$$

or

$$\frac{\text{Nu}}{\text{Re} \, \text{Pr}^{1/3}} = \frac{\text{Nu}_{A2}}{\text{Re} \, \text{Sc}_{A2}^{1/3}} \tag{E3.3-4}$$

At these conditions $\text{Pr} = 7.0$, $\text{Sc}_{A2} = 890$. Substituting the definitions of the respective Nusselt numbers, we obtain

$$\frac{hD}{k_2(7.0)^{1/3}} = \frac{{}^3k_{A2}D}{c_2 \mathscr{D}_{A2}(890)^{1/3}} \tag{E3.3-5}$$

Solving for h in Eq. E3.3-5 we obtain for $k_2 = 0.59 \text{J/m} \cdot \text{s} \cdot \text{K}$

$$h = \left(\frac{7.0}{890} \right)^{1/3} \left(\frac{36 \text{ m}}{\text{yr}} \right) \left(\frac{0.59 \text{J}}{\text{m} \cdot \text{s} \cdot \text{K}} \right) \left(\frac{\text{s}}{1E-5 \text{ cm}^2} \right) \left(\frac{1E+4 \text{ cm}^2}{\text{m}^2} \right) \left(\frac{\text{yr}}{3.51E+7 \text{ s}} \right)$$

$$h = 120 \text{ J/m}^2 \cdot \text{s} \cdot \text{K}$$

Problems

3.3A. ESTIMATING HEAT AND MASS TRANSFER FLUX RATES FROM MICROMETEOROLOGICAL DATA

Brooks and Pruitt[25] measured wind velocity, air temperature, and moisture near the ground. The following data are a part of that work.

Height (cm)	Air Velocity (cm/s)	Additional Data
16	185.4	
39	252.6	temperature at 90 cm was 14.96° C
90	319.5	temperature at 0 cm was 13.72° C
139	356.4	humidity at 90 cm was 8.98 g/m³
189	383.7	humidity at 0 cm was 9.12 g/m³
390	437.6	volume measured at STP
590	469.6	

1. Estimate the heat flux rate at the ground in $J/m^2 \cdot s$.
2. Estimate the mass flux rate at the ground in $g/m^2 \cdot s$.

3.3B. FAILURE OF THE REYNOLDS ANALOGY FOR LIQUIDS

Determine the error (%) in a particular mass transfer coefficient calculation (use $v_* = 0.1$ cm/s and $v_b = 5$ cm/s) if the Reynolds analogy is used. $Sc = 3500$.

1. Compare to the von Kármán equation.
2. Compare to the j-factor equation.

3.3C. A SIMPLE RELATIONSHIP BETWEEN THE LAMINAR SUBLAYER THICKNESS, THE THERMAL SUBLAYER THICKNESS, AND THE DIFFUSIVE SUBLAYER THICKNESS

As momentum, heat, and mass are transferred across a fluid–solid interface, profiles of velocity, temperature, and concentration develop in the boundary layer next to the solid surface. Figure 3.3C shows the case for momentum transfer in the negative y direction, heat and mass transfer in the negative y direction. The sublayer thicknesses of velocity, temperature, and concentration are usually different, as shown.

1. Using the result of the film theory for mass transfer (Eq. 3.1-13) and Eq. 3.2-18, show that the Chilton-Colburn j-factor analogy (Eq. 3.3-30) predicts that the diffusive sublayer thickness δ_{A2} is related to the laminar sublayer δ_v by

$$\delta_{A2} = \frac{\delta_v}{Sc_{A2}^{1/3}} \qquad (3.3C-1)$$

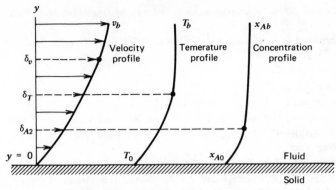

Figure 3.3C. Sublayer thickness and profiles.

2. Using the film theory for heat transfer:

$$q_o = \left(\frac{k_2}{\delta_T} \right) (T_o - T_b) \tag{3.3C-2}$$

and Eq. 3.2-18, show that the Chilton-Colburn j-factor analogy (Eq. 3.3-29b) predicts that the thermal sublayer thickness δ_T is related to the laminar sublayer thickness δ_v by

$$\delta_T = \frac{\delta_v}{Pr^{1/3}} \tag{3.3C-3}$$

REFERENCES

1. R. C. Reid and T. K. Sherwood, *The Properties of Gases and Liquids*, 2nd ed., McGraw-Hill, New York, 1966.

2. A. Lerman, "Time to Chemical Steady-States in Lakes and Oceans," in J. D. Hem, Ed., *Nonequilibrium Systems in Natural Water Chemistry*, Advances in Chemistry Series 106, American Chemical Society, Washington, D.C., 1971.

3. R. E. Bird, W. E. Steward, and E. N. Lightfoot, *Transport Phenomena*, Wiley, New York, 1960, p. 656f.

4. W. Nernst, *Z. Phys. Chem.*, **47**, 52 (1904).

5. W. G. Whitman, *Chem. Met. Eng.*, **29** 146 (1923).

6. R. Higbie, "The Rate of Absorption of a Pure Gas into a Still Liquid during Short Periods of Exposure," *Trans. Am. Inst. Chem. Eng.*, **31** , p. 365, (1935).

7. P. V. Dankwerts, "Significance of Liquid-Film Coefficients in Gas Absorption," *Ind. Eng. Chem.*, **43** (6), 1469 (1951).

8. R. E. Treybal, *Mass-Transfer Operations*, 2nd ed., McGraw-Hill, New York, 1968, pp. 46–60.

9. T. K. Sherwood, R. L. Pigford, and Charles R. Wilke, *Mass Transfer*, McGraw-Hill, New York, 1975, pp. 150–159.

10. Bird, op. cit., p. 617.

11. Treybal, op. cit., p. 63.

12. C. R. Wilke and P. Chang, Am. Inst. Chem. Eng. J., **1**, 1955, pp. 204–270.

13. *Chem. Eng. News*, "Japanese Oil Spill Has Wide Repercussions," June 2, 1975, p. 13.

14. J. O. Hirschfelder, C. E. Curtis, and R. B. Bird, *Molecular Theory of Gases and Liquids*, Wiley, New York, 1954.

15. J. T. Davies, *Turbulence Phenomena*, Academic, New York, 1972.

16. D. H. Slade, Ed., *Meteorology and Atomic Energy*, U.S. Atomic Energy Commission Technical Information Center, Oak Ridge, Tenn., 1969, pp. 66–73.

17. M. Wimbush, "The Physics of the Benthic Boundary Layer," in I. N. McCave, Ed., *The Benthic Boundary Layer*, Plenum, New York, 1976, p. 8.

18. Committee on Sedimentation, "Nomenclature for Bed Forms in Alluvial Channels," *Proc. Am. Soc. Chem. Eng.*, **92**, HY3 (May 1966).

19. G. Newmann and W. J. Pierson, Jr., *Principles of Physical Oceanography*, Prentice-Hall, Englewood Cliffs, N. J., 1966, pp. 29, 30, 327.

20. C. Y. Koh and L.-N. Fan, "Mathematical Models for the Prediction of Temperature Distributions Resulting from the Discharge of Heated Water in Large Bodies of Water, U.S. Environmental Protection Agency, Washington, D.C. Water Quality Office Report 16130DW010/70, October 1970, p. 127.

21. F. L. Parker and P. A. Krenkel, *Thermal Pollution: Status of the Art*, Report 3, School of Engineering, Vanderbilt University, Nashville, Tenn., 1969.

22. A. Defant, *Physical Oceanography*, Vol. 1, Macmillan, New York, 1961, p. 104.

23. C. O. Bennett and J. E. Myers, *Momentum, Heat and Mass Transfer*, 2nd ed., McGraw-Hill, New York, 1974, pp. 351–352.

24. T. H. Chilton and A. P. Colburn, *Ind. Eng. Chem.*, **26**, 1183 (1934).

25. F. A. Brooks and W. O. Pruitt, *Investigation of Energy, Momentum, and Mass Transfer Near the Ground*, U.S. Army Electronics Command, Atm. Sci. Lab., Res. Div., Fort Huachuca, Ariz., Final Report, 1965.

CHAPTER 4

CHEMICAL EXCHANGE RATES BETWEEN AIR AND WATER

There are many times when a knowledge of the flux of an organic or inorganic chemical between water and air is valuable. For example, in estimating the movement of halocarbons in the biosphere it is necessary to know the rate at which the material moves between the atmosphere and the hydrosphere. In a different situation such as a spill of a chemical into a river a knowledge of the rate of evaporation through the water–air interface is fundamental in estimating the resulting concentration of the chemical as it moves downstream. The ability to predict rates of chemical transfer between various regions of the ocean surface and atmosphere is necessary to the understanding of the cycles of a number of trace atmospheric chemicals. These chemicals include naturally occurring gases such as carbon dioxide and synthetic trace chemicals such as chloroform and carbon tetrachloride. For some of these chemicals the ocean is a potential sink; for others the ocean has been proposed as a natural source. If we are to develop the capability to make long-range forecasts of the effect of pollution on concentrations of the trace chemicals, we must be able to establish their air–sea transfer rates. To do this it is necessary to measure not only the concentration differences across the air–sea interface but also the mass transfer coefficients governing the exchange.

The processes that control chemical movements in the real world are extremely complex when taken all together. A system of this degree of complexity cannot be described exactly by a theoretical development.

Some information about the behavior of the system can, however, be obtained by analysis of highly simplified models. The reader may feel that the applications are too simple to be of interest; it is certainly true that they represent highly idealized situations, but the results find considerable use in the understanding of numerous topics in environmental chemistry and engineering.

In this chapter various aspects of the chemical transport at the air–water interface are presented in the context of specific problems and applications of theory. The first section considers the problem of desorption (or evaporation) of waste gases and liquids from aerated basins. In this case the desorption process is aided by mechanical means and involves air–water interfaces of the size of small ponds and lakes. The topics progress through the evaporation of pesticides from the surface of lakes, the flux of gases in both directions across the air–sea interface, and the weathering of hydrocarbon components from sea-surface slicks.

4.1. DESORPTION OF GASES AND LIQUIDS FROM AERATED BASINS AND RIVERS

Natural and artificial basins (ponds, lagoons) are often used as aerobic biochemical oxidation reactors for neutralizing organic wastewater. Carbonaceous compounds in the wastewater are converted to carbon dioxide and water by a mixed culture of microorganisms. The organisms use the carbon compounds as an energy source. Oxygen must be available and mechanical means are frequently employed to increase the rate of transfer and supplement the natural absorption processes.

A common mechanical device employed is a surface aerator. Impeller blades connected by a shaft to a motor beat the water surface into an agitated state. Oxygen absorption in water is liquid phase controlled (see Problem 4.1G), so that the agitation increases the absorption rate. Figure 4.1-1 shows a typical installation.

Industrial wastewater is varied in nature and contains a collection of chemicals used in the manufacturing process. These waste chemicals, in small amounts, originate from raw material handling, intermediates, final products, and so on. Aerobic biochemical oxidation is a common means of removing these chemicals prior to discharging the water to a receiving stream or lake. It is not uncommon for the wastewater entering the basin to contain components that are volatile. These volatile components, both gases and liquids, can desorb from the water directly into the air. The combination of a large interfacial area between water and air, mechanical

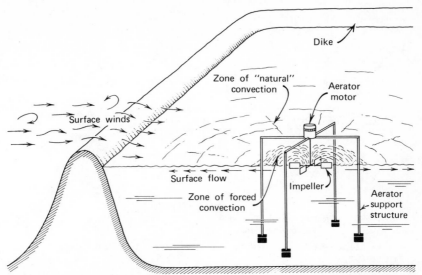

Figure 4.1-1. Natural and forced phenomena for desorption in aerated stabilization basins. (Reprinted with permission from Reference 1.)

agitation, and high relative volatility can result in significant quantities being desorbed.[1]

A necessary condition for the transfer of a volatile species from a liquid phase to a gas phase is a favorable chemical potential. The thermodynamic vapor–liquid equilibrium for dilute solutions is conveniently expressed by Henry's law (in mole fraction form):

$$y_A^* = H_{XA} x_A \tag{4.1-1}$$

where H_{XA} is the Henry's law constant for species A in water. Table 2.1-5 contains typical values of H_{XA} for common industrial waste components. Values of the relative volatility are also given in the table.

The desorption of component A occurs by interphase mass transfer of A from the aqueous phase to the air phase. An adequate rate equation for transfer at low mass transfer rates and dilute solutions is

$$\mathbf{W_{A0}} = {}^1K_{A2} a_v V(x_A - x_A^*) \tag{4.1-2}$$

where ${}^1K_{A2}$ is the overall liquid phase coefficient for A (mol/$L^2 \cdot t$), a_v is the interfacial area per unit volume (L^{-1}), V is volume of the desorbing system (L^3), x_A is the mole fraction concentration of A in the water, and

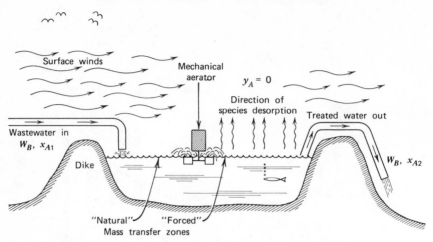

Figure 4.1-2. Idealized desorption wastewater treatment system.

x_A^* is water mole fraction in equilibrium with the surrounding air concentration of A. For a complex gas–liquid interfacial area only the volume is known with certainty and the coefficient area product $^1K_{A2}a_v$ is conveniently measured and reported as a single quantity. In this development it is assumed that the air above the basin contains no quantity of component A so that $x_A^* = 0$. It is also assumed that no biochemical reaction is occurring within the basin water, no settling or adsorption of species A by solids, and so on; only desorption through the air–water interface.

Figure 4.1-2 illustrates the idealized, steady-state desorption system. The concentration of A in the basin water x_A is important in determining the desorption rate. There are two idealized mixing models that need to be considered. The completely mixed basin model reasonably depicts basins of small volume in which the mechanical device stirs the liquid so that a uniform concentration of A is maintained at all points. The plug flow basin model assumes no mixing and is approached by real world basins made in the shape of long, narrow channels with flow in one end and out the other. Both models are highly idealized and represent the extreme in mixing and the extreme in unmixing, respectively. Real world basins rarely approach either case but usually involve some intermediate mixing regime.

Case A: Completely Mixed Basin

This idealized mixing concept assumes the liquid portion of the basin is in a state of mixing such that the concentration of A is the same everywhere.

The inlet water x_{A1} is immediately mixed with the water in the basin to yield a uniform concentration x_{A2}, which is also the concentration in the exit water. A component balance for A over the entire basin volume V according to Eq. 1.1-8 yields

$$W_B x_{A1} = W_B x_{A2} + {}^1K_{A2} a_v V(x_{A2} - 0) + 0 + 0 \qquad (4.1\text{-}3)$$

where W_B is the molar flow rate of water through the basin. Solving Eq. 4.1-3 yields the exit concentration of A:

$$x_{A2} = \frac{x_{A1}}{\left[1 + {}^1K_{A2} a_v V / W_B \right]} \qquad (4.1\text{-}4)$$

The relative quantity of species A lost from the basin by desorption is of interest. The fraction desorbed in the completely mixed basin is $F_M \equiv (x_{A1} - x_{A2})/x_{A1}$, assuming no water evaporation or seepage from the basin. This definition along with Eq. 4.1-4 yields

$$F_M = 1 - \frac{1}{1 + {}^1K_{A2} a_v V / W_B} \qquad (4.1\text{-}5)$$

which is independent of concentration. Fractional removal, decay, treatment, and so on in any first-order process such as mass transfer is concentration independent. The desorption rate is concentration dependent, however, and can be expressed in terms of fractional removal by

$$W_{A0} = W_B x_{A1} F_M \qquad (4.1\text{-}6)$$

These final two equations are useful in estimating the maximum fraction and quantity of component A desorbed from a completely mixed basin.

Case B: Plug Flow Basin

A plug flow basin is another name for a completely unmixed basin. In this idealized mixing model the liquid portion of the basin flows through as a slug of fluid. The elements of fluid hold their position and do not mix with elements in front or back. The concentration of A falls continuously from the high of x_{A1} at the inlet to the low of x_{A2} at the outlet.

A differential equation describes the material balance on component A (see Problem 4.1A) over an elemental volume of basin. The integrated form of this equation is

$$x_{A2} = x_{A1} \exp\left(\frac{-{}^1K_{A2}a_v V}{W_B}\right) \qquad (4.1\text{-}7)$$

The relative quantity desorbed from the basin is

$$F_P = 1 - \exp\left(\frac{-{}^1K_{A2}a_v V}{W_B}\right) \qquad (4.1\text{-}8)$$

and the desorption rate is

$$W_{A0} = W_B x_{A1} F_P \qquad (4.1\text{-}9)$$

These final two equations are useful to estimate the maximum fraction and quantity of component A desorbing from a treatment basin. Any real world basin should lie somewhere between a state of completely mixed to completely unmixed so that use of the preceding expressions will give the upper and lower bounds of chemical desorption without biochemical reaction and other competing mechanisms. Both of the preceding mixing models assume the mass transfer coefficient is uniformly distributed throughout the basin volume. The following section is devoted to obtaining realistic estimates of the overall mass transfer coefficient (i.e., ${}^1K_{A2}a_v$) for use in the preceding equations.

It is convenient to divide the gas–liquid interface of a typical basin into two zones. One zone, located near the surface agitator, is called the *zone of forced convection*. In this zone the mechanically driven impeller forces the liquid and the gas phases to move about and come into intimate contact. This forced convection zone extends to a radius of 10 to 12 ft from the impeller shaft. The other zone is located beyond the immediate influences of the agitator and assumes characteristics of the natural environment. In this zone water flows outward from the aerator not unlike a natural flowing stream. Surface winds dominate the movement of air over the water surface in this zone. (See Fig. 4.1-1.) The creation of these two zones reflects the physical state of the basin and guides the search for appropriate mass transfer coefficient correlations.

The two-resistance theory is a sufficient quantitative tool for describing the interphase desorption of volatile species in an aerated stabilization basin (ASB). Desorption is occurring in parallel from each of the zones so that the total rate is the sum of the "natural" (n) and "forced" (f) zones: $W_{A0} = W_{A0}^{(n)} + W_{A0}^{(f)}$. Applying Eq. 4.1-2 to each zone

$$W_{A0} = \left({}^1K_{A2}a_v\right)^{(n)} V(x_A - 0) + \left({}^1K_{A2}a_v\right)^{(f)} V(x_A - 0) \qquad (4.1\text{-}10)$$

where $({}^1K_{A2}a_v)^{(n)}$ is the "natural" overall liquid phase desorption mass transfer coefficient $(\text{mol}/L^3 \cdot t)$ that accounts for the basin surface stream flow (impeller-induced) type behavior of the liquid phase and the wind (air) surface behavior of the gas phase. The product $({}^1K_{A2}a_v)^{(f)}$ accounts for the pumping and phase-dispersing action occurring with the gas and liquid immediately adjacent to the surface agitator. As noted earlier, the "forced" convection zone occupies a region near each agitator and extends outward about 10 to 12 ft from the shaft. In reality, even the "natural zone" is influenced by the aerator.

The overall coefficients in Eq. 4.1-10 must be related to individual coefficients in each phase. The two-resistance theory developed in Chapter 3 is used (see Eq. 3.1-33) to give

$$\frac{1}{\left({}^1K_{A2}a_v\right)^{(n)}} = \frac{1}{\left({}^1k_{A2}a_v\right)^{(n)}} + \frac{1}{H_{XA}\left({}^2k_{A1}a_v\right)^{(n)}} \qquad (4.1\text{-}11)$$

for the natural coefficient and

$$\frac{1}{\left({}^1K_{A2}a_v\right)^{(f)}} = \frac{1}{\left({}^1k_{A2}a_v\right)^{(f)}} + \frac{1}{H_{XA}\left({}^2k_{A1}a_v\right)^{(f)}} \qquad (4.1\text{-}12)$$

for the forced coefficient. Therefore, a quantitative description of the rate of desorption of chemical A requires information on four individual phase mass transfer coefficients, the natural zone interfacial area and the forced zone interfacial area, plus the Henry's law constant. In the following section techniques used to obtain the individual coefficients and interfacial areas are presented in detail. Each coefficient area product will be developed as a product or individually as a coefficient and an area.

ESTIMATING MASS TRANSFER COEFFICIENTS FOR AERATED STABILIZATION BASINS

$({}^1k_{A2}a_v)^{(f)}$—MECHANICAL SURFACE AGITATOR LIQUID PHASE COEFFICIENT

The rate of absorption of oxygen from air into water by mechanically agitating the water surface has been studied extensively. Since oxygen absorption into water is liquid phase controlling, this coefficient provides a very good measure of the forced desorption coefficient. Many experimental observations on scale model and prototype units have resulted in a reliable correlating equation.[2] The reported oxygen coefficient transformed to yield a liquid phase mass transfer coefficient applicable to any chemical species absorbing or desorbing in the basin area under the immediate influence of the agitator is

$$(^1k_{A2}a_v)^{(f)} = 6.06\text{E }3\sqrt{\frac{\mathcal{D}_{A2}}{\mathcal{D}_{B2}}} \ \frac{n_{B0}^1 E\omega\alpha}{V}(1.024)^{T-20°C} \qquad (4.1\text{-}13)$$

where the subscript A denotes species A, B denotes molecular oxygen, and 2 denotes water. The coefficient area product has dimensions of lb mol/hr $\cdot$ft^3, V is in ft^3 and the dimensions of the remaining terms may be found in Problem 4.1C. The square-root dependence of the coefficient on liquid diffusivity assumes a penetration theory mechanism, that is, short liquid exposure times for the liquid phase transfer.

$(^2k_{A1}a_v)^{(f)}$—MECHANICAL SURFACE AGITATOR GAS PHASE COEFFICIENT

A minimum amount of information on this coefficient is available in the open literature and because of this it is necessary to synthesize a coefficient area product from an available correlation and minimal experimental observations. A preliminary experimental investigation on a low speed surface turbine aerator employing high and low speed still photographic techniques revealed the agitated volume to be made of droplets, spouts, and ebullient volumes.[3] This investigation led to the use of a correlation for liquid spherical droplets ejected into a gas phase.[4]

$$\text{Sh} = 2.0 + 0.347 \ (\text{Re Sc}^{1/2})^{0.62} \qquad (4.1\text{-}14)$$

where the Sherwood number $\text{Sh} \equiv {}^2k_{A1}^{(f)}d/c_1\mathcal{D}_{A1}$, the Reynolds number $\text{Re} \equiv dv\rho_1/\mu_1$, and the Schmidt number $\text{Sc} \equiv \mu_1/\rho_1\mathcal{D}_{A1}$. Measurements taken by Boundurant et al. from photographs yielded water droplet diameters $d = 0.25$ to 1.0 cm and droplet velocity $v = 0.5$ of impeller tip speed. The interfacial area generated per horsepower was found to be $a_{hp} = 1.2$ to 217 ft^2/hp with an average of 56.2. The interfacial area per unit volume may then be computed by

$$a_v^{(f)} = \frac{a_{hp}\omega}{V} \qquad (4.1\text{-}15)$$

The product $(^2k_{A1}a_v)^{(f)}$ can now be estimated from Eqs. 4.1-14 and 4.1-15. Reinhardt[5] has completed an investigation of this coefficient that provides a significantly improved estimate.

$(^1k_{A2}a_v)^{(n)}$—NATURAL SURFACE LIQUID PHASE COEFFICIENT

Operative transport mechanisms in the basin away from the immediate splashing and influence of the agitator are not unlike that which is operative in flowing streams and provides the aeration there. The stream aeration equation developed by Owens et al.[6] was transformed to yield a liquid phase mass transfer coefficient for any chemical absorbing or desorbing in the zone of natural convection

$$^1k_{A2}^{(n)} = 5.78\left(\frac{\mathcal{D}_{A2}}{\mathcal{D}_{B2}}\right)v^{0.67}h^{-0.85}(1.024)^{T-20°C} \qquad (4.1\text{-}16)$$

with $^1k_{A2}$ in lb mol/hr·ft^2, the average stream velocity v in cm/s, and the average depth h in cm. Details on obtaining realistic values of velocity and depth are contained in the original paper by Thibodeaux and Parker. The interfacial area per unit volume for the natural surface of the basin is obtained by

$$a_v^{(n)} = \frac{A}{V} \qquad (4.1\text{-}17)$$

where A is the plane surface area and V is the volume underneath. The product $(^1k_{A2}a_v)^{(n)}$ is obtained from Eqs. 4.1-16 and 4.1-17.

$(^2k_{A1}a_v)^{(n)}$—NATURAL SURFACE GAS PHASE COEFFICIENT

In the zone of natural convection the gas phase resistance is related to the movement of air across the plane surface of the basin. Water evaporation is gas phase controlling and gives a measure of this coefficient. Experimental field measurements on the evaporation of water from reservoirs and a resulting correlation developed by Harbeck[7] results in

$$^2k_{A1}^{(n)} = 3.0\text{E}-2\left(\frac{\mathcal{D}_{A1}}{\mathcal{D}_{C1}}\right)v_8A_r^{-0.05} \qquad (4.1\text{-}18)$$

where $^2k_{A1}^{(n)}$ is in lb mol/hr·ft^2, v_8 is the wind velocity at 8 m above the water surface in mi/hr, and A_r is the surface area of the reservoir in acres. The subscript C denotes water vapor. The film theory model was used in both Eqs. 4.1-16 and 4.1-18 to correct the original correlation for chemical species A. Equation 4.1-17 combined with 4.1-18 yields the product $(^2k_{A1}a_v)^{(n)}$.

With sufficient operating details on an existing (or planned) aerated (or unaerated) basin the individual mass transfer coefficients of species A can be estimated by Eqs. 4.1-13 through 4.1-18. Using the two-resistance theory, Eqs. 4.1-11 and 4.1-12 and assuming parallel desorption, we obtain the overall coefficient

$$^1K_{A2}a_v = \left(^1K_{A2}a_v\right)^{(n)} + \left(^1K_{A2}a_v\right)^{(f)} \qquad (4.1\text{-}19)$$

The completely mixed model equations or the plug flow model equations allows computation of desorption rates and percent removals. (See Problem 4.1F.)

A DESORPTION STUDY INVOLVING 13 BASINS AND 11 CHEMICALS

A simulation study of 13 existing aerated basins employed in the pulp and paper industry is summarized in Tables 4.1-1 through 4.1-3. Eleven common industrial chemicals were chosen for the study. A summary of the individual mass transfer coefficients computed in the study is shown in Table 4.1-1. The variation reflects the variation in chemical species and basin properties. Table 4.1-2 gives the interfacial areas in the basins and compares the size of the natural and generated surfaces. The surface agitators *do not* increase the interfacial area significantly; however, the product $(^1k_{A2}a_v)^{(f)}$ is approximately 100 times the product $(^1k_{A2}a_v)^{(n)}$.

Table 4.1-1. Mass Transfer Coefficients for Aerated Basins
(lb mol/hr·ft^3)

Coefficient	Average	Range
$(^1k_{A2}a_v)^{(f)}$	1.37	6.2–0.1
$(^2k_{A1}a_v)^{(f)a}$	0.00225^b	0.0092–0.00020^b
$(^1k_{A2}a_v)^{(n)a}$	0.0113	0.020–0.0051
$(^2k_{A1}a_v)^{(n)}$	0.00474	0.00847–.00173

(Reprinted by permission, *Source*.   Reference 1.)

^{a}Denotes the controlling coefficients.
^{b}Techniques used to estimate this coefficient are extremely rough.

Table 4.1-2. Interfacial Mass Transfer Area in Aerated Basins, Natural and Generated Area

Model Input Generated Area (ft^2/hp)	$a_v^{(f)}$ (ft^{-1})		Relative Quantity (%) of Natural Area^a (Based on Average)
	Range	Average	
1.2	0.0000086–0.00044	0.00011	99+
56.2	0.00040–0.021	0.00495	95
217.0	0.0016–0.079	0.0191	84

(Reprinted by permission, *Source*.   Reference 1.)

^{a}The natural area average is 0.0980 with a range of 0.056 to 0.14 ft^{-1}.

Table 4.1-3. Relative Quantities Desorbed in Aerated Basins
@ 25°C
(Average and Range)

Component	H_{XA}	$F_M(\%)$	$F_P(\%)$
Acetaldehyde	5.88	46.9 (20–82)	58.2 (22–99)
Acetone	1.99	28.8 (9.7–61)	34.3 (10–79)
Isopropanol	1.19	16.6 (4.8–41)	18.6 (4.9–50)
n-Propanol	0.417	9.35 (2.5–24)	10.0 (2.6–27)
Ethanol	0.363	9.10 (2.5–24)	9.72 (2.5–27)
Methanol	0.300	10.0 (2.7–26)	10.7 (2.8–29)
n-Butanol	0.182	3.60 (0.93–9.6)	3.70 (0.93–10)
Acetic acid	0.0627	2.07 (0.52–5.7)	2.11 (0.52–5.9)
Formic acid	0.0247	1.06 (0.26–3.0)	1.06 (0.26–3.0)
Propionic acid	0.0130	0.347 (0.086–.96)	0.348 (0.087–0.96)
Phenol	0.0102	0.241 (0.061–.64)	0.241 (0.061–0.64)

(Reprinted by permission, *Source.* Reference 1.)

Table 4.1-3 indicates that significant desorption of common industrial chemicals can occur from aerated basins.

No field observations on desorption rates are available to corroborate or discredit the range of predicted relative removals reported here. Laboratory stripping experiments provide the only experimental data to verify the significance of the preceding simulation model. Gaudy[8] reports 50% butanone removal in 4 hr by diffused air in a batch process and 50% propionaldehyde removal in 3 hr, 50% butyraldehyde in 2 hr, 50% valeraldehyde in 1 hr, and 50% acetone in 4.5 hr, all at 25°C. Eckenfelder[9] reports 20% removal of acetone in 4 hr in a similar apparatus at 20°C. The variability of sparged vessel results is due to the variations in gas rate–liquid volume ratios.

Goswami[10] reports that significant removals can occur under quiescent conditions. He observed the following removals in 12 hr from a 1 L reactor at 25°C: 17.2% removal of acetone, 18% methylethylketone, 27% propionaldehyde, and 35.8% valeraldehyde.

Wachs, Folkman, and Shemesh[11] report on the use of surface stirrers to promote the transfer of ammonia to the atmosphere. In all experiments a cylindrical tank of diameter 1.42 m was used. Ammonium chloride solution was introduced in the tank along with sufficient lime slurry to reach a pH of 11. The resulting liquid depth was 70 cm. The initial ammonia concentration was 80 mg/L. To study the effect of wind velocities a series of experiments were carried out in a wind tunnel where the wind velocity could be regulated by the use of screens and measured by a network of thermoanemometers.

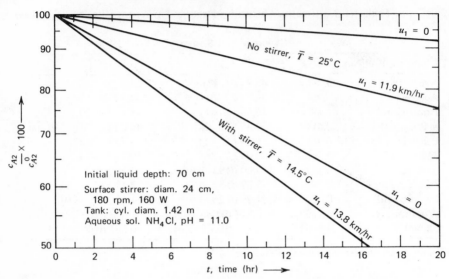

Figure 4.1-3. Ammonia desorption using surface stirrers. (Reprinted with permission from Reference 11.)

The surface agitators were of a type used in the biological treatment of wastewater, involving a turbine, made of a disk of perspex to which curved blades were attached. Two impellers 16 cm and 24 cm in diameter were employed.

Results of runs in the wind tunnel with and without surface stirring are shown in Fig. 4.1-3. There is a significant improvement in desorption resulting from the use of surface stirrers. Even when the surface stirrers were used, wind velocity has an appreciable effect on the desorption rates. Ammonia desorption rate is controlled by significant resistance in both phases.

Laboratory Simulations of Natural and People-Influenced Desorption (or Absorption) of Chemicals

As noted earlier, it is common practice to perform laboratory-scale and pilot-scale experiments with finite batches of water. Many environmental simulation experiments such as ammonia desorption and oxygen absorption are performed much easier in batch experiments than in continuous flow experiments. Time becomes the independent variable; however, with the proper mathematical description the important environmental coefficient can be extracted and used in flow situations with a degree of confidence.

Mass transfer involving absorption or desorption across a water–air interface is a first-order equation. (See Eqs. 3.1-30 and 3.1-31.) The chemical of study, for example ammonia, is placed in the batch of water in a concentration corresponding to typical environmental situations. In this case the process is unsteady state and the concentration of ammonia decreases with time just as does the desorption rate. The proper expression describing this transient behavior is

$$\ln\left(\frac{c_{A2}}{c_{A2}^{0}}\right) = -K_{\text{des}}t \tag{4.1-20}$$

where K_{des} is the experimentally observed desorption rate constant (t^{-1}). Figure 4.1-4 shows a typical apparatus arrangement.

For the case of the absorption of oxygen into water from air, assuming sufficient is available so that oxygen concentration in the air remains unchanged, and for trace chemicals in air, the proper expression is

$$\ln\left(\frac{c_{A2}^{*} - c_{A2}}{c_{A2}^{*} - c_{A2}^{0}}\right) = -K_{\text{abs}}t \tag{4.1-21}$$

where K_{abs} is the experimentally observed absorption rate constant. A denotes oxygen or any chemical being absorbed into the water.

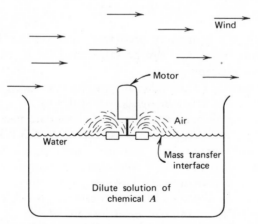

Figure 4.1-4. Laboratory simulation of batch desorption or absorption.

Experimental investigations performed in the laboratory as described earlier yield concentration-time data that can be used in Eqs. 4.1-20 and 4.1-21 to obtain the specific rate constants. Since water readily desorbs into air corrections for water losses may need to be made if significant evaporation occurs. Once the proper constant is obtained an additional transformation yields the overall mass transfer coefficient compatible with conventional rate equations. Using Eq. 3.1-31 as a guide, the experimental desorption coefficient is related to the overall liquid phase mass transfer coefficient by

$$^1K_{A2} = \frac{\mathfrak{M}}{A} K_{\mathrm{des}} = c_B h K_{\mathrm{des}} \tag{4.1-22}$$

where $\mathfrak{M}$ is the total moles of mixture in the batch, A is the plane area of the gas–liquid interface, and h is the depth of the liquid mixture in the uniform cross-sectioned vessel. Usually the chemical of study is present as a very dilute solution so that c_B is essentially the molar density of water, which is 0.0556 mol/cm^3 at 20°C.

Occasionally the product $^1K_{A2}a_v$ is useful, where a_v is the plane interfacial area per unit volume. In the case of uniform cross-sectioned vessels $a_v = 1/h$. For use in the rate equation

$$N_{A0} = {}^1K'_{A2}\Delta c_{A2} \tag{4.1-23}$$

where $^1K'_{A2}$ is the overall coefficient in L/t, the following relationship is applicable:

$$^1K'_{A2} = h K_{\mathrm{des}} \tag{4.1-24}$$

Equation 4.1-23 is a commonly used rate equation that employs a concentration driving force. K_{abs} can be substituted for K_{des} in the preceding equations for the absorption case.

Coefficients obtained from laboratory apparatus assembled to simulate environmental effects must be used with caution. In general, observed coefficients are overall mass transfer coefficients and measure the combination $^1k_{A2}$, $^2k_{A1}$, and H_{XA}. (See Eqs. 3.1-32 and 3.1-33.) Coefficients measured for one chemical do not necessarily apply to another even though identical environmental conditions of temperature, wind, and liquid motion are known to exist. However, when the proper chemical is chosen, it is possible that the observed coefficient is one of the individual phase coefficients. For example, measurement of the oxygen absorption rate into water (i.e., aeration) is in fact a measure of the individual liquid phase mass transfer coefficient.

Gas Exchange Rates between the Atmosphere and the Surface of Rivers

Mass transfer flux rates at the gas–liquid interface of rivers and streams are in general more rapid than in large bodies such as lakes and oceans. More rapid exchange rates are due primarily to enhanced liquid phase coefficients. Vigorous liquid side turbulence is present and is generated by the flowing water. Natural streamflow, a form of open channel flow, is normally turbulent. The Reynolds number for open channel flow is

$$\mathrm{Re} \equiv \frac{4r_h\rho\bar{v}}{\mu} \qquad (4.1\text{-}25)$$

where r_h is the hydraulic radius, defined as the area of the stream cross section divided by the wetted perimeter. Values of the hydraulic radius for some common cross sections are given in Table 4.1-4. The transition from laminar to turbulent flow in open channels occurs at Reynolds numbers between 2000 and 4000. Most natural streams are in turbulent flow.

AMMONIA DESORPTION FROM RIVERS AND STREAMS

The desorption of ammonia from natural streams is a specific case of some importance; however, the concepts developed may be applied to the desorption of any component. Free ammonia in water behaves as a dissolved gas and therefore exerts a vapor pressure in the surrounding air. Ammonia is frequently present in streams, its source being domestic wastewater, industrial wastewater, or agriculture runoff.

The equilibrium vapor pressure of free ammonia above an aqueous solution is a strong function of pH and temperature. Gases such as H_2S and NH_3 in water are partially dissolved as gases and partially ionized (dissociate) in water. Other important factors affecting reactive gas evolution (or solubility) are

1. The solubilities of the undissociated gases themselves.
2. The pH of the solution.
3. The dissociation constant of the reactants.
4. The concentration of other dissolved substances in solution.

Substances that react in aqueous solution to accept or donate electrons (and consequently donate or accept protons) dissolve to a degree dependent on solution pH and their dissociation constant K. The pH affects gas

Shape and/or Cross Section		r_h^a	a_v^b
Rectangle, depth h, width b		$\dfrac{bh}{b+2h}$	$\dfrac{1}{h}$
Semicircle, free surface on a diameter d		$\dfrac{d}{4}$	$\dfrac{8}{\pi d}$
Wide shallow stream rectangle with $h \ll b$		h	$\dfrac{1}{h}$
Triangle trough, angle $= 90°$		$\dfrac{b}{4} = \dfrac{h}{2\sqrt{2}}$	$\dfrac{2\sqrt{2}}{b} = \dfrac{2}{h}$
Trapezoid, depth h, bottom width b, angle $60°$		$h\left(\dfrac{b+h/\sqrt{3}}{b+4h/\sqrt{3}}\right)$	$\dfrac{1}{h}\left(\dfrac{b+h}{b+h/2}\right)$
Angle $45°$		$h\left(\dfrac{b+h}{b+2\sqrt{2}\,h}\right)$	$\dfrac{1}{h}\left(\dfrac{b+hy}{b+h}\right)$

$^a r_h \equiv$ cross-sectional area ÷ wetted perimeter.
$^b a_v \equiv$ air–water interfacial area ÷ volume of water.

evolution through the interaction of hydrogen ions with gases that dissociate. The equilibrium reaction between ammonium ion and gaseous ammonia is

$$NH_4^+ \rightleftharpoons NH_3 + H^+ \qquad (4.1\text{-}26)$$

Ionization removes the reactive gas from solubility considerations. The equilibrium constant for the ammonium ion dissociation reaction is

$$K \equiv \frac{c_{NH_3} c_{H^+}}{c_{NH_4^+}} \qquad (4.1\text{-}27)$$

This last equation shows the relationship between pH (i.e., C_{H^+}) and free ammonia concentration, C_{NH_3}. Temperature affects both solubility and the dissociation constant. Click and Reed[12] have developed tables of equilibrium concentrations of hydrogen sulfide and ammonia over aqueous solutions as a function of temperature and pH. Table 4.1-5 contains Henry's law constant (mole fraction form) data for ammonia between water and air.

Table 4.1-5. Henry's Law Constant (Mole Fraction Form: $y_A = H_{XA} x_A$) for Ammonia Gas in Water

pH	Temperature (°F)			
	40°	60°	80°	100°
6	0.000266	0.000754	0.00198	0.00486
7	0.00266	0.00753	0.0197	0.0480
8	0.0263	0.0734	0.186	0.428
9	0.238	0.586	1.20	2.05
10	1.20	1.94	2.65	3.31

Source. Reference 12.

Bulk water motion moves dissolved ammonia from the stream depths and makes it available at the air–water interface. Desorption occurs at the interface and surface winds carry the ammonia molecules away from the air space near the interface. The flow of small streams approaches the plug flow model closely while short stretches of large streams and pools may more nearly follow the completely mixed flow model. Mixing in the vertical and longitudinal directions is usually complete except for the case of some streams receiving warm water discharges. In this case thermal stratification may be present for a certain distance below the discharge.

Relations developed earlier for treatment basins may be applied directly to describe the desorption of ammonia from natural streams. These relations are also valid for the desorption (or absorption with proper modification) of any chemical from a flowing body of water. Equation 4.1-8 in a slightly altered form (see Problem 4.1H) relates the fraction desorbed

$$F_P = 1 - \exp\left(-\, {}^1K'_{A2}a_v\tau\right) \tag{4.1-28}$$

where ${}^1K'_{A2}$ = overall liquid phase mass transfer coefficient, L/t
a_v = stream air–water interfacial area per unit volume, L^2/L^3
τ = mean water residence time, t
The completely mixed stream model is essentially Eq. 4.1-5:

$$F_M = 1 - \frac{1}{(K'_{A2}a_v\tau + 1)} \tag{4.1-29}$$

Equations 4.1-9 and 4.1-6 give the desorption rates of ammonia from the stream for the plug and mixed flow models, respectively. Equation 4.1-11 gives the overall coefficient in terms of the individual phase coefficients.

Equations appearing previously may be used to estimate the individual phase coefficients. The liquid phase coefficient can be determined from Eq. 4.1-16 or the oxygen transfer coefficients in Table 1.2-1 properly modified. The gas phase coefficient can be estimated from Eq. 4.1-18 and 4.2D-1 (if corrected). No specific work on stream gas phase coefficients is available.

Tributaries often intercept the stream of interest and dilute the ammonia concentration. The desorption rate from water is concentration dependent as shown by Eqs. 4.1-9 and 4.1-6; however, the fraction desorbed is concentration independent as shown by Eqs. 4.1-28 and 4.1-29. Consider the subsectioned stream shown in Fig. 4.1-5a, which has three tributaries. The four subsections between the discharge point and Highway No. 7 bridge can be modeled simply for the plug flow case. For the first subsection the fraction desorbed is

$$F_P(1) = 1 - \exp\left[-\left({}^1K'_{A2}a_v\tau\right)_1\right] \tag{4.1-30}$$

The fraction desorbed through the second and succeeding subsections are calculated by

$$F_P(2) = 1 - \exp\left[-\left({}^1K'_{A2}a_v\tau\right)_1 - \left({}^1K'_{A2}a_v\tau\right)_2\right], \tag{4.1-31}$$

$$F_P(3) = 1 - \exp\left[-\left({}^1K'_{A2}a_v\tau\right)_1 - \left({}^1K'_{A2}a_v\tau\right)_2 - \left({}^1K'_{A2}a_v\tau\right)_3\right] \tag{4.1-32}$$

$$F_P(4) = 1 - \exp\left[\left({}^1K'_{A2}a_v\tau\right)_1 - \left({}^1K'_{A2}a_v\tau\right)_2 - \left({}^1K'_{A2}a_v\tau\right)_3 - \left({}^1K'_{A2}a_v\tau\right)_4\right]$$

$$\tag{4.1-33}$$

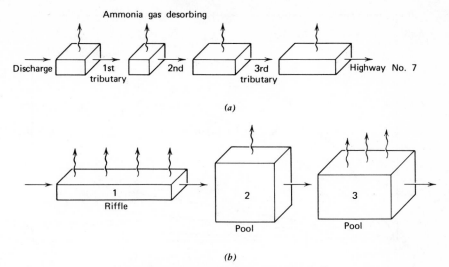

(a)

(b)

Figure 4.1-5. Stream subsectioning.

where $^{1}K'_{A2}a_v\tau$ is characteristic of the stream section between tributaries. The same subsectioning technique may be extended to quantify desorption in long streams accurately without significant tributaries but where landforms result in vastly different $^{1}K'_{A2}$, a_v, and τ values. Terrain resulting in pool-and-riffle-type streams may require this type of subsectioning, as shown in Fig. 4.1-5b. Because of the nonlinearity of the relations (i.e., Eq. 4.1-28 and 4.1-29) the use of average values can produce significant errors, usually yielding lower desorptions estimates than what is actually occurring.

The stream depth is reflected for the most part by the a_v term. Table 4.1-4 gives a_v values for various stream cross sections. For the case of wide streams a_v is essentially the reciprocal of the stream depth. Accurate data on stream depth, which are particularly difficult to obtain for small streams, are essential if realistic desorption predictions are to be made. Figure 4.1-6 shows the sensitivity of the fraction of ammonia desorbed with stream depth for a small stream in south Arkansas. Desorption in deep rivers is significantly smaller than in small streams (see Problem 4.1M) mainly owing to depth.

ESTIMATING MASS TRANSFER COEFFICIENTS FOR PONDS, RIVERS, AND SMALL WATER BODIES

In the absence of specific studies made to determine mass transfer coefficients in various environmental settings it is necessary to employ

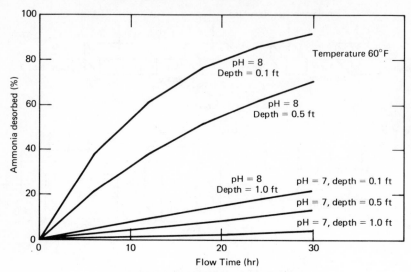

Figure 4.1-6. Predicted ammonia desorption.

existing correlations and a certain amount of inventiveness to obtain reasonable estimates. A host of correlations exists in the literature based on both laboratory and field measurements. With the proper mass transfer theory and fluid dynamic interpretations this existing information can be transformed to fit new environmental situations. This tack was employed in estimating the coefficients used for the desorption of chemicals from aerated stabilization basins.

Blosser et al.[13] reported velocities, zone-of-aerator influence, and pumping characteristics observed in a field study involving numerous full-scale surface aerators used in oxygenating paper industry wastewater. Based on the results in that report it was recognized that the hydraulics behavior of the water flow in the basin zone beyond the immediate influence of the mechanical device was not unlike the flow in a natural stream. The pumping action of the impellers caused the basin surface water to have a definite velocity component outward and away from the aerator. At some point this outward bound flow blended with that of adjacent aerators. Continuity was maintained by a backflow in the opposite direction toward the aerator. The return flow to the aerator was detected underneath the outward surface flow, as shown in Fig. 4.1-7. From reported values of surface flow velocity and flow depth (D in Fig. 4.1-7) reasonable estimates of v and h for Eq. 4.1-16 were available and a value of the liquid phase coefficient was obtained. The range of velocities and effective water depths

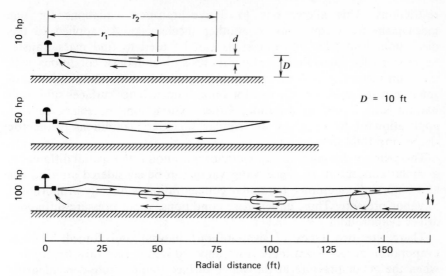

Figure 4.1-7. Zone-of-aerator influence in mixing basin water. (Reprinted with permission from Reference 13.)

reported by Blosser et al. were 0.3 to 72 cm/s and 30 to 210 cm, respectively.

For estimating the gas phase coefficient a simple correlation developed for the evaporation of water from lakes was transformed and used for each chemical species (Eq. 4.1-18). That particular equation is possibly adequate for the specific application of the aerated stabilization basin (ASB) environment; however, a more sophisticated equation is needed for general use in estimating the gas phase mass transfer coefficient for small water bodies.

The mass transfer coefficient for the movement of a chemical species in the air boundary layer above a body of water is influenced by the wind velocity, wind fetch, and air stability. The latter parameter is related to the direction of water vapor movement and accounts for free convection buoyancy above highly heated ponds and sections of lakes and rivers receiving thermal discharges. Goodling et al.[14] have reviewed existing water evaporation equations for open bodies exposed to the atmosphere and have converted standard horizontal flat plate heat transfer relationships to a mass transfer or evaporation equation.

As shown in a previous section, where the analogy theories were presented, the similarity of the governing equations for heat, mass, and momentum make possible similar solutions for all three under several restrictions. This allows one to derive transport equations for one mechanism from equations of another mechanism. As applied to this discussion, heat transfer expressions can be used to find mass transfer expressions for equivalent configurations and boundary conditions with the main restriction that the mass transfer rate be small. The restriction of small transfer rates is easily met for smooth open water surfaces subject to natural and forced evaporation. Other assumptions necessary for the application of this analogy are similar boundary conditions, geometric shape, and fluid flow.

The primary driving force for water evaporation is the spatial differences in molar concentration. Since water vapor can be considered an ideal gas under normal conditions, the molar concentration is directly proportional to vapor pressure. Thus, under these conditions, vapor pressure difference drives evaporation.

There are two types of evaporation: natural and forced. Natural evaporation occurs when there is no imposed free stream flow (wind) and when the vapor pressure at the water surface (temperature-dependent) is greater than that in the free stream (dependent on relative humidity and temperature). Since the molecular weight of water vapor ($M = 18$) is less than that of air ($M \sim 29$), the diffusion of this lighter gas (water vapor) into the surrounding water–air mixture leads to a positive buoyant force and therefore upward macroscopic movement. The phenomena of natural evaporation is not unlike the problem of heat transfer from a hot plate facing upward, that is, buoyancy-induced motion. When the vapor pressure of the water at the surface is less than that in the air above, the situation is stable and the resulting condensation is governed by the slow process of molecular diffusion and laminar flow.

Water vapor movement away from the zone of "natural" convection shown in Fig. 4.1-1 occurs by the parallel mechanisms of natural and forced evaporation. Since water ($H_2O \equiv B$, air $\equiv 1$) is gas phase controlled the rate equation for evaporation is

$$N_{B0} = \left({}^2k_{B1N} + {}^2k_{B1F} \right) \left(y_{Bi} - y_B \right) \qquad (4.1\text{-}34)$$

The effective gas phase coefficient is the sum of the natural (subscript N) and forced (subscript F) components. As noted earlier, in the absence of a mass transport equation, one may borrow from a known heat transport equation and convert. Using the experimental correlation recommended by McAdams[15] based on the work of W. Kraus for turbulent natural convection over a horizontal flat plate of length L, transformed with dimensionless parameters, we obtain

$$\text{Sh}_{B1N} = 0.14 [\text{Gr}_{B1} \text{Sc}_{B1}]^{1/3} \qquad (4.1\text{-}35)$$

where Sh, Gr, and Sc are the Sherwood, Grashof, and Schmidt numbers, respectively. The numbers are defined as follows:

$$\text{Sh}_{B1N} \equiv \frac{^2 k_{B1N} L}{c \mathcal{D}_{B1}}, \qquad \text{Gr} \equiv \frac{g \zeta_{B1} L^3 (y_{Bi} - y_B)}{\nu^2}, \qquad \text{Sc} \equiv \frac{\nu}{\mathcal{D}_{B1}}$$

$$(4.1\text{-}36a, b, c)$$

The Grashof number contains the parameter ζ_{B1}, termed the concentration coefficient of volume expansion, the definition of which is

$$\zeta_{B1} \equiv -\frac{1}{\rho_1} \left(\frac{\partial \rho_1}{\partial y_B} \right)_{p,T} \qquad (4.1\text{-}37)$$

where ρ_1 is the mass density of the air–water vapor mixture. For ideal gas mixtures this coefficient may be obtained by

$$\zeta_{B1} = \frac{-1}{y_{Bi} + M_1/(M_B - M_1)} \qquad (4.1\text{-}38)$$

For dilute solutions $y_{Bi} \ll 1$, Eq. 4.1-38 becomes $\zeta_{B1} = 1 - M_B/M_1$. Equation 4.1-35 is applicable only when $y_{Ai} > y_A$. Should $y_{Ai} < y_A$, the resulting heat and mass transfer additions due to condensation are so small as to be considered negligible and in this case the natural convection portion of the coefficient is omitted.

When air flows horizontally over a water surface, the resulting evaporation is termed *forced*. Again, we draw from known heat transfer equations and convert them to mass transfer equations for the reasons stated earlier. For even mild winds above an open span of water, the flow is turbulent and fully developed. For this situation the turbulent equation recommended by McAdams is used:

$$\text{Sh}_{B1F} = 0.036 \, \text{Re}^{0.8} \text{Sc}_{B1}^{1/3} \qquad (3.1\text{-}21)$$

where Re is the Reynolds number. The new numbers are defined as follows:

$$\text{Sh}_{B1F} \equiv \frac{^2 k_{B1F} L}{c \mathcal{D}_{B1}}, \qquad \text{Re} \equiv \frac{v_\infty L}{\nu} \qquad (4.1\text{-}39a, b)$$

where L is the fetch of the water body and v_∞ is the wind speed measured 8 to 10 m above the surface.

The water evaporation coefficient sum in Eq. 4.1-34 may be obtained from Eqs. 4.1-35 and 3.1-21:

$$^2k_{B1N} + {}^2k_{B1F} = \frac{(0.14\,\mathrm{Gr}_{B1}^{1/3} + 0.036\,\mathrm{Re}^{0.8})c\mathcal{D}_{B1}\,\mathrm{Sc}_{B1}^{1/3}}{L} \qquad (4.1\text{-}40)$$

Goodling et al.[14] compared the natural convection and forced convection portions of Eq. 4.1-40 with empirical equations based on field observations and evaporative pan studies. He concluded that the natural convection portion gives results "between and within the range of" the empirical equations, and for the forced convection portion Eq. 4.1-40 gives results that "fall within those found by previous investigators but display a dependency on total fetch of the water body."

Equation 4.1-40 provides the means for specifying an individual gas phase coefficient for the movement of chemical species A in either direction (i.e., absorption or desorption) through the gas boundary layer. For arbitrary chemical A, with the coefficient definition $^2k_{A1} \equiv {}^2k_{A1N} + {}^2k_{A1F}$, Eq. 4.1-40 becomes

$$^2k_{A1} = \frac{\left[0.14\,\mathrm{Gr}_{B1}^{1/3} + 0.036\,\mathrm{Re}^{0.8}\right]c\mathcal{D}_{A1}\,\mathrm{Sc}_{A1}^{1/3}}{L} \qquad (4.1\text{-}41)$$

where $B \equiv H_2O$ and $1 \equiv$ air. The subscript B is retained on the Grashof number because water vapor buoyancy provides the air mixture with macroscopic movement when $y_{B1} > y_B$. In essence, water evaporation can have a decided effect on the individual gas phase mass transfer coefficient of chemical A for a small water body in the absence of wind.

DeWalle[16] used the Shulyakovoskiy equation, which is similar to Eq. 4.1-40, for the prediction of downstream water temperature and compared the results with field measurements from a thermally loaded reach of the west branch of the Susquehanna River near Shawville, Pennsylvania. The equation tended to overestimate the heat loss and consequently positive prediction errors were obtained. The average error was only $0.89 \pm 0.25\,°C$. A preliminary study was conducted to determine the effects of wind direction on water evaporation prediction.

Wind direction can affect predictions of evaporation by controlling the fetch as well as the velocity of the air relative to the direction of motion of the stream. Fetch is greatest for upstream and downstream winds and lowest for cross-stream winds. Equation 4.1-41 accounts for fetch through the variable L. The horizontal wind velocity used in any equation to compute the forced convection contribution should be the velocity of the stream, v'_1, according to

$$v'_1 = |v_1 - v_2 \cos\theta| \qquad (4.1\text{-}42)$$

where v_1 is the wind velocity measured from a stationary mast, θ is the angle between the stream direction and the wind direction in degrees, and v_2 is the stream velocity. Thus if a 1 ms^{-1} upstream wind ($\theta = 180°$) is measured and the stream velocity is 1 ms^{-1}, v'_1 is 2 ms^{-1}. But if a 1 ms^{-1} wind is in the downstream direction ($\theta = 0°$), the relative velocity is zero. Only when the wind is blowing perpendicular to the stream direction is correction for relative motion unnecessary. Correction of wind velocity data for relative motion improved the predictions of water temperature in the Susquehanna River slightly.

Problems

4.1A. MODEL FOR CHEMICAL DESORPTION FROM A PLUG FLOW BASIN

1. Derive the differential equation that describes the desorption in a plug flow basin. (*Hint:* Using Eq. 1.1-8 and Fig. 4.1A as a guide write a mass balance on component A for an elemental volume ΔV.)
2. Integrate the result of part 1 with the proper boundary conditions and use the definition of F to obtain Eq. 4.1-8.

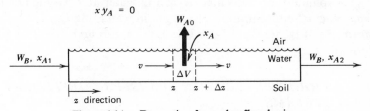

Figure 4.1A. Desorption from plug flow basin.

4.1B. DESORPTION OF ETHANOL AND PHENOL FROM AN AERATED STABILIZATION BASIN

1. Compute the fraction (F_M and F_P) of ethanol and phenol desorbed at 25°C from a basin containing 50E 6 gal with a flow of 10E 6 gal/d. Use the average coefficients given in Table 4.1-1. Report the fraction desorbed as percentages.

2. Compute the desorption rate of each chemical for each basin type in k mol/d. The inlet concentrations are 100 mg/L and 10 mg/L for ethanol and phenol respectively.

3. In the cases of ethanol and phenol, which phase resistance controls the rate of mass transfer?

4.1C. A LIQUID PHASE MASS TRANSFER COEFFICIENT FOR MECHANICAL SURFACE AGITATORS

As given in Section 3.1, individual phase mass transfer coefficients are scalars and are therefore independent of the direction of the concentration driving force. This means that a coefficient measured under absorption conditions can be transformed and used in desorption applications.

Experimental observations on the absorption of molecular oxygen into water by mechanical surface aerators has resulted in the correlation:[2]

$$W_B' = \frac{n_{B0}' E \omega \alpha}{9.17} (1.024)^{T-20°C} [\rho_{B2}^* - \rho_{B2}] \text{ lb } O_2/\text{hr} \qquad (4.1C-1)$$

where $n_{B0}' = 2$ to 4 lb O_2/hr·hp depending on the specific aerator

E = specific aerator power delivery efficiency, 0.65 to 0.9, dimensionless

ω = nameplate horsepower, hp

α = correction factor to clean water–oxygen mass transfer for specific wastewater, 0.8 to 0.85, dimensionless

T = water temperature, °C

ρ_{B2}^* = solubility of molecular oxygen in water at T°C, mg O_2/L

ρ_{B2} = actual concentration of oxygen in water, mg O_2/L

Transform Eq. 4.1C to Eq. 4.1-13. Use Eq. 3.1-12b and 3.1-18 to aid the transformation.

4.1D. A LIQUID PHASE MASS TRANSFER COEFFICIENT FOR NATURAL STREAMS

See the introduction given in Problem 4.1C. The following equation is given for the absorption of oxygen from air by natural streams[6]:

$$^1k_{A2}' = 27.5 v_2^{0.67} h^{-0.85} (1.024)^{T-20°C}, \text{cm}/\text{hr} \qquad (4.1D-1)$$

where v_2 is the average water velocity in ft/s, and h is the average stream depth in ft. Transform Eq. 4.1D to Eq. 4.1-16. Use Eq. 3.1-11b and 3.1-18 to aid the transformation.

4.1E. A GAS PHASE MASS TRANSFER COEFFICIENT FOR LAKES AND RESERVOIRS

Experimental field measurements on the evaporation of water from reservoirs has been correlated by Harbeck[7] to give the equation

$$n'_{B0} = 3.38 E - 3 v_2 A_r^{-0.05} (p^*_{B1} - p_{B1})$$ (4.1E-1)

where n'_{B0} = water evaporation rate in in./d
  v_2 = wind speed at 2 m ($v_2 = v_8/1.375$) in mi/hr
  A_r = reservoir surface in acres
  p^*_{B1} = saturated vapor pressure of water at the surface temperature in millibars
  p_{B1} = actual vapor pressure of water at 2 m above the reservoir surface in millibars

Transform this equation to give the mass transfer Eq. 4.1-18. Use Eq. 3.1-11a and 3.1-18 to aid the transformation. (Note that 1 bar = 0.9869 atm.)

4.1F. DESORPTION OF AMMONIA FROM AN ASB

Compute the fractional removal and rate (kg/d) of ammonia by desorption using both the completely mixed and plug flow models. The following data are available ($A \equiv NH_3$, $B \equiv O_2$, $1 \equiv$ air, $2 \equiv H_2O$): $v_8 = 4$ mi/hr, $T_1 = T_2 = 25°C$, $v_2 = 0.1$ ft/s, $h = 7$ ft, $\alpha = 1.0$, $n'_{B0} = 3$ lb/hr·hp, $E = 0.83$, $d = 1.0$ cm, $a_{hp} = 56.2$ ft^2/NPHP, $A = 65.5$ acres, $Q_2 = 27E\ 6$ gal/d, $V = 384E\ 6$ gal, 13 low speed (50 rpm) aerators of 100 NPHP each, impeller diameter = 3 ft, $\rho_{A21} = 15$ mg A/L, Sc = 0.61 for A, $\mathcal{D}_{A1} = 0.1$ cm^2/s, $\mathcal{D}_{A21} = 2E-5$ cm^2/s, $\mu_2 = 1.2E-5$ lbm/ft·s

4.1G. COEFFICIENT CONTROLLING THE MASS TRANSFER RATE

Show that the liquid phase coefficient dominates the mass transfer rate for the absorption (or desorption) of oxygen near the agitator in an ASB. Which phase controls the desorption of H_2S in this same region? Use the forced coefficients in Table 4.1-1 in your calculations.

4.1H. DESORPTION AS A FUNCTION OF RESIDENCE TIME

It is often convenient to use residence time (or detention time) as a desorption parameter. The group of terms $^1K_{A2}a_v V/W_B$ appearing in Eqs. 4.1-5 and 4.1-8 is dimensionless. $W_B = Q_B c_B$, where Q_B is the volumetric flow rate through the basin and c_B is the molar density of water. V/Q_B is the residence time (or detention time) of the water in the basin, defined as τ, and $c_B/{}^1K_{A2}a_v$ is a characteristic mass transfer time for the component in the particular basin. Equations 4.1-5 and 4.1-8 can be written as

$$F_M = 1 - \cfrac{1}{1 + \cfrac{{}^1K_{A2}a_v}{c_B}\tau_m} \qquad (4.1\text{H-1})$$

and

$$F_p = 1 - \exp\left(-\frac{{}^1K_{A2}a_v}{c_B}\tau_p\right) \qquad (4.1\text{H-2})$$

1. If $F \leqslant 0.05$ is an insignificant amount of desorption show that the basin-water residence times must be

$$\tau_m \text{ and } \tau_p > \frac{0.05c_B}{{}^1K_{A2}a_v}$$

for a significant amount of desorption to occur.
2. Compute the minimum significant residence time for ethanol and phenol. Use the average coefficients given in Table 4.1-1.

4.1I. PILOT PLANT STUDY OF AMMONIA DESORPTION[11]

The results of a "batch" run, carried out for 12 hr in a pilot plant desorption of ammonia from lime-treated waste effluent of a stabilization pond are given in Table 4.1I. The experimental conditions were pH range

Table 4.1I. Pilot Plant Desorption of Ammonia

Time (hr)	Concentration (mg/L)
0	23
1	22.5
2	22.5
3	18.9
4	17.9
5	15.9
7	13.6
9	11.5
11	10.4
12	8.4

Source. Reference 11.

10.4 to 10.9, aerator impeller diameter 24 cm, 160 rpm, liquid depth 65 cm, temperature range 25.8 to 28.5°C, wind velocity 2 to 13.5 km/hr.

1. Compute the ammonia desorption coefficient (hr^{-1}).
2. Discuss the general utility of the coefficient determined in part 1 for
   (a) Ammonia desorption in aerated vessels operated under different experimental conditions of pH, temperature, impeller diameter and speed, and wind speed.
   (b) Hydrogen sulfide desorption under identical experimental conditions.

4.1J. SIMULTANEOUS ACETONE DESORPTION AND OXYGEN ABSORPTION

Acetone desorption and oxygen absorption were studied simultaneously in a pilot-scale surface-agitated vessel. Experimental conditions: 0.075 hp electric motor, single-blade 3.0 in. diameter impeller, vessel 50 in. in diameter, water depth 8.75 in., floor fan to simulate wind at 3 to 4 mph, temperature 18 to 24.5°C, impeller speed 465 to 545 rpm, 35.3 g Na_2SO_3 and 1.08 g CoO were used to create an oxygen deficit in the vessel. The experimental results are shown in Table 4.1J.

Table 4.1J. Simultaneous Absorption and Desorption

Acetone Desorption Data		Oxygen Absorption Data	
time (mins)	Concentration (mg C/L)	time (mins)	Concentration (mg O_2/L)
0	3170	0.0	1.95
18	3140	1.5	2.50
35	2920	3.87	2.80
52	2480	5.83	3.20
67	2170	8.13	3.70
90	2050	10.9	4.10
127	1780	14.3	4.60
157	1490	20.1	5.20
363	660	26.6	5.80

1. Determine the experimental desorption coefficient for acetone (hr^{-1}).
2. Determine the experimental absorption coefficient for oxygen (hr^{-1}).
3. Explain the difference, if any, in the coefficients for desorption and desorption.

4. The physical measurements from the vessel, impeller, and so on, and the techniques developed in this chapter were used to calculate a coefficient of 0.701 lb mol/hr·ft^3. Convert the acetone experimental coefficient in part 1 to these dimensions to compare predicted and experimental values.

4.1K. REYNOLDS NUMBER OF NATURAL STREAMS

Compute the Reynolds numbers for the following streams:
1. The Mississippi River near Baton Rouge, La. See Problem 5.1C for data.
2. Haynes creek in south-central Arkansas. See Table 4.1K for data.

Table 4.1K. Haynes Creek, South-Central Arkansas

Parameter	Section			
	1	2	3	4
Velocity, ft/s	0.28	0.31	0.34	0.86
Depth, ft	0.35	0.44	0.32	0.52
pH	8.4	8.0	8.0	7.5
Temperature, °C	20	15	15	15
Detention time				
(τ), hr	3.27	1.68	3.55	2.5
Flow, ft^3/s	1.78	2.00	2.22	5.58

4.1L. HYDRAULIC RADIUS AND INTERFACIAL AREA— VOLUME OF STREAMS

Compute hydraulic radius (r_H) and interfacial area–volume (a_v) for the following:
1. A stream of 2 ft depth (maximum) and unknown cross-sectional shape.
2. A stream of 2 ft depth (maximum) and bottom width of 4 ft.

4.1M. SENSITIVITY OF DESORPTION TO STREAM DEPTH

1. Show by calculation that the fraction of ammonia desorbed is a strong function of depth for wide, shallow streams. For calculations use $\tau = 24$ hr, $^1K_{A2} = 0.04$ lb mol/hr·ft^2, $h = 0.5$ ft ± 0.25 ft.
2. Show by calculation that the fractional quantity of ammonia desorbed in a deep river is small. For calculation use $\tau = 24$ hr, $^1K_{A2} = 0.04$ lb mol/hr·ft^2, $h = 30$ ft.

4.1N. A COMPREHENSIVE AMMONIA DESORPTION PROBLEM

A small creek (Haynes data in Problem 4.1K) receives ammonia from the discharge of an ammonia nitrate manufacturing facility. Compute the fraction desorbed and the quantity desorbed after each subsection. For mass transfer coefficients use $^2k_{A1}=0.4$ lb mol/hr·ft^2 and $^1k_{A2}=0.43$ lb mol/hr·ft^2. The ammonia concentration into subsection 1 is 50 mg/L.

4.1O. ANOTHER COMPREHENSIVE AMMONIA DESORPTION PROBLEM

A small stream receives the effluent from a secondary wastewater treatment facility that purifies (i.e., removes carbon and suspended solids) the domestic waste from the town of Faysprings, Ark. The purified discharge water contains 0.5 mg/L ammonia and constitutes the bulk of the water flow in the stream. Stream conditions: $Q=1.0$ ft^3/s, depth$=0.5$ ft, velocity 0.3 ft/s, temperature 20°C, pH$=8.0$, wind speed 10 mph.
Calculate the following:

1. The fraction of ammonia desorbed in flow times of 10, 20, and 30 hr for a stream with a "plug" flow regime.
2. Repeat for a "completely mixed" flow regime.

4.1P. CARBON DIOXIDE DESORBING FROM WAIKATO RIVER

Carbon dioxide is present in geothermal effluents and possibly enhances growth of *Largorosiphon major* (a bottom-growing weed commonly used to supply oxygen in aquariums) by providing additional carbon for photosynthesis.[17] Pure water in equilibrium with the atmosphere contains approximately 0.5 ppm (by weight) of CO_2. Mixing of effluent with the river water increases the CO_2 concentration to 3 ppm.

Lake Aratiatia (New Zealand) is about 4 km downstream, on the Waikato River, from the Wairabei geothermal power station wastewater discharge. Predict the percent of the CO_2 excess that will reach the Lake Aratiatia. Data: River flow 127 m^3/s, river depth 10 m, river flow time to lake 40 min, and overall mass transfer coefficient between 1.5 and 2.0 cm/hr.

4.1Q. CODISTILLATION OF CHEMICALS WITH WATER

One mechanism by which chlorinated hydrocarbons dissipate from aquatic environments is termed the *codistillation* phenomenon. Table 4.1Q shows

Table 4.1Q.   Loss of Insecticides from Water by Codistillation

Insecticide	Original Concentration	Percent Codistilled 20 hr @ 26.5°C
Aldrin	0.024 ppm	93
Dieldrin	0.024	55
Heptachlor	0.21	91
Heptachlor epoxide	0.25	42
p,p'-DDT	0.0056	30
γ-Chlordane	0.20	70
Lindane	0.023	30

(Reprinted by permission, from "Organic Pesticides in the Environment." *Source.* Reference 18)

the loss of various insecticides by codistillation 20 hr after they were introduced into a jar of water. The codistillation experiments are based on a "still" system.

1. Compute the desorption coefficient for each insecticide (s^{-1}).
2. If the "still" system consists of seven identical jars at constant temperature and placed in a laboratory so that each jar is under identical conditions of temperature, humidity, wind currents, and so on, explain how it is possible to obtain such variation in the observed desorption coefficient.

4.1R.   EVAPORATION RATES OF LOW MOLECULAR WEIGHT CHLORINATED HYDROCARBON IN LABORATORY STUDIES[19]

Solutions that contain 1.0 ppm (weight) of each of five chlorinated compounds were prepared. The solution of the chlorinated compounds in water (200 mL, solution depth 65 mm before stirring) was placed into a beaker and after starting the stirrer (200 rpm stainless steel shallow pitch propeller in 250 mL Pyrex beaker) samples were withdrawn and mass spectra scanned after 1 min and periodically thereafter. Peak heights were assumed to be proportional to concentration. Figure 4.1R shows typical evaporation results. Table 4.1R contains solubility and pure component vapor pressure data for the five chemicals.

1. Which phase resistance controls the evaporation?
2. Will stirrer speed affect the evaporation rate?
3. Compute $^1K'_{A2}$ (cm/hr) average for the five compounds based on the data in Fig. 4.1R. Evaporation temperature was 25°C.

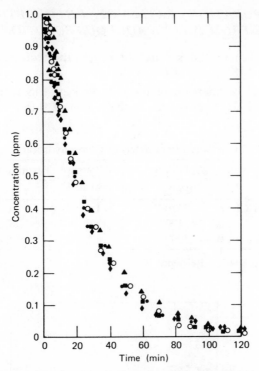

Figure 4.1R. Evaporation of rates of CH_2Cl_2(◆), $CHCl_3$ (●), CH_3CCl_3 (○), $CHCl=CCl_2$ (■), and $CCl_2=CCl_2$ (▲) from water. (Reprinted with permission from Reference 19. Copyright by the American Chemical Society.)

Table 4.1R. Chlorinated Hydrocarbon Data

Compound	Solubility (ppm)	Vapor Press (mm Hg)
CH_2Cl_2	19,800	426
$CHCl_3$	7,950	200
CH_3CCl_3	1,300	123
$CHCl=CCl_2$	1,100	74
$CCl_2=CCl_2$	400	19

Source. Reference 19.

4.1S. EVAPORATION RATES OF CHLOROETHANES AND PROPYLENES FROM DILUTE AQUEOUS SOLUTIONS[20]

Laboratory studies of the kinetics of evaporation of three compounds shown in Table 4.1S appear in Fig. 4.1S.

1. Make a visual comparison of the data in Fig. 4.1S with those in Fig. 4.1R. Ignoring the data scatter in the region of 0 to 5 min explain the

Table 4.1S. Evaporation Parameters of Chlorohydrocarbons in Water

Compound		Solubility (ppm)	Vapor Pressure (mm Hg) 20°C	Partition Coefficient H^*
$CH_2 = CHCH_2Cl$	■	3370	361	0.44
CH_2ClCCl_3	▲	1100	13.9	0.11
$CHCl_2CHCl_2$	●	3000	6.5	0.019

$H^* \equiv c_1 H_{AX}/c_2$　Source.　Reference 20.

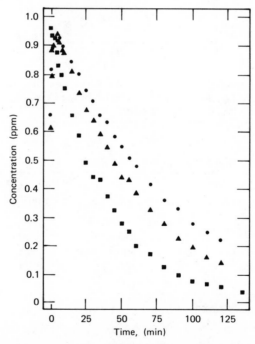

Figure 4.1S. Evaporation rates of $CH_2=CHCH_2Cl$ (■), CH_2ClCCl_3 (▲), and $CHCl_2CHCl_2$ (●) from water. (Reprinted with permission from Reference 20. Copyright by the American Chemical Society.)

wide variation of evaporation rates observed in Fig. 4.1S. Note that the experimental technique for obtaining the data in both figures is identical.

2. Compute the overall liquid phase mass transfer coefficient (cm/min) for $CHCl_2CHCl_2$ from the experimental data in Fig. 4.1S.

3. Compute the overall liquid phase mass transfer coefficient (cm/min) for $CHCl_2CHCl_2$ from the two-resistance theory, given

$$\frac{1}{{}^1K'_{A2}} = \frac{1}{{}^1k'_{A2}} + \frac{1}{H^2k'_{A1}}$$

$$^2k'_{A1} = 50\left(\frac{M_{H_2O}}{M_A}\right)^{1/2} \quad cm/min$$

$$^1k'_{A2} = 0.33\left(\frac{M_{CO_2}}{M_A}\right)^{1/2} \quad cm/min$$

4.2 EXCHANGE OF CHEMICALS ACROSS THE AIR–WATER INTERFACE OF LAKES AND OCEANS

Lakes and oceans differ from rivers and small basins in more ways than just physical size. Chemical exchanges at the air–water interface are dominated by the effect of the wind on the water. The flow of water through the system resulting in a detention time is not pertinent for large lakes and oceans; however, density stratification is important. An idealized picture of a stratified body of water is a well-mixed layer at the surface, a layer with a more or less pronounced density gradient (pycnocline) below it, and a well-mixed layer below the pycnocline. In many freshwater lakes the density stratification is thermal in origin. See Fig. 7.1-1 for an example of thermal stratification in lakes.

Rates of Loss of Low Solubility Contaminants from Water Bodies Such as Lakes and Oceans.[21,22]

Chlorinated hydrocarbons, such as pesticides and polychlorinated biphenyls (PCBs), have been transported widely throughout the global environment, even to remote Arctic and Antarctic regions. The major route by which these contaminants are transported is apparently through the atmosphere. Analysis of rainwater in England has shown concentrations of total pesticide residues of 104 to 229 ppt DDT; concentrations of 40 ppt have been reported in meltwater from Antarctic ice. The residues are presumably vapor or adsorbed on dust particles and may be carried many thousands of miles from the original source. Some of these materials are applied by spraying techniques, which allow the possibility of direct evaporation; however, most are used as solids, liquids, or wettable powders in which transport to the atmosphere can take place only by natural evaporative processes when exposed to the atmosphere.

PHASE EQUILIBRIUM OF LOW SOLUBILITY CONTAMINANTS.

As presented in Chapter 2, the partial pressure of chemical A and its concentration in the aqueous solution may be linked by equating the fugacities in both phases (e.g., Eq. 2.1-2). For this case

$$y_A \gamma_{A1} f_{A1}^0 = x_A \gamma_{A2} f_{A2}^0 \tag{4.2-1}$$

and it can be assumed throughout that the vapor phase fugacity coefficient and the activity coefficient are unity, thus equating the fugacity to the partial pressure of chemcial A. Under these conditions Eq. 4.2-1 becomes

$$p_A = x_A \gamma_{A2} f_{A2}^0 \tag{4.2-2}$$

in which x_A, the mole fraction, is typically about $1E-7$. Here γ_{A2}, the activity coefficient, is about $1E7$ and f_{A2}^0 is the fugacity of the pure liquid or solid A at the same temperature and pressure. In the case where pure A is a liquid at the system pressure and temperature, the reference fugacity can be equated to the saturation vapor pressure. If, as in the case of most pesticides, it is a solid, then the appropriate reference fugacity is the vapor pressure of the hypothetical supercooled liquid. Prausnitz[23] has discussed this problem and methods of estimating the reference fugacity and activity coefficient in such cases.

For the case of pure chemical A (solid or liquid) in equilibrium with water Eq. 4.2-2 becomes

$$p_A^0 = x_A^* \gamma_{A2} f_{A2}^0 \tag{4.2-3}$$

where p_A^0 is the vapor pressure of the pure solid or liquid and x_A^* is the solubility in water. The product $\gamma_{A2} f_{A2}^0$ must be equal to p_A^0/x_A^*. It is then possible to obtain a phase equilibrium relationship from accessible properties p_A^0 and x_A^* or ρ_{A2}^* (mg/L) by Eqs. 4.2-2 and 4.2-3,

$$p_A (\equiv y_A p_T) = \left(\frac{p_A^0}{x_A^*} \right) x_A \tag{4.2-4}$$

Some pure component vapor pressure and solubility data for important environmental chemicals appears in Table 4.2-1.

Table 4.2-1. Evaporation Parameters for Various Compounds @ 25°C

Compound	M_A Molecular Weight	ρ_{A2}^* Solubility (mg/L)	p_A^0 Vapor Pressure (mm Hg)	$^1K_{A2}'$ (m/h)
Alkanes				
n-Octane	114	0.66	14.1	0.124
2,2,4-Trimethyl pentane	114	2.44	49.3	0.124
Aromatics				
Benzene	78	1780	95.2	0.144
Toluene	92	515	28.4	0.133
o-Xylene	106	175	6.6	0.123
Cumene	120	50	4.6	0.119
Naphthalene	128	33	0.23	0.096
Biphenyl	154	7.48	0.057	0.092
Pesticides				
DDT($C_{14}H_9Cl_5$)	355	0.0012	1×10^{-7}	9.34×10^{-3}
Lindane	291	7.3	9.4×10^{-6}	1.5×10^{-4}
Dieldrin	381	0.25	1×10^{-7}	5.33×10^{-5}
Aldrin	365	0.2	6×10^{-6}	3.72×10^{-3}
Polychlorinated biphenyls (PCBs)				
Aroclor 1242	258	0.24	4.06×10^{-4}	0.057
Aroclor 1248 ($C_{12}H_6Cl_4$)		5.4×10^{-2}	4.94×10^{-4}	0.072
Aroclor 1254 ($C_{12}H_5Cl_5$)		1.2×10^{-2}	7.71×10^{-5}	0.067
Aroclor 1260 ($C_{12}H_4Cl_6$)		2.7×10^{-3}	4.05×10^{-5}	0.067
Other				
Mercury	201	3×10^{-2}	1.3×10^{-3}	0.092

(Reprinted with permission from reference 22. Copyright by the American Chemical Society)

QUANTIFICATION OF THE RATES OF EVAPORATION FROM AQUEOUS SOLUTION OR SUSPENSION[21,22]

In deriving the rate equations it is convenient to consider a column of water $1 \, m^2$ in cross section of depth h m containing $h \, m^3$ of water as shown in Fig. 4.2-1. If the concentration, c_{A2}, of the evaporating compound A is representative of the depth the quantity in the column is $c_{A2}h$ moles. Equation 4.2-4 is suitable for phase equilibrium. The rate of mass transfer across the phase boundary can be expressed in terms of an overall mass transfer coefficient obtained by combining two individual phase mass transfer coefficients (e.g., Eq. 3.1-33). The mass flux across the phase boundary can be expressed in terms of the bulk liquid concentration and

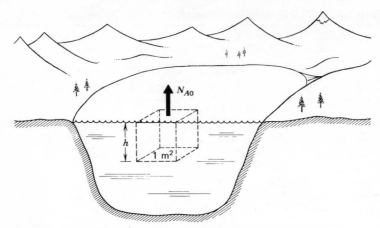

Figure 4.2-1. Chemical evaporation from lake ecosystem.

the partial pressure in the atmosphere converted to an equivalent liquid phase concentration, c_{A2}^*

$$N_{A0} = {}^1K'_{A2}(c_{A2} - c_{A2}^*) \tag{4.2-5}$$

assuming there is a nonnegligible background atmospheric level of the contaminant.

UNSTEADY-STATE MODEL

A mass balance on A, for the volume of water shown in Fig. 4.2-1 assuming no other methods of loss, yields

$$\frac{dc_{A2}}{dt} = -\frac{{}^1K'_{A2}(c_{A2} - c_{A2}^*)}{h} \tag{4.2-6}$$

Integrating this equation with c_{A2}^0 the concentration at zero time results in

$$c_{A2} = c_{A2}^* + \left(c_{A2}^0 - c_{A2}^*\right) \exp\left(\frac{-{}^1K'_{A2}t}{h}\right) \tag{4.2-7}$$

If the background atmospheric level of the contaminant is low compared to the local level and a half-life $\tau_{A\frac{1}{2}}$ is defined as the time required for the concentration to drop to half its original value, then Eq. 4.2-7 becomes

$$\tau_{A\frac{1}{2}} = \frac{0.69h}{{}^1K'_{A2}} \tag{4.2-8}$$

The unsteady-state model represents a situation in which a water body becomes depleted of a compound introduced at a point in time, for example from an accidental spill. It also represents a situation of agricultural runoff containing a pesticide in which the warmer river water "floats" on the surface of the receiving lake and becomes depleted of the compound.

STEADY-STATE MODEL

Situations arise where there is a fairly constant influx from other sources and the evaporation rate of the contaminant is exactly balanced by this influx. The concentration c_{A2} adjusts to a value such that these rates are equal. If the influx rate is I_A mol/m²·hr, then equating this to N_{A0} in Eq. 4.2-5 yields

$$c_{A2} = \frac{I_A}{^1K'_{A2}} + c^*_{A2} \tag{4.2-9}$$

If the mean residence of time of A in the volume is defined

$$\tau_A \equiv \frac{c_{A2}h}{I_A} \tag{4.2-10}$$

it can then be calculated from Eq. 4.2-9

$$\tau_A = h\left(\frac{1}{^1K'_{A2}} + \frac{c^*_{A2}}{I_A}\right) \tag{4.2-11}$$

Equations 4.2-10 and 4.2-11 are valid for $I_A > 0$. If c^*_{A2} is negligible the mean residence time reduces to $h/^1K'_{A2}$ and differs from the unsteady-state model half-life only by the factor 0.69. Table 4.2-1 gives the overall liquid phase mass transfer coefficient for various compounds at 25°C. The half-lives vary from 4.81 hours for benzene to 12,940 hours for dieldrin. Most of the resistance to mass transfer resides in the liquid phase. Calculations of half-life and resistances are left as an exercise for the student. (See Problem 4.2C.)

The half-lives for a water depth of 1 m show that most of these compounds evaporate rapidly from solution. In situations where the water body is turbulent with frequent exchange between the surface water layer and the bulk (e.g., in a fast-flowing shallow river or during whitecapping on a lake or ocean), the liquid phase mass transfer coefficient may be considerably increased and the evaporation rate increased accordingly. For depth greater than 1 m the half-life is correspondingly increased assuming that the rate of eddy diffusion is substantial.

The model equations developed above are valid for any body of water that is well mixed through the depth h. Clearly, this is not the depth of the water column from the sediment–water interface to the air–water interface in many lakes and the ocean. For lakes, a reasonable value of h is the depth of the epilimnion. The epilimnion is mainly surface waters above the thermocline characterized by uniform mixing and near isothermal conditions. A similar region exists in ocean waters. Other rate-limiting diffusion processes at depths in water bodies, for example, through thermoclines, were not considered in the preceding treatment. The topic of intraphase diffusional processes is presented in Section 7.2. A brief introduction and some consequences of intraphase diffusion follows next.

Times to Chemical Steady States in Lakes and Oceans[24]

In natural systems of large dimensions many chemical processes are controlled by the transport of the species through the system. The distribution of chemical species in natural systems is only too often not homogeneous; concentration gradients and more or less abrupt changes in abundance from one part of an environment to another are commonplace. With reference to vertical migration of chemical species through the water column of lakes and oceans, the relevant diffusional process is eddy diffusivity.

The difference of several orders of magnitude between the molecular and eddy diffusion coefficients reflects the much more rapid dispersal by turbulent eddies in natural bodies of water. The much higher values of the eddy diffusivities in surface waters are owing to the greater effect of the wind-generated turbulence, as compared with the deeper parts of the basin. (See Table 3.1-4.)

It is possible to choose "reasonable" lower and upper limits of the diffusion coefficients and thereby to bracket the model in short and long time estimates. Lerman[24] considers an idealized three-layer water column with a mixed upper layer, a less mixed pycnocline, and a well-mixed layer below the pycnocline. When a three-layer system remains closed and the dimensions of the water layers do not change, a conservative chemical species in one of the mixed layers redistributes itself between the two layers because of the diffusional flux down the concentration gradient from one mixed layer into the other.

For the case of transport from the lower into the upper mixed layer Lerman calculated the change in concentration in the upper layer as a function of time. Calculations were made for a 60 m deep water column with lower layer 25 m, pycnocline 10 m, and upper layer 25 m for three different eddy diffusion coefficients in the pycnocline ($\mathcal{D}_{A2}^{(l)}$ = 5E–3, 1E–2, and 5E–2 cm ^{2}/s). These values of the eddy diffusion coefficients are in the range reported for pycnoclines in stratified lakes. The calculations show the concentration of a chemical species in the two mixed layers would equalize in a period of 10 to 40 years.

The time required to attain equal concentrations in any given lake depends on the eddy diffusivity in the pycnocline and on the vertical dimensions of the individual layers characteristic of the particular lake. The characteristics in the preceding example reasonably represent the time to chemical steady-state in many lakes. In light of the period of time for intraphase movement it appears that the evaporation half-lives of 48.1 to 12,940 hours reported by Mackay et al. are very rapid. It appears that if a chemical species is uniformly distributed in a water column but calculations are made using a 1 m mixing depth, then the result will be a minimum estimate of the half-life.

Flux Rate of Gases across the Air–Sea Interface

Liss and Slater[25] describe the use of the two-resistance model of interphase mass transfer to estimate the flux of eight common gases across the air–sea interface. Using reasonable estimates of the individual phase mass transfer coefficients plus a knowledge of Henry's law constant in conjunction with Eqs. 4.2-5 and 3.1-33, together with observed concentrations differences of each gas across the interface, the flux of the gases can be calculated. Mean concentrations of these gases in oceanic air and seawater are required. The concentration measures for many chemicals have been made by direct sampling from research vessels and are available in the literature.

FACTORS AFFECTING INTERFACE RESISTANCE

A review of the published literature established that reasonable values of the gas and liquid phase mass transfer coefficients appropriate to the sea surface for unreactive gases are

$$^2k'_{A1} = 3000 \, \text{cm/hr}; \qquad ^1k'_{B2} = 20 \, \text{cm/hr}$$

where $A \equiv H_2O$ and $B \equiv CO_2$. To obtain the gas phase coefficient for gases other than water vapor the preceding value must be multiplied by the ratio of the square roots of the molecular weights of H_2O and the other gas. The value of the liquid phase coefficient is based largely on measurements of CO_2 exchange and is probably valid for gases of molecular weight 40 ± 25. For gases outside this range this value should be multiplied by the ratio of the square roots of the molecular weights of CO_2 and the other gas.

In chemically reactive gases, transport in the liquid phase may be more complex than by straightforward diffusion processes. For example, with CO_2 as well as the usual concentration gradient for gas molecules in physical solution at a suitable pH there is also a similar gradient for HCO_3^- and CO_3^- ions. It has been shown in the laboratory that for $pH > 5$, and under moderately calm conditions, the ionic species gradient can contribute significantly to the exchange of CO_2 across the interface. The chemical enhancement of oceanic CO_2 gas exchange results from the chemical reaction

$$CO_2 + H_2O + CO_3^- \rightleftharpoons 2HCO_3^- \qquad (4.2\text{-}12)$$

Hoover and Berkshire[26] have used a one-layer (liquid) film model to derive the following equation, which successfully predicts the exchange enhancement of CO_2 found in laboratory experiments:

$$\alpha_B = \frac{K_{B2}^* K \delta_{B2}}{\left[(K_{B2}^* - 1)(K\delta_{B2}) + \tanh(K\delta_{B2})\right]} \qquad (4.2\text{-}13)$$

where $K \equiv \sqrt{K_{B2} K_{B2}^* / \mathcal{D}_{B2}}$. Here α_B is the fractional increase in the liquid phase mass transfer coefficient due to chemical reaction Eq. 4.2-12, K_{B2}^* is the ratio of total to ionic forms of inorganic carbon, K_{B2} is the hydration reaction rate constant for CO_2 in water, and δ_{B2} is the thickness of liquid film. Therefore, for a chemically reactive gas it is appropriate that the liquid phase mass transfer coefficient be corrected for chemical enhancement by multiplying it by the value of α_B calculated from Eq. 4.2-13.

Table 4.2-2 contains exchange constants and Henry's law constants for a number of gases crossing the air–sea interface. For all gases considered other than SO_2, liquid phase resistance controls the exchange. For SO_2, the gas phase resistance is all-important. This is in part due to the high solubility of SO_2, but also because of its extremely rapid reaction rate constant in seawater. Exchange of other gases, with high solubilities and for rapid aqueous phase reaction such as NH_3, SO_3, and HCl, is probably also controlled by the gas phase resistance.

Based on the exchange constants in Table 4.2-2 Liss and Slater computed the flux directions and rates across the air–sea interface for the gases listed. The results of these calculations appear in Table 4.2-3. Further details appear in Problem 4.2B.

Table 4.2-2.   Exchange Constants for Selected Gases Across the Air–Sea Interface

Gas	$^2k'_{A1}$ (cm/hr)	α_A	$^1k'_{A2}$ (cm/hr)	H_{AX}	$^1K'_{A2}$ (cm/hr)
SO_2	1600	1721	34,420	47	1600
N_2O	1900	1.0	20	2,000	20
CO	2400	1.0	20	62,000	20
CH_4	3180	1.0	20	52,000	20
CCl_4	1030	1.0	10.7	1,300	10.7
CCl_3F	1085	1.0	11.3	6,200	11.3
MeI	1070	1.0	11.1	300	10.6
$(Me)_2S$	1620	1.0	20.0	370	19.2

(Printed by permission, Macmillan Journals Ltd. **Source**. Reference 25.)

Table 4.2-3.   Calculated Fluxes of Selected Gases Crossing the Air–Sea Interface in g/yr[a]

SO_2	N_2O	CO	CH_4	CCl_4	CCl_3F	MeI	$(Me)_2S$
1.5E 14	1.2E 14	4.5E 13	3.2E 12	1.3E 10	5.3E 9	2.8E 11	7.2E 12
(+)	(−)	(−)	(−)	(+)	(+)	(−)	(−)

(Printed by permission, Macmillan Journals Ltd. **Source**. Reference 25.)

[a] The sign (+) denotes into the sea and (−) out of the sea.

Because of the chemical reactivity of SO_2 (rapid hydration and oxidation), it seems reasonable to assume that the SO_2 concentration in oceanic surface waters is zero and that the ocean acts as a perfect sink for SO_2. The calculated flux rate of SO_2 closely estimates the total world input to the atmosphere from the burning of fossil fuels (1E 14 g/yr). The air-to-sea flux of N_2O may be of very considerable importance in the oceanic nitrogen budget. The flux of N based on the N_2O calculation is large enough to account for all the nitrogen inflow to the oceans not removed to the marine sediments. The magnitude of the calculated CO flux is about 20% of the amount of CO injected into the atmosphere by human activities (2E 14 g/yr). The calculated flux of CCl_4 into the oceans is about 30% of the industrial production of this chemical that is not converted to CCl_3F and similar compounds (5E 10 g/yr). The total world production of Freon-11 (CCl_3F) is about 0.3E 12 g/yr. As there are no known natural sources of this compound the calculated flux from the atmosphere to the ocean seems to be about 2% of the world production. The flux of iodomethane from the ocean to the atmosphere accounts for about half the total amount of iodine needed to balance the budget for this element. The dimethyl sulfide calculation corresponds to 4% of the amount needed to balance the sulfur budget. Apparently terrestrial sources account for most of the release of $(Me)_2S$.

The preceding account makes readily apparent the role interphase flux calculations can play in assessing the fate of natural and synthetic chemicals within the global environment. Obviously the results are only as good as the data used in the calculations. There is much uncertainty in choosing mean values for the gas and liquid exchange constants. It is not known from measurements at the sea surface how $^1k_{A2}$ varies with wind speed. Even greater uncertainties arise involving the air and water concentrations of some of these gases. As further data become available it will be a relatively simple matter to recalculate the fluxes and to extend the approach to other gases.

ESTIMATING MASS TRANSFER COEFFICIENTS OF THE AIR–SEA INTERFACE

Liss and Slater employed the reasonable values of the gas and liquid phase of 3000 and 20 cm/hr, respectively, for calculations of the flux of selected chemicals. Several investigators have undertaken the study of mass transfer coefficients in the ocean environment. In general, these investigations commence with laboratory experimentations employing wind tunnel–water tank apparatus to simulate conditions in the ocean. The exchange of simple chemicals such as oxygen, carbon dioxide, and water vapor is studied to gain a basic understanding of the importance of wind speed, fetch, waves, salinity, surface contamination, and so on, on the individual phase coefficients. From observation aboard research vessels at sea it is possible to obtain actual gas exchange rates. The concentrations of radon gas (half-life 3.85 days) and radiocarbon data (carbon 14) measured at sea provide this information. The following is a brief review of laboratory and oceanic studies.

Downing and Truesdale[27] studied the effects of a number of factors on the rate of solution of oxygen in fresh and saline water to provide information about reaeration in the polluted Thames Estuary. In an estuary the water is subjected to considerable agitation both at and below the surface, owing to the motion of currents caused by tides and the flow of fresh water, but the chief agent causing disturbances at the surface is the wind. To simulate the former effect, the water in the laboratory experiments was stirred at varying speeds with impellers. To simulate the effects of wind directly was impossible since a very long fetch is required to work up waves of the magnitude of those observed in the estuary. The local action of wind in producing ripples and wavelets on the surface was studied in a small experimental tank and larger waves were generated mechanically in a wave tank. Figure 4.2-2 shows a schematic of a typical

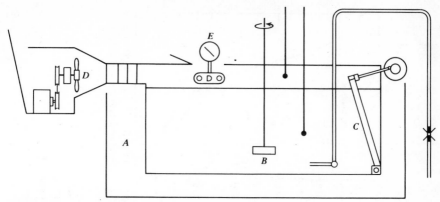

Figure 4.2-2. Apparatus for studying the effects of surface agitation on rate of solution of oxygen in water. (Reprinted with permission from Reference 27.)

gas exchange simulation apparatus. Tank A is 90 cm long, 30 cm wide, and 38 cm deep; B is a stirrer; C is a wave generator; D is a fan; and E is a small cup anemometer.

EFFECT OF SURFACE WIND

The rate of solution of oxygen in distilled water and seawater at several wind speeds between 0 and 9 m/s measured at 5 cm above the water surface yielded estimates of the liquid phase mass transfer coefficient $^1k'_{A2}$. The results appear in Fig. 4.2-3a. There was little change in the appearance of the water surface until the wind speed exceeded 3 m/s. As the wind velocity increased above this value, the surface became increasingly ruffled by small wavelets and ripples although these rarely exceeded 2 cm in height. The rate of solution of oxygen increased with increase in the disturbances of the surface. There was little change from the initial rate of 1.5 cm/hr until the wind velocity exceeded 3 m/s, but above this velocity there was a steady increase up to 35.5 cm/hr at a velocity of 9 m/s.

The authors report that the laboratory results are not in good agreement with estimates of the effect of wind obtained from the results of direct measurement of the rate of reaeration in an estuary. These indicate a gradual steady increase in rate of solution with increasing wind velocity up to values of the order of 25 cm/hr at a velocity of 10 m/s.

EFFECT OF WAVES

To measure the effect of waves, progressive waves of different heights and the same wavelength of 99.3 cm were generated at a constant frequency of 75 per minute. The rate of solution of oxygen increased almost linearly

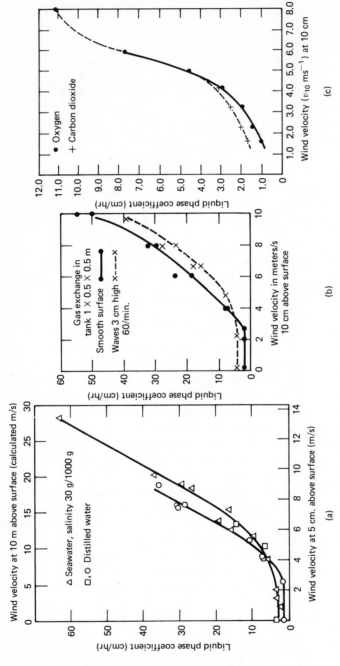

Figure 4.2-3. Sea-surface liquid phase coefficient variation with wind speed. [(a) Reprinted with permission from Reference 27. (b) Reprinted with permission from Reference 28. Copyright 1963, Pergamon Press Ltd. (c) Reprinted with permission from Reference 29. Copyright 1963, Pergamon Press Ltd.]

184

from 9.6 to 37 cm/hr as the height of the waves increased from 2.8 to 10.8 cm. Varying the frequency of waves 8 cm in height from 36 to 75 waves per minute increased the rate of solution from 14.6 to 27.2 cm/hr. The results are to be compared with those obtained from direct determinations of the rate of entry in the Thames Estuary, which indicate a roughly linear increase in rate of solution with increasing wave height from 5 cm/hr for calm conditions to 25 to 30 cm/hr for wave heights of 50 cm. For waves 10 to 13 cm high the rates of solution were 10 to 12 cm/hr, or less than half those in the laboratory tank.

Downing and Truesdale also report results of the effect of temperature, oil films, and soluble surface-active agents on the oxygen mass transfer coefficient. In most cases the coefficient increase is approximately linear with increasing temperatures in the range 0 to 35°C. In terms of the change in the coefficient at 20° the values are correlated by

$$^1k_{A2}(@T, °C) = {}^1k_{A2}(@20°C)\alpha^{T-20°C} \qquad (4.2\text{-}14)$$

with α between 1.015 and 1.035 ($A \equiv O_2$).

Oil films had little effect until the thickness was greater than 1 μ; films of greater thickness tended to reduce the coefficient. No similar experiments appear to have been carried out at sea. It would seem reasonable to assume that, except in a region of accidental oil spillage, the amount of oil likely to be found at the sea surface would be insufficient to have any measurable effect on the transfer coefficient.

There is some evidence that monomolecular surface layers composed of organic molecules can decrease the rate of the exchange of dissolved gases, such as oxygen and carbon dioxide. However, the film has to be in the close-packed condition to offer appreciable resistance to the passage of carbon dioxide. The same limitation almost certainly applies to transfer of oxygen and other dissolved gases. It seems unlikely that such a close-packed monomolecular film could exist over more than a very small part of the ocean surface.

Kanwisher[28] addressed the exchange of gases between the atmosphere and the sea. He observed that the liquid phase coefficient is influenced by the turbulent structure of the underlying water of the surface layer. Since the wind is ultimately responsible for most of this turbulence through the shearing stress it exerts on the water surface, it is not surprising to find that this coefficient depends critically on the wind velocity. Figure 4.2-3b shows the effect of wind velocity on gas exchange rate. The results are similar to those of Downing and Truesdale.

It appears from these laboratory experiments that the liquid phase coefficient at the ocean surface at any place will be proportional to the average of the quantity (wind velocity)^2. To obtain the average of (wind)^2 it is necessary to know the degree of variability of the wind. In the trades, where it is fairly constant, the two are not very different. In mid-latitudes, however, where much of the gas exchange rate may be due to short periods of high wind velocities (storms) the use of average wind speed may cause large error. The result is that the liquid phase coefficient at sea can vary considerably with latitude.

The role of breaking waves and bubbles in enhancing the transport of atmospheric gases across the air–sea interface has been discussed by Kanwisher. He concludes that when conditions are sufficiently rough for bubbles to be produced this may contribute significantly to the gas exchange.

Liss[29] measured the transfer coefficients of O_2, CO_2, and water vapor across an air–water interface using both a laboratory tank filled with water and a wind–water tunnel. When the water pH is less than 5, CO_2 exists in water only as the physically dissolved species. The measured coefficient for the desorption of CO_2 was the same as for the absorption of oxygen. The results appear in Fig. 4.2-3c. Under calm conditions and with water pH greater than 5 there is a measurable enhancement of the exchange of CO_2 relative to O_2. This enhancement is thought to be due to the ionic species gradient present at the liquid surface under these conditions. Results indicate that the exchange of water vapor is controlled by resistance in the air. The wind tunnel experiments demonstrate that the mass transfer coefficients for both CO_2 and O_2 increase approximately as the square of the wind velocity, whereas the coefficient for water vapor increases linearly with wind velocity. (See Fig. 4.2-4.)

Liss has compiled the results of liquid phase mass transfer coefficients measured at sea. The range of values reflects the difficulty of making these measurements. The rather large spread may be due to different degrees of turbulence at the interface or to experimental error. The compiled values appear in Table 4.2-4.

Cohen, Cocchio, and Mackay[30] have studied the volatilization of toluene and benzene in an effort to measure the rate of air–water mass transfer of nonionizing low solubility pollutants. Measurements of the liquid phase mass transfer coefficients in a laboratory wind–water tank are presented and the effects of waves on mass transfer rate is discussed.

Waves and drift current contribute to turbulence near the surface and hence increase the rate of transfer. Waves also increase the interfacial area but this effect cannot account for more than 4% of the increase in the transfer rate. In the presence of wind-induced waves the turbulence and

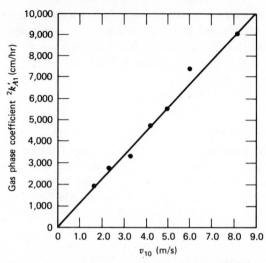

Figure 4.2-4. Graph of the variation of the exchange constant for water vapor ($^2k'_{A1}$), with the wind velocity measured at a height of 10 cm above the water surface (v_{10}). (Reprinted with permission from Reference 29. Copyright 1963, Pergamon Press Ltd.)

Table 4.2-4. Published Liquid Phase Mass Transfer Coefficients Measured in Situ at the Sea Surface

Gas	Transfer Process	$^1k'_{A2}$ (cm/hr)	Comments
Rn	Desorption	7	average wind speed 8 m/s
Rn	Desorption	25	speed <8 m/s
O_2	Absorption/desorption	22	whole year average
CO_2	Absorption	29	speed <4.2 m/s, waves <30 cm
CO_2	Absorption/desorption	36	three experiments
CO_2	Desorption	4–14	method unknown
CO_2	Absorption	11	7 yr average

(Reprinted with permission from Reference 29. Copyright (1973), Pergamon Press, Ltd.)

wind velocity profile near the interface are highly irregular and difficult to quantify. A mean logarithmic velocity profile is usually assumed:

$$v = \left(\frac{v_*}{\kappa_1}\right)\ln\left(\frac{y}{y_0}\right) \tag{4.2-15}$$

where v_* and y_0 are the friction velocity and effective roughness height and κ_1 is the von Kármán constant.

The transfer rate is influenced by the drift current, the velocity profiles, and the surface roughness. It is believed that the parameters v_* and y_0 are potentially useful in correlating the exchange process. Both of these terms can be combined into a modified Reynolds number

$$\text{Re}_* = \frac{y_0 v_*}{\nu_1} \tag{4.2-16}$$

Both v_* and y_0 change with wind velocity and fetch.

Laboratory experiments on the volatilization of dilute solutions of benzene and toluene in a wind–water tank gave the same variation of the liquid phase coefficient with velocity as shown in Fig. 4.2-3. Because the dependence of ${}^1k'_{A2}$ on wind velocity is not simple and the claim that it is proportional to the square of wind speed is incorrect, based on both theory and experimental facts, an alternative correlation of ${}^1k'_{A2}$ with v_* and y_0 combined as Re is suggested. A correlation based on the results of both compounds is proposed for $0.11 < \text{Re}_* < 102$:

$$^1k'_{A2} = 11.4\,\text{Re}^{*(0.195)} - 5 \tag{4.2-17}$$

The coefficient is in cm/hr. The uncertainty in applying the preceding correlation to natural bodies lies in the substantial difference between laboratory wind waves and those occuring in nature. The coefficients are believed to be lower than the actual environmental values owing to the greater water surface roughness of natural water bodies when compared to the experimental conditions at the same Re_*. Reported ${}^1k'_{A2}$ data suggest that the factor may not exceed 2.

Another approach is to obtain Re_* by using the drag coefficient C_d of wind over water. C_d can be expressed in terms of the shear stress τ_0 as $(\tau_0/\rho_1 v_{10}^2)$, where v_{10} is the velocity at a height of 10 m. Since τ_0 is equivalent to $\rho_1 v_*^2$, the relationship v_* and v_{10} is readily obtained

$$v_* = v_{10}\sqrt{C_d} \tag{4.2-18}$$

Using the logarithmic velocity profile, it can be shown that

$$\text{Re}_* = \frac{10 v_{10}\sqrt{C_d}}{\left[\nu_1 \exp\left(0.4/\sqrt{C_d}\,\right)\right]} \tag{4.2-19}$$

For the range $1\ \text{m/s} \leqslant v_{10} \leqslant 15\ \text{m/s}$ the following correlation based on oceanic determinations of the wind stress coefficient is applicable:

$$C_d = 5\text{E} - 5 v_{10}^{1/2}$$

with v_{10} in cm/s. Using Eq. 4.2-19 the expression for Re. becomes

$$\text{Re.} = \frac{7.07\text{E}{-}2v_{10}^{5/4}}{\nu_1 \exp(56.6/v_{10}^{1/4})} \qquad (4.2\text{-}20)$$

where v_{10} is in cm/s. This final result is more convenient to use in obtaining environmental Re. values for Eq. 4.2-17.

MEASURING MASS TRANSFER COEFFICIENTS AT THE AIR–SEA INTERFACE

The foregoing laboratory results are important to the task of estimating mass transfer coefficients at the air–sea interface. The question of how applicable these results are to the actual interface is open. Table 4.2-4 gives some values of the liquid phase coefficient ${}^1k'_{A2}$ obtained at sea. The following is a demonstration of how such numerical results are obtained.

Measurements of the vertical distribution of radon (${}^{222}\text{Rn}$) in surface seawater offers a means of determining the gas exchange rate between the ocean and the atmosphere.[31] Radon (half-life 3.85 days) is generated within the sea by the decay of dissolved radium

$$^{226}\text{Ra} \rightarrow {}^{222}\text{Rn} + \alpha$$

The partial pressure of radon in the seawater produced in this manner greatly exceeds the radon pressure in oceanic air. Radon therefore desorbs, leaving the surface waters with a lower concentration than that accounted for by the decay of the parent radium. By measuring the difference between the radium and radon activity as a function of depth, it is possible to determine the rate of radon loss, and in turn the liquid phase mass transfer coefficient.

The assumption is made that the main barrier to radon desorption is the resistance on the sea side of the oceanic interface. The net loss of radon atoms through each unit area of the sea surface must exactly balance the integrated deficiency of radon activity in the water column. Figure 4.2-5 shows a typical profile of radon content for various depths below the surface. The integrated deficiency of radon content m_A (where $A \equiv {}^{222}\text{Rn}$) is:

$$m_A = A \int_0^\infty (\rho_{A2}^{**} - \rho_{A2}) \, dy \qquad (4.2\text{-}21)$$

where ρ_A^{**} is the equilibrium mass concentration of radon supported by the radium present in the water and y is the distance below the sea surface.

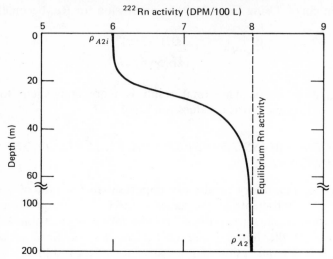

Figure 4.2-5. Radon content below the sea surface.

For convenience, the value of the integral can be expressed in terms of ρ_{A2i}, the radon concentration at the surface, and h, the depth to which the surface anomaly would have to be carried such that

$$A \int_0^\infty (\rho_{A2}^{**} - \rho_{A2})\, dy = A(\rho_{A2}^{**} - \rho_{A2i})h \qquad (4.2\text{-}22)$$

A steady-state mass balance on radon for a volume Ah of seawater of uniform concentration ρ_{A2i} near the surface yields that the production rate of radon must equal the decay rate plus the rate of loss at the air–water interface:

$$(-r_B)h = \rho_{A2i}hk_A''' + {}^1k_{A2}'(\rho_{A2i} - \rho_{A2}^*) \qquad (4.2\text{-}23)$$

where $-r_B$ is the rate of disappearance of radium ($B \equiv {}^{226}\text{Ra}$), k_A''' is the fraction of radon atoms decaying per unit time, and ρ_{A2}^* is the equilibrium solubility of atmospheric radon. A similar mass balance on a volume V of water located well below the surface where no radon gradients exist yields that $(-r_B)V = \rho_{A2}^{**} V k_A'''$, from which $-r_B$ can be obtained:

$$(-r_B) = \rho_{A2}^{**} k_A''' \qquad (4.2\text{-}24)$$

Equations 4.2-23 and 24 can be combined and rearranged to yield an expression for the liquid phase coefficient. If, as is almost always the case for radon, $\rho_{A2i} \gg \rho_{A2}^*$

$$^1k'_{A2} = k'''_A h\left(\frac{\rho_{A2}^{**}}{\rho_{A2i}} - 1\right) \tag{4.2-25}$$

This result is useful in estimating $^1k'_{A2}$ values at sea from measurements of the vertical distribution of radon. Equation 4.2-22 is used to compute h. Implied assumptions necessary for the development of this simple expression are that the radium content of seawater is constant and there is no turbulent diffusion of radon in the lateral and vertical directions. See Problem 4.2E for an application of Eq. 4.2-25 to data obtained from the Coast Guard vessel *R. V. Rockaway* in the Bomex area.

EVAPORATION FROM THE SURFACE OF THE SEA

Defant[32] presents H. U. Sverdrup's simple mass transfer model for the evaporation of water from the surface of the sea. He assumed the air space above the sea surface was composed of two zones. The lower zone is immediately above the water surface and consists of a thin boundary layer through which water vapor transport proceeds only by molecular diffusion. Above this boundary layer, the water vapor transport proceeds through turbulent exchange. The two zones are illustrated in Fig. 4.2-6.

The thickness of the boundary layer immediately above the water surface depends on the wind velocity. The layer itself can hardly be regarded as invariably composed of the same air particles. Since the turbulent eddies penetrate down to and into the boundary layer, it must be clearly understood that this layer occasionally disappears completely; however, after some time it is always reformed so that a mean thickness $\delta_{\mathcal{D}}$ of this layer can be introduced.

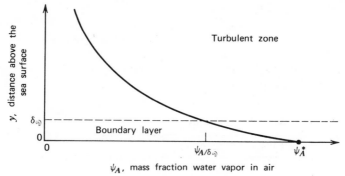

Figure 4.2-6. Two-layer surface at the air–sea interface.

The turbulent diffusivity is a linear function of the height above the water surface and depends on the roughness of the surface, as shown in Eq. 3.2H-5

$$\mathcal{D}_{A1}^{(t)} = \kappa_1(y+y_0)v_*$$ (4.2-26)

The flux of water vapor ($A \equiv H_2O$) in the turbulent zone is

$$j_A = -\kappa_1(y+y_0)v_*\rho_1\frac{d\psi_A}{dy}$$ (4.2-27)

Since j_A is constant for steady-state evaporation from the sea surface, Eq. 4.2-27 may be integrated from the upper edge of the boundary layer, $\psi_A|_{\delta_\mathcal{D}}$ at $y = \delta_\mathcal{D}$ to an arbitrary distance y at which ψ_A is $\psi_A|_y$ to give

$$j_A = \frac{\rho_1\kappa_1 v_*}{\ln[(y+y_0)/(\delta_\mathcal{D}+y_0)]}\left(\psi_A|_{\delta_\mathcal{D}} - \psi_A|_y\right)$$ (4.2-28)

Within the boundary layer, mass transfer occurs by molecular diffusion for which Eq. 3.1-13 can be used in the form

$$j_A = \frac{\mathcal{D}_{A1}\rho_1}{\delta_\mathcal{D}}\left(\psi_A^* - \psi_A|_{\delta_\mathcal{D}}\right)$$ (4.2-29)

where ψ_A^* is the mass fraction of water in equilibrium with the salinity and temperature of the sea surface. By combining Eqs. 4.2-28 and 4.2-29, the unknown vapor concentration at the junction between the boundary layer and the turbulent zone can be obtained and is

$$\psi_A|_{\delta_\mathcal{D}} = \frac{\dfrac{\mathcal{D}_{A1}\psi_A^*}{\delta_\mathcal{D}} + \dfrac{\kappa_1 v_*\psi_A|_y}{\ln[(y+y_0)/(\delta_\mathcal{D}+y_0)]}}{\dfrac{\mathcal{D}_{A1}}{\delta_\mathcal{D}} + \dfrac{\kappa_1 v_*}{\ln[(y+y_0)/(\delta_\mathcal{D}+y_0)]}}$$ (4.2-30)

This equation for $\psi_A|_{\delta_\mathcal{D}}$ may then be used in Eq. 4.2-29 to yield an expression for the vapor flux

$$j_A = \left[\frac{v_*}{\dfrac{\ln[(y+y_0)/(\delta_\mathcal{D}+y_0)]}{\rho_1\kappa_1} + (v_*\delta_\mathcal{D})/(\rho_1\mathcal{D}_{A1})}\right](\psi_A^* - \psi_A|_y)$$

(4.2-31)

Now using Eq. 3.2-42 to replace v_* yields the final desired vapor flux expression obtained by Sverdrup

$$j_A = \left(\frac{\kappa_1/\ln[(y+y_0)/y_0]}{\ln[(y+y_0)/(\delta_\mathscr{D}+y_0)]/\rho_1\kappa_1 + (v_*\delta_\mathscr{D}/\rho_1\mathscr{D}_{A1})} \right)(\psi_A^* - \psi_A|_y)v_x|_y$$

(4.2-32)

Observations at sea were necessary to establish y_0 and $\delta_\mathscr{D}$. From data gathered aboard the research vessel Atlantis, a value of y_0 of 0.6 cm was obtained. From values of wind velocity above the water surface, tempera-ture, humidity, salinity, and evaporation rates $\delta_\mathscr{D}$ can be determined from Eq. 4.2-32. It was found that the mass transfer boundary layer thickness could be roughly correlated with v_* from which the simple equation $\delta_\mathscr{D} = 4.12/v_*$ was obtained. The term $\delta_\mathscr{D}$ is in cm, and v_* is in cm/s.

Employing the relation between $\delta_\mathscr{D}$ and v_* and a reasonable range of v_* values observed at sea ($13.2 \leqslant v_* \leqslant 36.3$ cm/s), it is possible to simplify Eq. 4.2-32. If one chooses to measure ψ_A and v_x at $y = 10$ m, Eq. 4.2-32 becomes

$$j_A = k_{10}(\psi_A^* - \psi_A|_{10})v_x|_{10}$$

(4.2-33)

and k_{10} lies between 1.86×10^{-6} and 1.89×10^{-6} g/cm^3.

Equation 4.2-32 is quite remarkable since it is based solely on theoretical considerations. The theory of evaporation discussed earlier involves a hydrodynamically smooth surface with a laminar boundary layer with a turbulent layer of air above it. The preceding development is essentially the Prandtl-Taylor analysis. In the Prandtl-Taylor analysis, molecular transport of mass is assumed to be the only mechanism of importance in the laminar sublayer. In the turbulent core, the Reynolds analogy is applied. The linear dependence of velocity on the gas phase mass transfer coefficient is supported by field observations. See, for example, Harbeck's result for evaporation from the surface of lakes, Eq. 4.1-18, and Schooley's data, Eq. 4.2D-1, and Fig. 4.2-4.

Disappearance of Selected Hydrocarbons from Sea-Surface Slicks[33]

When oil is spilled in an aqueous environment (see Fig. 4.2-7), it is altered from its original composition by a series of processes that influence its weathering. Processes that contribute to weathering are evaporation, pho-tochemical and oxidative reactions, dissolution of individual components

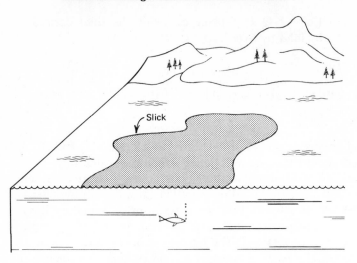

Figure 4.2-7. Oil slick at sea.

and emulsion formation, as well as the action of microorganisms. In a system where all these processes are occurring, quantification and modeling are extremely difficult, compounded by the multitude of components in the oil. The following is a simple transport model study of the disappearance of aromatic and aliphatic components from small sea-surface slicks.

Water-soluble aromatic and aliphatic hydrocarbons may have sublethal effects on marine organisms at concentrations of 10 to 100 ppb, lethal toxicity at 0.1 to 1.0 ppm for most larval stages, and lethal effects at 1 to 100 ppm for most adult organisms. It is thus of interest to investigate the rates of disappearance of specific aromatic and aliphatic components from crude oil slicks under conditions of essentially constant water temperature and nearly constant air temperature. Studies of the fate of slick components during the early stages of slick aging are crucial because the lower boiling fractions contain most of the lethal components of the slick.

Although a complete mathematical model of the fate of hydrocarbon components from an oil spill is beyond the scope of this book and probably not possible, it is instructive to develop a simple model that quantifies the relative importance of evaporation and dissolution. Previous investigations have demonstrated that all the lower boiling components evaporate or dissolve within a few hours of slick initiation. Although little is known of the relative percentages of loss of these slick components due to evaporation and dissolution, it is assumed that they are mainly lost by evaporation, at least under conditions of low sea-surface roughness.

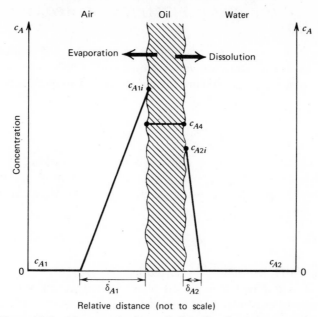

Figure 4.2-8. Mass transfer from oil slick, film theory interpretation.

Lower boiling components leave the slick by two routes: mass transfer to the air and to the seawater. Figure 4.2-8 shows the directions of chemical movements and concentration of a typical low boiling component within the slick and in the adjoining phases. The total rate of mass transfer for component A is

$$N_{A0} = {}^{4}k'_{A1}(c_{A1i} - c_{A1}) + {}^{4}k'_{A2}(c_{A2i} - c_{A2}) \tag{4.2-34}$$

where ${}^{4}k'_{A1}$ is the gas phase mass transfer coefficient above the slick surface, c_{A1i} is the air concentration of component A at the slick interface in equilibrium with the concentration of A in the slick, c_{A4}. The slick is denoted as phase 4. On the water side of the slick, ${}^{4}k'_{A2}$ is the mass transfer coefficient, and c_{A2i} is the seawater concentration of chemical A at the slick interface in equilibrium with c_{A4}. Far removed from the interfaces, the concentrations in air and water are c_{A1} and c_{A2}, respectively. The liquid phase mass transfer coefficients in the oil (i.e., ${}^{1}k'_{A4}$ and ${}^{2}k'_{A4}$) are assumed to be large and do not limit the rate of mass transfer. This assumption is likely invalid for the highly volatile components such as the n-alkanes from c_1 to c_{10}. See exercise problem 4.2G.

SINGLE COMPONENT BEHAVIOR, EVAPORATION, AND DISSOLUTION

Consider a slick containing a component mole fraction of chemical A, x_{A4}. This chemical has a saturation vapor pressure p_A^*, activity coefficient γ_{A4} so that the air phase partial pressure for component A immediately above the interface is

$$p_{A1i} \equiv p_T y_{Ai} = x_{A4} \gamma_{A4} p_A^* \qquad (4.2\text{-}35)$$

Partial pressure in air can be converted to concentrations by using the ideal gas law relationship $c_{A1} = p_{A1}/RT_1$. If component A has an aqueous phase solubility of c_{A2}^* the water–slick interface concentration can be approximated by

$$c_{A2i} = c_{A2}^* x_{A4} \gamma_{A4} \qquad (4.2\text{-}36)$$

See Problem 2.1L for details on developing Eq. 4.2-36.

Now if it is assumed that the oil phase is well mixed, the background concentration of chemical A in air and seawater is zero, a differential mass balance on a square meter of slick with a "thickness" corresponding to $\mathfrak{M}/A$ mol/m^2 and containing component A yields

$$\frac{d(x_{A4} \mathfrak{M}/A)}{dt} = -{}^4k_{A1}' c_{A1i} - {}^4k_{A2} c_{A2i} \qquad (4.2\text{-}37)$$

where $\mathfrak{M}$ is the total moles of oil in the slick and A is the area of the sea surface covered. Employing Eq. 4.2-35, the ideal gas law, and Eq. 4.2-36 in the preceding mass balance yields a simple differential equation with x_{A4} as the dependent variable:

$$\frac{d(x_{A4} \mathfrak{M}/A)}{dt} = \frac{-{}^4k_{A1}' x_{A4} \gamma_{A4} p_A^*}{RT} - {}^4k_{A2}' c_{A2}^* x_{A4} \gamma_{A4} \qquad (4.2\text{-}38)$$

Rearranging and assuming that $\mathfrak{M}/A$ is constant and integrating between limits of $x_{A4} = x_{A4}^0$ at $t = 0$ to x_{A4} at t gives

$$x_{A4} = x_{A4}^0 \exp(-Kt) \qquad (4.2\text{-}39)$$

where

$$K \equiv \frac{\{{}^4k_{A1}' \gamma_{A4} p_A^*/RT + {}^4k_{A2}' \gamma_{A4} c_{A2}^*\}}{\mathfrak{M}/A} \qquad (4.2\text{-}40)$$

The preceding description follows closely that of Harrison et al.[33]

The rate at which chemical A leaves the slick and enters the water decreases as the slick ages and the concentration decreases. The rate at which chemical A enters the air behaves similarly. In both cases the rate is proportional to the individual phase mass transfer coefficients and concentration of the chemical. The rate of evaporation is

$$N_{A0}(\text{evap}) = {}^4k'_{A1}\left[\frac{\gamma_{A4}p_A^*}{RT}\right]x_{A4}^0\exp(-Kt) \qquad (4.2\text{-}41)$$

and the dissolution rate is

$$N_{A0}(\text{diss}) = {}^4k'_{A2}c_{A2}^*\gamma_{A4}x_{A4}^0\exp(-Kt) \qquad (4.2\text{-}42)$$

In the interest of assessing the fate of chemical A, it is important to know the quantity entering each phase. The quantity of A that has moved into the air per unit of sea surface covered by the slick of age t is

$$\mathfrak{M}_A(\text{evap}) \equiv \int_0^t N_{A0}(\text{evap})\,dt$$

$$= {}^4k'_{A1}\left[\frac{\gamma_{A4}p_A^*}{RT}\right]\left(\frac{x_{A4}^0}{K}\right)\{1-\exp(-Kt)\} \qquad (4.2\text{-}43)$$

The quantity that transferred to the water during a similar period is

$$\mathfrak{M}_A(\text{diss}) \equiv \int_0^t N_{A0}(\text{diss})\,dt \qquad (4.2\text{-}44)$$

$$= {}^4k'_{A2}c_{A2}^*\gamma_{A4}\left(\frac{x_{A4}^0}{K}\right)\{1-\exp(-Kt)\} \qquad (4.2\text{-}45)$$

The last four expressions show that the movement of chemical A into the adjoining phases occurs rapidly immediately after the slick is formed but that the rate decreases in an exponential manner.

Harrison et al. report on the dynamic behavior of two components in small experimental sea-surface slicks formed of south Louisiana crude. The experiments were performed off the south shore of Grand Bahama Island. The crude oil was spiked with cumene (isopropyl benzene) to yield a solution of 4.2% cumene by weight. The concentrations of cumene, nonane, and other hydrocarbons within the oil phase were determined by gas chromatograph for the first few hours of slick aging.

Cumene and all lower boiling aromatics disappeared within the first 90 minutes. In general, cumene disappeared faster than nonane. The time to achieve 63% loss in spill 1 was 13 minutes for cumene and 28 minutes for nonane; 25 and 35 minutes, respectively, in spill 2; and 20 and 27 minutes, respectively, for spill 3. It was concluded from seawater samples that the dissolution rate was considerably slower than the evaporation rate and that only organisms in water that are or have been in close proximity to a spill for an extended time are likely to suffer toxic effects. (See Problem 4.2F for further details.)

MULTICOMPONENT EVAPORATION

The assumption that the "oil lay," $\mathfrak{M}/A$, is constant is valid only for a very short period of time. In general, this assumption is invalid since the spill spreads. (See problem 3.1F.) $\mathfrak{M}/A$ decreases in proportion to the spill thickness. Also as the oil evaporates and dissolves, the amount of oil remaining decreases, and this causes a further decrease in $\mathfrak{M}/A$ and causes a change in the oil composition. It is very difficult to quantify these effects accurately since information on the spill area as a function of time and on the oil composition is required.

Crude oil and its distilled fuel products consist of a host of individual hydrocarbon compounds. Regnier and Scott[34] performed evaporation rate studies of n-alkane components of Arctic diesel 40, a No. 2 fuel, to clarify the multicomponent behavior of slicks. The work reports the determination of evaporation rate constants of selected oil components. It is desirable to obtain the evaporation rate constants of the components as they occur in the oil matrix. If a sufficient number of these constants is known, it should be possible to predict the evaporative loss of oils spilled in the environment. Such values would be applicable to spills on ice and water where the oil lay is more than just a few molecules thick.

A mathematical description of a multicomponent hydrocarbon mixture consisting of N components undergoing evaporation commences with a component mass balance for constituent i:

$$d\left(\frac{x_{i4}\mathfrak{M}}{A}\right)/dt = \frac{-^{4}k'_{i1}x_{i4}\gamma_{i4}p_i^*}{RT} \tag{4.2-46}$$

Expanding the left-hand side yields

$$x_{i4}\frac{d(\mathfrak{M}/A)}{dt} + \frac{\mathfrak{M}}{A}\frac{dx_{i4}}{dt} = \frac{-^{4}k'_{i1}x_{i4}\gamma_{i4}p_i^*}{RT} \tag{4.2-47}$$

A total mass balance that includes all components is

$$\frac{d(\mathfrak{M}/A)}{dt} = \frac{-\sum_{i=1}^{N} {}^{4}k'_{i1}x_{i4}\gamma_{i4}p_i^*}{RT} \tag{4.2-48}$$

A description of this simple "nonspreading" slick evaporation model requires the simultaneous solution of $N-1$ component balances and the total balance. The necessary initial condition at $t=0$ is

$$x_{14} = x_{14}^0, x_{24} = x_{24}^0, \dots, x_{N-1,4} = x_{N-1,4}^0, \mathfrak{M} = \mathfrak{M}^0 \qquad (4.2\text{-}49)$$

The solution of the preceding system using a differential equation computer software package such as CSMP is not a difficult task and can yield slick composition and evaporation rate as a function of spill lifetime.

Regnier and Scott placed 12 g samples of Arctic diesel 40 in 90 mm diameter Petri dishes in an environmental chamber to perform evaporation studies. The chamber provided a constant wind speed of 21 km/hr over the samples and was set to operate at temperatures of 5, 10, 20, and 30°C. The oil thickness was roughly 3 mm. The air movement over the samples produced small wavelets that resulted in a stirring action. Samples were taken for gas chromatographic analysis.

The evaporation study involved following the kinetic behavior of 10 n-alkanes. The n-alkanes and the fractions in the Arctic diesel appear in Table 4.2-5. Figure 4.2-9 is the evaporation profile of the refined oil as a function of the remaining oil. Although a smooth curve was obtained at each temperature, no simple order for evaporation of the oil could be determined. This might be expected since the oil is a multicomponent system, and each component would contribute to the total order of evaporation.

Table 4.2-5. Percentages of n-Alkanes in Oil Samples

Component	Percentage
n-C$_9$	0.71
n-C$_{10}$	1.88
n-C$_{11}$	2.74
n-C$_{12}$	2.81
n-C$_{13}$	2.83
n-C$_{14}$	2.88
n-C$_{15}$	1.95
n-C$_{16}$	1.33
n-C$_{17}$	0.80
n-C$_{18}$	0.34
Total	18.27

A detailed study of the 10 individual n-alkanes revealed that a first-order model was sufficient to quantify the evaporation rate of each. The evaporation rate constants are listed in Table 4.2-6. It was hoped that by

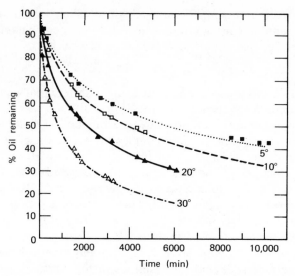

Figure 4.2-9. Evaporation of oil. (Indicators represent experimental points; curves are calculated.) (Reprinted with permission from Reference 34. Copyright by the American Chemical Society.)

Table 4.2-6. First-Order Evaporation Rate Constants of n-Alkane Components of Arctic Diesel 40

	Evaporation Rate Constants (min^{-1})			
Com-	Temperature °C			
ponent	5	10	20	30
$n\text{-}C_9$	3.49×10^{-3}			
$n\text{-}C_{10}$	1.19×10^{-3}	1.87×10^{-3}	3.44×10^{-3}	6.98×10^{-3}
$n\text{-}C_{11}$	4.15×10^{-4}	7.17×10^{-4}	1.31×10^{-3}	2.48×10^{-3}
$n\text{-}C_{12}$	1.57×10^{-4}	2.86×10^{-4}	5.25×10^{-4}	1.28×10^{-3}
$n\text{-}C_{13}$	1.57×10^{-4}	1.20×10^{-4}	2.46×10^{-4}	5.72×10^{-4}
$n\text{-}C_{14}$	2.21×10^{-5}	4.20×10^{-5}	1.14×10^{-4}	2.94×10^{-4}
$n\text{-}C_{15}$	5.61×10^{-6}	4.28×10^{-5}	5.24×10^{-5}	1.14×10^{-4}
$n\text{-}C_{16}$	1.08×10^{-6}	2.58×10^{-5}	3.99×10^{-5}	6.14×10^{-5}
$n\text{-}C_{17}$	4.00×10^{-7}	5.70×10^{-5}	4.08×10^{-5}	1.11×10^{-4}
$n\text{-}C_{18}$		2.20×10^{-5}	4.00×10^{-5}	

quantifying the kinetic behavior of the 10 n-alkanes, the evaporative behavior of the whole sample would be reflected. Since Eq. 4.2-39 reflects first-order kinetics, the evaporation rate constants (K) in Table 4.2-4 must be related to the gas phase mass transfer coefficient, vapor pressure, temperature, evaporation surface area, and total moles by

$$K = \frac{^4 k'_{A1} \gamma_{A4} p^*_A A}{RT \mathfrak{M}} \qquad (4.2\text{-}50)$$

It was found that a $1:1$ correspondence existed between K and p^*_A and a least squares fit of those data with $p^*_A > 1E-6$ atm was performed. The following equation with p^*_A in atm and K in $\min^{-1}$

$$\log p^*_A = 1.25 \log K + 0.160 \qquad (4.2\text{-}51)$$

makes it possible to calculate the evaporation rate constant of an n-alkane if its vapor pressure is known.

The evaporation rate constants and the initial concentrations of each n-alkane were used to ascertain, on a percentage basis, the amount of the n-alkane remaining at definite time intervals. This was then compared with the percentage of the total oil remaining at the same time intervals. The experimental values were in excellent agreement with the calculated results, as shown in Fig. 4.2-9.

Example 4.2-1.   Half-life of Normal Hexane in an Oil Slick

Estimate the half-life of n-hexane in an oil slick of temperature 75°F. The vapor pressure of n-hexane is 3.0 psia.

SOLUTION

Use Eq. 4.2-51 to estimate the evaporation rate constant, and use Eq. 4.2-39 for the time. This procedure assumes evaporation is the dominant mechanism of n-hexane disappearance from the slick. From Eq. 4.2-51

$$\log K = \frac{[\log(3/14.7) - 0.160]}{1.25}$$

$$K = 0.209 \text{ min}^{-1}$$

From Eq. 4.2-39 with $x_{A4} = x^0_{A4}/2$

$$t_{1/2} = \frac{-\ln(1/2)}{K} = 3.32 \text{ min}$$

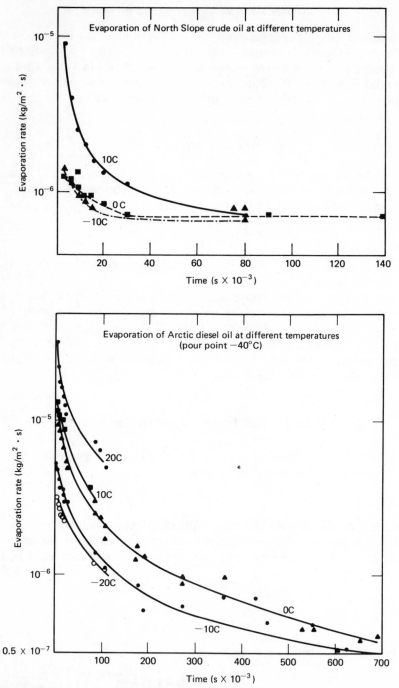

Figure 4.2-10. (*a*) Evaporation of North Slope crude oil as a function of time and temperature. (*b*) Evaporation rates of Arctic diesel as a function of time and temperature. (**Source.** Reference 35.)

CLOSURE

The simple models previously presented contain several essential aspects of transport processes associated with oil slicks or similar chemical slicks on the surface of water. It appears that evaporation is a dominant mechanism, and from the viewpoint of interphase chemical movement, the air receives the bulk of the transported contaminants. The evaporation process occupies only a brief period during the life of a crude oil slick. Figure 4.2-10 contains evaporation rate data of an Arctic diesel and a North Slope (Alaska) crude oil. It appears that the evaporation rate can decrease by as much as two orders of magnitude in 8 days.

Later during the lifetime of an oil slick, depending on the physical and chemical properties of the oil and on the environmental conditions, other factors become equally or more important. Increased viscosity of the mixed oil as the evaporation proceeds modifies the kinetic parameters of evaporation. As the more volatile components evaporate, the mixture becomes more viscous, and diffusion processes within the oil phase have progressively more influence on rates of evaporation. It has been observed that a type of film forms on the surface of unmixed crude oil that essentially stops evaporation of the lighter fractions.

In addition to evaporation, oil on water undergoes other processes, including spreading, emulsification, oxidation, bacterial action, and sinking. Emulsification occurs as a result of wave action. The turbulence produced during whitecapping results in the formation of oil-in-water and water-in-oil emulsifications. Heavy fractions accumulate and form tar balls. Tar balls are lighter than water and float; however, with age, sand and shell particles accumulate, and the ball sinks to the bottom. Hydrocarbon in the liquid phase is attacked by oxygen. The hydrocarbons that are more readily affected are paraffins and aromatic hydrocarbons with suitable side chains. Microorganisms such as pseudomonas, which are found in seawater and sand-contaminated beaches, can use hydrocarbons under aerobic conditions. Oil slicks may be destroyed by bacteria, but the process would take many months.

Problems

4.2A. TRITIUM LOSS FROM WATER EXPOSED TO THE ATMOSPHERE[36]

Laboratory and field observations were made on the fate of small quantities of tritium as HTO found in process water in a certain nuclear facility. The water is usually released into streams or open basins excavated in the ground.

In laboratory experiments, the transfer of HTO across an air–water interface was measured using stoppered 250 mL bottles initially containing 100 mL of liquid. The depth of the liquid was 4.6 cm. The space above the liquid was swept with 600 cm^3 of air per minute. H_2O loss was determined by weight and tritium analysis was by liquid-scintillation counting. Experiments were conducted at $24.5°C \pm 1°$. In these experiments the samples were hand-shaken twice a day. Tritium desorption into H_2O-saturated air resulted in as overall liquid phase mass transfer coefficient of 6.32E−6 cm/s. A desorption experiment into dry air with H_2O evaporation replenished with distilled H_2O resulted in a coefficient of 5.75E−6 cm/s. Absorption of HTO from H_2O-saturated air yielded a coefficient of 6.00E−6 cm/s.

1. The preceding experiments represent the same overall HTO mass transfer coefficients. If H_2O was vaporized into dry air in the experimental apparatus and the evaporation rate 5.5E−6 $g/cm^2 \cdot s$ measured, which phase resistance controls the HTO mass transfer rate? At 25°C the vapor pressure of HTO is 0.92 that of H_2O.
2. Use Eq. 3.1-16 to estimate the liquid phase mass transfer coefficient. Use 1E−5 cm^2/s for the molecular diffusivity of HTO in H_2O. Which phase resistance controls?

In the field study, an impermeable basin (13,000 m^2 in area) containing tritium was isolated from further waste discharges and studied for a 3 year period. Monthly, water was taken from a depth of 1 ft at one location and the tritium content determined. The field data are shown in Figs. 4.2A*a* and *b*. A slow mixing of surface water with underlying liquid was observed. The slow mixing is caused by the slight specific gravity difference between the waste containing a small amount of salt and rain. The value of the overall mass transfer coefficient depends on the rate of mixing and is different for each rate.

3. Compute an average overall liquid phase mass transfer coefficient for the basin using the field data. Answer in cm/s.
4. Confirm by calculation that tritium loss is liquid phase controlled, owing apparently to slow water maxing. Use 750 cm/s as a reasonable gas phase coefficient for earthen basins.

4.2B. AIR–SEA INTERFACE FLUX CALCULATION[25]

Calculate the flux rate in g/yr and direction of transfer for the gases given in Table 4.2B at the air–sea interface. Use 3.6E 18 cm^2 as the air–water surface area of the oceans. Compare your answers with the values given in Table 4.2-3.

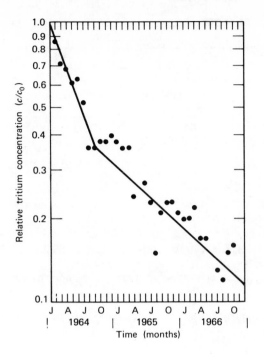

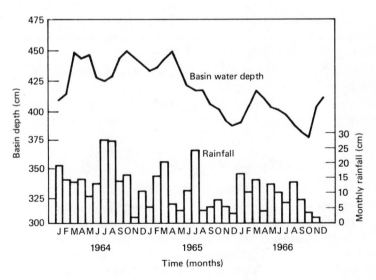

Figure 4.2A. (*a*) Tritium concentrations in an isolated impermeable basin from January 1964 through December 1966. (*b*) Depth of water in an isolated impermeable basin containing tritium and the rainfall received from January 1964 through December 1966. (Reprinted with permission from Reference 36. Copyright by the American Chemical Society.)

Table 4.2B. Concentrations of Gases in the Marine Environment

Gas	SO_2	N_2O	CO	CH_4	CCl_4
In air	3 $\mu g/m^3$	0.25[c]	0.13[c]	1.4[c]	71.2E−6[c]
In water	0	0.4 ppm[a]	6E−8[b]	4E−8[b]	60E−12[b]

Gas	CCl_3F	MeI	$(Me)_2S$
In air	50E−6[c]	1.2E−6[c]	~0
In water	7.6E−12[b]	135E−12[b]	1.2E−11 g/cm^3

Source. Reference 25.

[a]This is the N_2O concentration in air in equilibrium with the ocean surface water N_2O concentration in ppm (by volume).

[b]Units of concentration are cm^3 gas at STP + cm^3 liquid water.

[c]Units of concentration are parts per million (by volume).

4.2C. EVAPORATION OF LOW SOLUBILITY CONTAMINANTS FROM WATER SURFACES[21,22]

The group $\gamma_{A2} f_{A2}^0$ is of considerable importance because it quantifies the evaporating tendency of a chemical and its vapor liquid equilibrium. High values of $\gamma_{A2} f_{A2}^0$ denote chemicals that are less soluble in water and partition preferentially into the vapor phase.

1. Calculate $\gamma_{A2} f_{A2}^0$ for each class (i.e., alkanes, aromatics, etc.) of compounds in Table 4.2-1 for 25°C.
2. Calculate the vaporization half-life and mean residence time of representative compounds from each class. Use a depth h of 1 m.

4.2D. EVAPORATION IN A SHORT-FETCH WATER–WIND TUNNEL[37]

In a comparison of evaporation in the laboratory and at sea, Schooley employed a short-fetch water–wind tunnel to measure evaporation for wind velocities between 200 and 800 cm/s. The laboratory water–wind channel was 90 cm long, 7.5 cm wide, and 26 cm deep and the water depth was 16 cm. A 10 cm high column of air was drawn over the water surface with a blower. The average wind speed was measured with a small commercial wind velocity probe. A spring-balance arrangement weighed the tunnel before and after each experiment to determine the volume evaporated. The experimental data are shown in Fig. 4.2D.

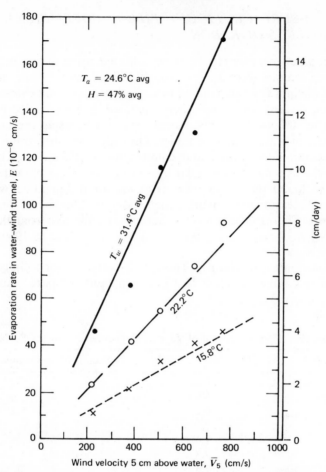

Figure 4.2D. Evaporation rate versus average wind speed at 5 cm above water for three different water temperatures in a short-fetch water-wind tunnel. Air temperature and relative humidity held constant. (Reprinted with permission from Reference 37.)

1. Show that the equation

$$n'_{B0} = Kv_{10}(p^*_{B1} - p_{B1}) \qquad (4.2D\text{-}1)$$

fits the experimental data. Here n'_{B0} is in inches of water per day, v_{10} is air speed at 10 m in mi/hr, and p_{B1} is water vapor pressure in mb. Determine the least squares value of K.

2. Compare the value of K with that calculated using Eq. 4.1E−1. Assume $v_{5\ cm} = 0.4\ v_{10\ m}$ for the laboratory data.

4.2E. SEA-SURFACE LIQUID PHASE COEFFICIENT FROM MEASUREMENTS OF RADON[31]

Subsurface water samples (10 L) were collected using a Niskin sampler while surface samples were drawn through tubing lowered from the fantail of the ship. Helium was recirculated through the sample and extracted 90% of the radon present, which was then separated in traps cooled by liquid nitrogen. The 90% yield was maintained constant by using the same flow rate and extraction time for all samples. The radon was then transferred to an alpha scintillation counter and counted for 4 to 8 hours. The ^{226}Ra content of the sample was obtained by storing the processed water in the bottle for 7 to 10 days. The new crop of radon generated was then extracted and measured in a manner identical to that for the original radon measurement. The samples reported here were collected from the Coast Guard vessel *Rockaway* during the period May 2 to June 9, 1969, from a position 15°N, 46°W.

The radon results are given in Table 4.2E corrected for salinity. Using the data in this table calculate $^1k'_{A2}$ in cm/hr and compare your results with values reported in Table 4.2-4.

Table 4.2E.   Radon Content at Various Depths below the Sea Surface

Depth (m)	^{222}Rn (dpm^a/100 L)	No. of Observations
1	6.1	16
10	5.9	8
20	6.2	4
25	7.1	5
35	7.4	6
50	7.8	6
75	7.5	2
100	8.2	6
200	8.0	4

Source.   Reference 31.

^{a}Disintegrations per minute.

4.2F. EVAPORATION AND DISSOLUTION OF CUMENE AND NONANE FROM SMALL SEA-SURFACE SLICKS[33]

A small sea-surface slick was created by placing 275 gal (1.04m³) of crude oil spiked with cumene ($c^0_{A4} = 0.34$ mol/L). Water temperature was essentially constant at 23.6°C, air temperature was 20.5 to 24.1°C, relative humidity was 60 to 79%, wind at 3 m was calm to 18 mph with gusts to 22 mph (9.8 m/s), and sea-surface conditions ranged from calm with gentle swell to extensively whitecap covered.

1. Compute the time (min) for 63% of the cumene to disappear from a slick 1 mm in thickness. Use the following data: $\mathfrak{M}/A = 5$ mol/m^2 (1 m^3 spread over 10^3 m^2); $^4k'_{A1} = 1E-2$ m/s, $^4k'_{A2} = 5.5E-5$ m/s; $p_A^* = 5.3E-3$ atm (4.1 mm Hg); $c_{A2}^* = 50$ mg/L; $\gamma_{A4} = 1.0$. Compare the computed time with the observed times reported in the text.

2. Compute the time (min) for 63% of the original nonane to disappear. Use the following data: $p_A^* = 3.0$ mm Hg; $c_{A2}^* = 0.22$ mg/L.

3. Equation 4.2-40 is useful in estimating the relative importance of evaporation and dissolution in the disappearance of a specific component from a slick. Using the numerator of Eq. 4.2-40 as the criterion, calculate the relative quantity of cumene transferred to the air and water. Repeat the calculation for the chemical nonane. Is the conclusion that dissolution is "considerably slower" justified?

4. Using Eq. 6.2-31, verify that the numerical value of the gas phase coefficient given in part (1) is a reasonable value.

4.2G. CONTROLLING RESISTANCE FOR EVAPORATION OF ALKANES FROM OIL SLICKS

1. It has been assumed in the development of model equations for the weathering of oil slick components that the oil phase resistance is not limiting. (See specifically Eq. 4.2-34.) Compute the overall coefficient for the n-alkanes given in Table 4.2G-1, assuming an oil of density 0.825 g/cm^3 and molecular weight 200. Give the answer in cm/hr.

   Based on the calculated result, which phase resistance controls the evaporation of the higher molecular weight hydrocarbons from the slick?

2. By using Eq. 4.2-50 it is possible to calculate the gas phase coefficient $^4k'_{A1}$ from the evaporation rate constant K, measured by Regnier and Scott[34] and presented in Table 4.2-6. Calculate the rate constants for all 10 n-alkanes at 10°C. Use the Antoine equation for estimating the vapor pressure:

$$\log p_{A1}^* = A - \frac{B}{C+T} \tag{4.2G-1}$$

where p_{A1}^* is the vapor pressure of component A in mm Hg and T is the temperature in °C. Constants for Eq. 4.2G-1 appear in Table 4.2G-2. Consult the text material concerning the experiment as a source of

Table 4.2G-1.   Henrys' Law Constants for Some Alkanes

Component	H_{AX} (Equilibrium Constant $H_{AX} \equiv y/x$ at 75°F)
Ethane	28.
Propane	7.8
n-Butane	2.0
n-Pentane	0.60

Table 4.2G-2.   Constants for the Antoine Equation

Component	A	B	C
$n\text{-}C_9$	7.26430	1607.12	217.54
$n\text{-}C_{10}$	7.31509	1705.60	212.59
$n\text{-}C_{11}$	7.3685	1803.90	208.32
$n\text{-}C_{12}$	7.35518	1867.55	202.59
$n\text{-}C_{13}$	7.5360	2016.19	203.02
$n\text{-}C_{14}$	7.6133	2133.75	200.8
$n\text{-}C_{15}$	7.6991	2242.42	198.72
$n\text{-}C_{16}$	7.03044	1831.317	154.528
$n\text{-}C_{17}$	7.8369	2440.20	194.59
$n\text{-}C_{18}$	7.9117	2542.00	193.4

Source.   Reference 38.

necessary data. Assume molecular weight of 200. Report the coefficient in cm/hr.

3. After reviewing the numerical values calculated, discuss the magnitude of the coefficients. Are the values reasonable?

4.2H.   n-NONANE CONCENTRATION UNDER AN OIL SLICK

Harrison et al.[33] suggest that an alternative modeling approach to the one represented by Eq. 4.2-39 is to assume unsteady-state penetration diffusion of the hydrocarbon into the air and water phases. This is invalid for the air phase since there is continual replacement of the air above the spill. It is more valid for the water phase, although there is some relative motion between the oil and the water.

Oil has a calming effect on the sea surface. Waves are dampened significantly. It has been reported that the center of slicks travels at less than 5% of the wind velocity. The water layers adjacent to the underside of a slick are imperfectly mixed, and it is therefore not too unreasonable to assume a penetration diffusion-type model for those calm sea-surface periods. This allows the calculation of water concentrations underneath a slick.

Calculate the average concentration of *n*-nonane in a 3 cm zone underneath an oil slick that contains 1 (mol) % *n*-nonane, which would develop in a 24 hour calm period. Assume a water temperature of 25°C. Report the concentration in mg/m^3 (ppb). (*Hint:* Review subsection in Chapter 3 entitled "Diffusion of Trace Chemicals in Stagnant Media," paying particular attention to Eqs. 3.1-3, 3.1-4, and 3.1-5. See Problem 3.1G.)

4.3. HEAT TRANSFER ACROSS THE AIR–WATER INTERFACE

Temperature is an important variable in assessing chemical movement rates across the air–water interface. The temperature at the water surface significantly affects the vapor pressure of the chemical species and it also affects the magnitude of the transfer coefficients. Fundamental concepts of heat transfer were presented in Section 3.3. In this section specific attention is devoted to the rate equations for heat transfer between water bodies and the air. The material presented is generally applicable at the surfaces of streams, lakes, and oceans.

A study of heat transfer in water bodies commences with the equation of energy. If one considers a fixed volume of water V containing an air–water interface of area A_{12} and a water–soil interface of area A_{23}, the equation of energy can be written simply as

$$\rho_2 \hat{c}_\rho \frac{dT}{dt} = \frac{1}{V}(q_{12}A_{12} + q_{23}A_{23}) \qquad (4.3\text{-}1)$$

where gradients of temperatures, convective flow, and heat generation within the volume are neglected. Equation 4.3-1 states that the change in enthalpy of a fixed water body is the result of heat flow through the interfaces. The sign convention adopted is that energy additions to a surface are positive, and energy losses are negative. All energy or heat flux relations are in J/m^2·s. (1 Btu/ft^2·hr = 3.15 J/m^2·s.)

If we now assume steady state and neglect heat losses through the bottom and sides of the lake, pond, or stream, then the heat (or energy) flux through the air–water interface is seen to consist of four major sources:

$$q_{12} = q_{lw} + q_e + q_c + q_s \qquad (4.3\text{-}2)$$

where q_{lw} is the longwave radiant energy flux, q_e is the evaporative heat flux, q_c is the conductive heat flux, and q_s is the shortwave radiant energy.

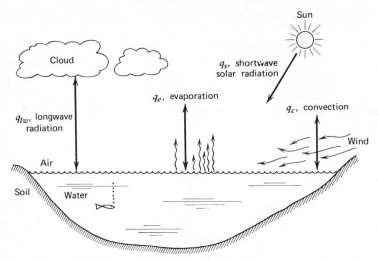

Figure 4.3-1.   Heat and energy transfers at the air–water interface.

Figure 4.3-1 illustrates these transfer mechanisms. As is shown, the magnitude of the fluxes depends on the water body, its geographic location, and time. All fluxes except q_s can be positive or negative.

Components of the Energy Balance Equation

SHORTWAVE RADIATION

Shortwave radiation originates directly from the sun, although the energy present at the outer edges of the atmosphere is depleted via absorption by O_3, scattering by dry air, absorption scattering by particulates, and absorption and scattering by water vapor. The electromagnetic wave spectrum from the sun consists of wavelengths of 0.2E−6 to 3.0E−6 m. The amount of solar radiation incident on a horizontal surface varies, depending on the geographic location, elevation, season, and meteorological conditions. Although q_s can be empirically calculated, it is much better to measure it using a Pyrheilometer, which is capable of giving the accuracy required for the energy budget.

Solar radiation intensity observations are made at a number of U.S. Weather Bureau stations. Table 4.3-1 gives the total daily solar radiation at the top of the atmosphere as given in the Smithsonian meteorological tables. The entries in this table are cal/cm²·d (1 cal = 4.1868 J). Table 4.3-2 gives the solar radiation observed at ground stations in 1971. The

Table 4.3-1. Total Daily Solar Radiation at the Top of the Atmosphere[a]

Latitude	Longitude of the Sun															
	0°	22½°	45°	67½°	90°	112½°	135°	157½°	180°	202½°	225°	247½°	270°	292½°	315°	337½°
	Mar. 21	Apr. 13	May 6	May 29	June 22	July 15	Aug. 8	Aug. 31	Sept. 23	Oct. 16	Nov. 8	Nov. 30	Dec. 22	Jan. 13	Feb. 4	Feb. 26
								Approximate Date								
								(cal/cm²)								
90°		423	772	999	1077	994	765	418								
−80	155	423	760	984	1060	980	754	418	153	7						7
−70	307	525	749	939	1012	934	742	519	303	129	24				24	131
−60	447	635	809	934	979	929	801	629	442	273	146	72	49	73	146	276
−50	575	732	867	958	989	954	859	725	568	414	286	204	176	205	289	419
−40	686	807	910	972	991	967	901	798	677	545	429	348	317	350	434	553
−30	775	865	929	967	975	960	921	856	765	663	564	492	466	494	568	670
−20	841	894	923	935	935	930	916	884	831	760	685	627	605	630	691	769
−10	882	897	893	881	873	877	886	887	871	835	789	748	733	752	795	845
− 0	895	873	837	804	790	800	830	863	885	886	870	851	843	855	878	896
−10	882	824	760	707	687	704	753	814	871	910	927	931	933	936	936	921
−20	841	750	660	593	567	590	654	741	831	907	959	988	999	993	968	918
−30	775	654	543	465	436	463	538	646	765	877	964	1020	1041	1025	973	888
−40	686	538	413	329	297	328	409	533	677	819	944	1027	1059	1032	953	828
−50	575	408	276	193	165	192	274	404	568	743	901	1014	1056	1018	909	752
−60	447	269	140	68	47	68	139	266	442	644	840	987	1046	992	847	652
−70	307	127	23				23	126	303	532	778	993	1081	998	785	539
−80	155	7						7	153	429	790	1041	1132	1046	796	434
−90										429	801	1056	1149	1062	809	434

[a]Values are in cal/cm² and apply to a horizontal surface.

Source: Smithsonian Meteorological Tables by Robert J. List, 6th revised edition, 1949.

213

Table 4.3-2. Solar Radiation Data, 1971

[Average Daily Values (Direct and Diffuse) Received on a Horizontal Surface, Tabulated in Langley]

Station	January	February	March	April	May	June	July	August	September	October	November	December	Annual
Albuquerque, N.M.	330	411	536	611	720	744	679	600	509	396	300	220	505
Ames, Iowa	167	204	373	499	507	625	569	545	393	307	207	–	–
Annette, Ak.	51	95	183	305	470	477	513	258	–	–	47	40	–
Apalachicola, Fla.	250	330	437	511	581	560	478	488	463	392	337	221	421
Argonne Nat. Lab.	177	179	306	452	505	569	526	476	328	250	162	94	335
Astoria, Ore.	61	158	248	403	426	418	458	484	433	234	–	41	–
Atlanta, Ga.	193	264	386	527	583	557	444	445	373	283	276	159	374
Barrow, Ak.[a]	2	34	178	374	549	665	541	313	121	54	3	0	236
Bethel, Ak.	–	139	314	491	432	431	358	251	206	102	55	22	–
Bismarck, N.Dak.	174	277	357	439	587	576	688	600	370	236	155	125	382
Blue Hill, Mass.	123	150	273	314	373	666	551	511	339	270	–	120	388
Boise, Idaho	145	255	347	507	592	637	700	605	475	322	164	128	406
Brookings, S.Dak.	169	216	305	437	471	516	531	506	328	208	150	129	331
Brownsville, Tex.	284	382	493	486	576	624	–	568	395	405	334	265	–
Burlington, Ver.	159	210	401	377	551	614	552	453	348	231	118	90	342
Cape Hatteras, N.C.	168	326	425	540	557	569	485	475	405	260	259	192	388
Caribou, Me.	140	–	348	333	424	499	443	359	297	192	132	103	–
Charleston, S.C.	218	282	416	489	527	554	442	393	406	248	275	184	370
China Lake, Calif.	–	–	–	–	–	567	–	–	–	396	286	222	–
Cleveland, Ohio	133	177	283	423	–	520	511	–	–	–	–	84	–
Columbia, Mo.	201	244	394	526	543	628	543	529	398	316	214	113	387
Davis, Calif.	191	331	427	–	565	704	706	658	538	400	262	182	–
Dodge City, Kan.	252	314	443	516	632	–	–	–	–	–	–	–	–
E. Lansing, Mich.	165	187	291	427	548	567	–	–	–	–	–	–	–
El Centro, Calif. NPF	296	384	498	591	640	663	614	516	483	397	312	254	471
El Paso, Tex.	325	433	565	623	691	709	665	586	500	398	339	269	509

Location													
Ely, Nev.	253	347	465	505	557	710	687	570	580	346	272	—	—
Eppley, Newport, R.I.	167	205	—	369	402	576	514	508	355	260	152	112	—
Fairbanks, Ak.	11	48	—	393	499	596	488	331	207	68	30	5	—
Flaming Gorge, Utah	210	259	410	546	539	591	—	—	—	320	260	191	—
Fort Worth, Tex.	285	318	456	480	520	582	537	457	399	297	260	161	396
Fresno, Calif.	199	301	460	560	577	749	701	613	519	388	251	167	457
Gainesville, Fla.	255	345	441	567	567	—	471	—	—	—	383	273	—
Geneva, N.Y.	142	163	261	325	396	452	450	390	293	192	116	75	277
Geneva, #2, N.Y.[c]	—	—	—	—	—	—	—	—	—	—	136	85	—
Glasgow, Mont.	132	246	371	439	548	586	649	532	373	240	132	117	364
Gloucester Pt., Va.	151	—	—	—	—	527	529	477	404	213	246	178	—
Grand Junction, Colo.	227	315	448	517	467	630	—	—	531	361	239	185	—
Great Falls, Mont.	129	221	372	437	520	593	664	541	381	273	155	119	367
Greensboro, N.C.	179	270	402	525	488	499	472	454	347	252	249	165	359
Indianapolis, Ind.	173	199	—	454	504	578	523	493	349	297	186	107	—
Inyokern, Calif.[b]	—	—	483	587	649	708	653	561	512	—	—	—	—
Ithaca, N.Y.	202	—	298	423	469	528	497	436	305	240	130	95	—
Lake Charles, La.	243	349	390	470	489	543	485	439	374	343	287	178	383
Lakeland, Fla.	312	372	484	495	581	478	499	457	419	358	320	289	422
Lander, Wyo.	192	299	448	509	495	660	598	556	437	282	243	182	408
Laramie, Wyo.	173	266	379	480	481	632	603	554	444	333	229	200	398
Las Vegas, Nev.	278	372	503	616	648	732	—	557	556	373	277	203	—
Lexington, Ky.	186	233	—	508	481	—	483	428	307	263	182	—	—
Little Rock, Ark.	199	288	377	463	508	567	511	463	405	324	234	139	373
Los Angeles, Calif.	248	349	459	570	557	591	657	615	492	382	281	210	451
Los Angeles, Calif. U	250	342	—	—	534	564	670	562	456	366	266	204	—
Madison, Wics.	190	244	366	467	546	588	557	513	388	252	160	117	366
Manhattan, Kan.	215	247	354	462	493	605	564	500	423	303	199	131	380
Matanuska, Ak.	46	74	258	338	—	457	389	—	210	85	43	381	—
Medford, Ore.	114	194	282	428	482	593	650	571	452	301	134	92	358

Table 4.3-2. (*Continued*)

Station	January	February	March	April	May	June	July	August	September	October	November	December	Annual
Miami, Fla.	341	–	459	519	529	432	539	452	427	309	290	278	–
Midland, Tex.	334	420	545	601	601	612	638	516	458	368	321	249	472
Nashville, Tenn.	157	276	361	504	523	543	494	462	375	282	215	128	360
New York, Central Park, N.Y.	162	200	324	430	419	537	530	504	299	241	168	129	329
North Omaha, Neb.	217	265	388	489	490	606	604	513	–	292	177	129	–
Oak Ridge, Tenn.	166	250	342	482	497	503	412	457	350	294	215	122	341
Oklahoma City, Okla.	250	287	458	460	521	581	555	470	399	297	232	168	391
Palmer Aaes, Ak.	39	63	235	320	406	401	356	255	183	78	42	12	199
Page, Ariz.	–	–	357	–	–	–	–	511	–	–	–	–	–
Phoenix, Ariz.	299	383	507	564	635	636	631	548	494	390	289	224	467
Portland, Me.	170	211	348	368	380	592	514	458	316	235	147	121	322
Prosser, Wash.	122	201	302	448	564	614	651	531	417	–	–	–	–
Rapid City, S.Dak.	159	258	344	409	498	555	559	468	371	253	157	140	350
Reno, Nev.	224	287	395	464	465	597	591	525	455	309	194	145	388
Richland 25 NW, Wash.	114	211	326	482	586	622	686	602	448	286	148	97	384
Riverside, Calif.	292	375	504	569	568	718	–	651	542	–	333	255	–
Ruston, La.	197	298	368	461	509	576	478	482	398	341	279	–	–
Saint Cloud, Minn.	179	251	375	453	541	514	560	506	335	181	124	104	344
Salt Lake City, Utah	170	277	402	494	578	710	718	583	479	296	214	146	422
San Antonio, Tex.	281	347	489	483	517	514	617	455	410	316	267	181	406
Santa Maria, Calif.	280	369	475	555	563	648	645	603	509	401	295	226	464
Sault Ste. Marie, Mich.	137	214	368	460	524	543	583	473	303	171	100	74	329
Seattle–Tacoma, Wash.	66	141	240	385	457	481	585	545	364	189	88	53	299
Seattle, Wash. Univ.	71	142	234	344	380	382	472	447	314	180	92	68	261

Spokane, Wash.	98	176	294	407	524	531	642	556	393	248	102	97	339
State College, Penn.	154	196	299	386	392	417	463	405	240	175	63	–	–
Sterling, Va.	179	224	365	528	467	532	547	544	364	217	198	138	359
Swan Island, W.I.	–	–	503	561	557	488	472	507	452	405	315	320	–
Tallahassee, Fla.	219	301	385	437	502	490	388	–	–	–	–	–	–
Tampa, Fla.	300	360	482	545	630	538	523	471	415	380	312	298	438
Tucson, Ariz.	315	391	523	590	676	649	678	591	535	414	330	244	50
Upper Marlboro, Md.	–	–	–	–	435	516	555	501	–	218	198	137	–
Wake Island, Pacific	428	446	594	612	677	647	638	591	437	512	435	400	535

Note: Langley is the unit to denote one gram calorie per square centimeter.

(U) Indicates urban sites.

[a]Sun below horizon November 19 through January 23, inclusive.

[b]Station name changed to China Lake, California, effective October 1971.

Geneva, N.Y., is a new pyranometer run concurrently with old equipment for 2 months for comparative purposes.

[c]Effective January 1972 Geneva #2 will be known as Geneva.

entries in Table 4.3-2 are monthly averages of the daily radiation (1 langley/d $= 4.1868$ J/cm$^2\cdot$d).

Cloud cover decreases the incoming radiation. This can be appreciated by comparing incident values from Tables 4.3-1 and 4.3-2 at the same latitude and month. Stations recording cloud cover are more numerous than those recording ground level radiation. A table giving average percentage of possible sunshine in selected cities is given in Appendix E. If this type of cloud cover information is available, the following equation may be used to estimate the shortwave radiation:[39]

$$q_s = q_{s0}(0.803 - 0.340n - 0.458n^2) \qquad (4.3\text{-}3)$$

where q_{s0} values are obtained from Table 4.3-1 and n is the fractional cloud cover.

LONGWAVE RADIATION

By virtue of a temperature difference, the water and its surrounding air canopy exchange radiant energy. Because the temperatures involved are relatively low compared to the sun, the electromagnetic waves are long, about $E-6$ m. The rate equation for the net exchange is

$$q_{lw} = \beta \sigma T_1^4 - e_2 \sigma T_2^4 \qquad (4.3\text{-}4)$$

where β is the cloud cover factor, σ is the Stefan-Boltzmann constant ($5.67E-12$ J/s$\cdot$cm$^2\cdot$K^4), T_1 is the air temperature in kelvin, e_2 is the emissivity of the water surface, and T_2 is the water surface temperature in kelvin.

The first term on the right-hand side of Eq. 4.3-4 is the longwave atmospheric radiation. It depends primarily on air temperature and humidity and increases as the air moisture content increases. It may be a major input on warm cloudy days when direct solar radiation approaches zero. It is actually a function of many variables, including CO_2 and O_3, although it can be fairly accurately calculated via a simple empirical formula. As noted earlier, β is a constant that is a function of the type of cloud cover given by

$$\beta = a + bp_{A1} \qquad (4.3\text{-}5)$$

where p_{A1} ($A \equiv H_2O$ vapor) is the vapor pressure of water in inches of mercury (1 in. of Hg $= 3.377E3$ Pa), a and b are constants. For each

Table 4.3-3. Cloud Cover Constants

Cloud Cover (tenths)	$\beta =$
0	$0.74 + 0.15\, p_{A1}$
1	$0.75 + 0.15\, p_{A1}$
2	$0.76 + 0.15\, p_{A1}$
3	$0.77 + 0.143\, p_{A1}$
4	$0.783 + 0.138\, p_{A1}$
5	$0.793 + 0.137\, p_{A1}$
6	$0.80 + 0.135\, p_{A1}$
7	$0.81 + 0.13\, p_{A1}$
8	$0.825 + 0.12\, p_{A1}$
9	$0.845 + 0.105\, p_{A1}$
10	$0.866 + 0.09\, p_{A1}$

value of cloud cover, these constants can be approximated as shown in Table 4.3-3. The vapor pressure of water can be calculated in inches of Hg by

$$p_{A1} = 0.0295 \left[\exp \left\{ 21.66 - \left(\frac{5431.3}{T} \right) \right\} \right] H_R \qquad (4.3\text{-}6)$$

where T is in kelvin and H_R is the fraction relative humidity, $0 \leqslant H_R \leqslant 1$.

The second term on the right-hand side of Eq. 4.3-4 is the longwave radiation originating at the water surface. A value of 0.97 is reasonable for the emissivity of water.

EVAPORATIVE ENERGY EXCHANGE

If the air above a body of water is less than 100% saturated with water vapor (i.e., $H_R < 1$), there is a potential for evaporation from the surface. The evaporated water requires energy, latent heat of evaporation, as it changes from a liquid to a vapor. Equation 3.3-5 relates the mass and energy flux rates. The Lake Hefner study (see Problem 4.1E) resulted in the following equation:

$$q_e = -30.5 v_1 \left(p_{A1}^0 - p_{A1} \right) \qquad (4.3\text{-}7)$$

where q_e is the evaporative heat loss in $J/m^2 \cdot s$, v_1 is wind speed in km/hr,

p_{A1}^0 is the vapor pressure of water in inches of mercury at T_2 (the water surface temperature), and p_{A1} is the vapor pressure of water vapor in the air far removed from the lake surface. Note that p_{A1}^0 can be obtained from Eq. 4.3-6 with $H_R = 1$ and $T = T_2$. Atmospheric conditions of temperature (T_1), relative humidity (H_R), and wind velocity (v_1) are usually measured at 8 to 10 m above the surface. These values can be used in Eq. 4.3-7.

It would be expected that coefficients for evaporation and heat lost would be much different for rivers and streams than for lakes and would be dependent on water velocity and turbulence. There appears to be a lack of data in this area. (See the subsection entitled "Estimating Mass Transfer Coefficients for Ponds, Rivers, and Small Water Bodies" in Section 4.1.)

SENSIBLE HEAT EXCHANGE

Heat enters or leaves water by conduction if the air temperature is greater or less than the water temperature. There is a transfer of sensible heat, and the rate expression is of the form of Eq. 3.3-3. The rate expression is

$$q_c = 0.27 v_1 \{ T_1 - T_2 \} \qquad (4.3\text{-}8)$$

where q_c is the sensible heat transfer rate in $J/m^2 \cdot s$, T_1 and T_2 are the air and water temperatures respectively in kelvin.

SUMMARY

The relative magnitudes of each of the heat transfer terms for a typical water surface are as follows:

solar radiation $= 50$ to 370 $J/m^2 \cdot s$
longwave atmospheric radiation $= \pm 320$ to 420 $J/m^2 \cdot s$
longwave back radiation $= \pm 320$ to 470 $J/m^2 \cdot s$
evaporative heat loss $= \pm 260$ to 1050 $J/m^2 \cdot s$
conductive heat losses or gain $= -40$ to $+50$ $J/m^2 \cdot s$

There is some reflectivity at the water surface. The reason solar reflectivity is more variable than atmospheric reflectivity is that the former is a function of sun altitude and cloud cover, whereas the latter is relatively constant. The Lake Hefner study indicated that the atmospheric reflectivity was approximately 0.03, whereas on an annual basis, the solar reflectivity of water was 0.06.

Example 4.3-1. Heat Loss Through the Bottom

At the beginning of this section it was assumed that the heat losses through the bottom and sides of the lake, pond, or stream are negligible. The temperature difference between the water at the bottom and the mud surface is of the order of a tenth degree Celsius. Using the heat transfer coefficient of Example 3.3-1, estimate the bottom heat flux, and decide whether it is negligible.

SOLUTION

The heat exchange across the water–mud interface is by a conductive mechanism, and Eq. 3.3-3 can be used.

$$q_{23} = h_{23} \Delta T = \frac{120. \text{ J}}{\text{m}^2 \cdot \text{s} \cdot \text{K}} \left| \frac{\pm 0.1 \text{ K}}{} \right| = \pm 12.0 \text{ J}/\text{m}^2 \cdot \text{s}$$

Even if ΔT was ± 1.0 K, the heat exchange would be small compared to the exchanges occurring at the air–water surface, which are of the order of ± 400 J/m$^2 \cdot$ s.

Problems

4.3A. MAXIMUM RADIANT ENERGY LOSS FROM LAKE SURFACE

Compute the rate of radiant energy lost from the surface of a lake at 20°C on a clear night. Assume the sky is black with no clouds (i.e., a perfect adsorber). Report your answer in J/m$^2 \cdot$ s.

4.3B. TYPICAL ENERGY TRANSFER RATES THROUGH THE SURFACE OF A RESERVOIR

Compute the energy transfer rate through the air–water interface of Beaver Reservoir. Account for all energy and heat transfer mechanisms. Note direction and magnitude. Compute the totals for each day. Use the following environmental conditions:

1. Julian day 123 (~May 3): Wind speed 10.4 mi/hr, relative humidity 71%, air temperature 66°F, water temperature 16°C.

2. Julian day 280 (~October 7): Wind speed 9.7 mi/hr, relative humidity 68%, air temperature 69°F, water temperature 21°C.

Report your answers in $J/m^2 \cdot s$.

REFERENCES

1. L. J. Thibodeaux and D. G. Parker, "Desorption Limits of Selected Industrial Gases and Liquids from Aerated Basins," in C. Rai and L. A. Spielman, Ed., *Air Pollution Control and Clean Energy*, Am. Inst. Chem. Eng. Symp. Ser., 72 156, (1976).

2. Technical Practice Committee, Subcommittee on Aeration in Wastewater Treatment, *Manual of Practice S*, Water Pollution Control Federation, Washington, D.C. 1972.

3. J. S. Boundurant, S. Luce, and H. Townsend, Unpublished chemical engineering senior project report, Department of Chemical Engineering, University of Arkansas, Fayetteville, 1973.

4. R. E. Treybal, *Mass Transfer Operations*, 2nd ed., McGraw-Hill, New York, 1968, pp. 63.

5. J. A. Reinhardt, "Gas-Side Mass-Transfer Coefficient and Interfacial Phenomena of Flat-Bladed Surface Agitators," unpublished doctoral dissertation, University of Arkansas, Fayetteville, 1977.

6. M. Owens, R. W. Edwards, and J. W. Gibbs, *Int. J. Air Water Pollut.*, 8, 469 (1964).

7. G. E. Harbeck, Jr., "A Practical Field Technique for Measuring Reservoir Evaporation Utilizing Mass-Transfer Theory," *Geological Survey Prof. Paper*, 272-E, U.S. Govt. Printing Office, Washington, D.C., 1962.

8. A. F. Gaudy et al., *J. Water Pollut. Control Fed.*, 33 (2) 1961; 33 (4) 1961; 35 (1) 1963.

9. W. W. Eckenfelder, *Industrial Water Pollution Control*, McGraw-Hill, New York, 1966.

10. S. R. Goswami, unpublished Ph.D. thesis, Oklahoma State University, Stillwater, O K 1969.

11. A. M. Wachs, Y. Folkman, and D. Shemesh, "Use of Surface Stirrers for Ammonia Desorption from Ponds," Application of New Concepts of Physical-Chemical Wastewater Treatment, September 18–22, 1972.

12. C. N. Click and J. C. Reed, "Atmospheric Release of Hydrogen Sulfide and Ammonia from Wet Sludges and Wastewater," in L. K. Cecil, Ed., *Proc. 2nd National Conference on Complete Water Reuse*, Am. Inst. Chem. Eng. Symp. Ser., New York, 1975, pp. 426.

13. R. B. Blosser, J. J. McKeown, and D. Buckley, *A Study of the Mixing Characteristics of Aerated Stabilization Basins*, National Council on Air and Stream Improvement, Tech. Bul. 245, New York, 1971.

14. J. S. Goodling, B. L. Sill, and W. J. McCabe, "An Evaporation Equation for an Open Body of Water Exposed to the Atmosphere," *Water Res. Bull.*, 12 (4), 843–853 (1976).

15. W. H. McAdams, *Heat Transmission*, 3rd ed., McGraw-Hill, New York, 1954.

16. D. R. DeWalle, "Effect of Atmospheric Stability and Wind Direction of Water Temperature Predictions for a Thermally-Loaded Stream," School of Forest Products, Pennsylvania State University, University Park, 1975.

17. R. C. Axtmann, "Environmental Impact of a Geothermal Power Plant," *Science*, 187 (4179), 1975.

18. L. E. Mitchell, "Pesticides: Properties and Prognosis," in A. A. Rosen and H. F. Kraybill, Eds., *Organic Pesticides in the Environment*, Advances in Chemistry Series 60, American Chemical Society, Washington, D.C., 1966, pp. 1–22.

19. W. L. Dilling, N. B. Tefertiller, and G. J. Kallos, "Evaporation Rates and Reactivities of Methylene Chloride, Chloroform, 1,1,1-Trichloroethane, Trichloroethylene, Tetrachloroethylene, and Other Chlorinated Compounds in Dilute Aqueous Solutions," *Environ. Sci. Tech.*, **9** (9), 833–838 (1975).

20. W. L. Dilling, "Interphase Transfer Processes, II. Evaporation Rates of Chloro Methanes, Ethanes, Ethylenes, Propanes, and Propylenes from Dilute Aqueous Solutions: Comparison with Theoretical Predictions," *Environ. Sci. Tech.*, **11** (4) 405, 1977.

21. D. Mackay and Q. W. Wolkoff, "Rate of Evaporation of Low Solubility Contaminants from Water Bodies to Atmosphere," *Environ. Sci. Tech.*, **7** (7), 611–614 (1973).

22. D. Mackay and P. J. Leinonen, "Rate of Evaporation of Low-Solubility Contaminants from Water Bodies to Atmosphere" *Environ. Sci. Tech.*, **9** (19), 1178–1180 (1975).

23. J. M. Prausnitz, *Molecular Thermodynamics and Fluid Phase Equilibria*, Prentice-Hall, Englewood Cliffs, N.J., 1969.

24. A. Lerman, "Time to Chemical Steady-States in Lakes and Oceans," in J. D. Hern, Ed., *Nonequilibrium Systems in Natural Water Chemistry*, Advances in Chemistry Series 106, American Chemical Society, Washington, D.C., 1971. pp. 31–76.

25. P. S. Liss and P. G. Slater, "Flux of Gases Across the Air–Sea Interface," *Nature*, **247**, 181–184 (1974).

26. T. E. Hoover and D. C. Berkshire, "Effect of Hydration on Carbon Dioxide Exchange Across an Air–Water Interface," *J. Geophys. Res.*, **74**, 456 (1969).

27. A. L. Downing and G. A. Truesdale, "Some Factors Affecting the Rate of Solution of Oxygen in Water," *J. Appl. Chem.*, **5**, 570–581 (October 1955).

28. J. Kanwisher, "On the Exchange of Gases between the Atmosphere and the Sea," *Deep-Sea Res.*, **10**, 195–207 (1963).

29. P. S. Liss, "Processes of Gas Exchange Across an Air–Water Interface," *Deep-Sea Res.*, **20**, 221–228 (1973).

30. Y. Cohen, W. Cocchio, and D. Mackay, "Laboratory Study of Liquid-Phase Controlled Volatilization Rates in Presence of Wind Waves," *Environ. Sci. Technol.*, **12** (5), 553–558 (1978).

31. W. S. Broecker and T.-H. Peng, "Gas Exchange Rates between Air and Sea," *Tellus*, **26** 21–35 (1974).

32. A. Defant, *Physical Oceanography*, Vol. 1, Macmillan, New York, 1961, pp. 226–231.

33. W. Harrison, M. A. Winnik, P. T. Y. Kwong, and D. Mackay, "Crude Oil Spills: Disappearance of Aromatic and Aliphatic Components from Small Sea-Surface Slicks," *Environ. Sci. Technol.*, **9** (3), 231–234 (1975).

34. Z. R. Regnier and B. F. Scott, "Evaporation Rates of Oil Components," *Environ. Sci. Technol.*, **9** (5), 469–472 (1975).

35. R. O. Ramseier, "Oil Pollution in Ice-Infested Waters," in I. Hoffman, Ed., International Symposium on the Identification and Measurement of Environmental Pollutants, Ottawa, Ontario, Canada, 1971, pp. 273–276.

36. J. H. Horton, J. C. Corey, and R. M. Wallace, "Tritium Loss from Water Exposed to the Atmosphere," *Environ. Sci. Technol.*, **5** (4), 338–343 (1971).

37. A. H. Schooley, "Evaporation in the Laboratory and at Sea," *J. Mar. Res.*, **27** 335, (1969).

38. J. A. Dean, Ed., *Lang's Handbook of Chemistry*, 11th ed., McGraw-Hill, New York, 1973, pp. 10–31.

39. D. M. Gates, "Radiant Energy, Its Receipt and Disposal," *Meteorol. Monogr.*, **6**, 28 (1965).

40. F. L. Parker, "Thermal Pollution and the Environment," in N. Irving Sax, Ed., *Industrial Pollution*, Van Nostrand Reinhold, New York, 1974, pp. 160.

CHAPTER 5   # CHEMICAL EXCHANGE RATES BETWEEN WATER AND THE ADJOINING EARTHEN MATERIAL

The interfaces at the bottom of water bodies such as streams, lakes, estuaries, and the oceans are unfamiliar to most humans except for some oceanographers, professional divers, other underwater specialists, and recreation divers. We spend most of our time on the Earth at the other two interfaces and in general have little sense experience of the basic happenings at the water–earthen material interface. Although much can be derived from the analogy with the air–earthen material interface, this does not replace feeling the bottom currents, seeing and treading on the bottom geometric forms, and directly sampling either phase at the interface. In tracking chemical movements at this interface, we must by necessity draw on a body of knowledge that depends on remote sensing and remote sampling plus laboratory simulation for most of the data base.

There are a host of biogenic and anthropogenic chemicals whose movement at the water–earthen material interface is important from the viewpoint of both pollutants and ecosystem balance. This chapter investigates chemical movements at the interface of interest by focusing on specific applications. Just as in the previous chapter, two goals are accomplished. First, the principles presented in Chapter 2 on equilibrium and in Chapter 3 on transport are demonstrated, and second, specific relevant problems are presented and studied in detail. This approach should demonstrate to the student the process of translation from basic principles to specific application. The student should be able to attack new and different chemodynamic problems.

The specific applications chosen cover the diverse nature of this chemodynamic topic. The first topic concerns the spill of heavy ($\rho > 1$) chemicals in flowing streams and the dissolution process. Attention is given to the geometric forms of the bottom-residing liquid chemicals, in-stream concentrations, and lifetimes. The next application is concerned with the movement of chemicals from lake bottom muds into the water column. Nitrogen- and phosphorus-containing chemicals, thought to accelerate the natural eutrophic processes in lakes, are emphasized. The final application concerns processes at the bottom of the sea and chemical movement there.

5.1. DISSOLUTION OF CHEMICALS ON THE BOTTOM OF FLOWING STREAMS

Forced Convection Dissolution

Accidental spills of liquid materials into rivers, lakes, estuaries, and so on, is occurring and will undoubtedly increase as waterborne traffic increases. These spills are mainly caused by transportation accidents but can also result from inadvertent releases from production facilities located near a water body or from routine disposal procedures.

Materials and chemicals, both solid and liquid, heavier than water ($\rho > 1$) move toward the bottom immediately on being spilled or released. Natural flow, chemical processes, and physical processes operate to transport, disperse, cover, dissolve, adsorb, and transform the material. Translocation occurs because of the bulk flow of the aqueous body, and dispersion occurs by flow-induced fluid turbulence. While the material resides on the bottom, it can be covered by sediments, can be adsorbed by the natural bottom materials, and can undergo microbial attack and dissolution due to solubility and the bottom transport processes. Table 5.1-1 contains a list of "sinker" chemicals.

Some of the natural processes that occur immediately after a spill of a slightly soluble, high density, immiscible liquid substance can be anticipated. Figure 5.1-1 depicts the process for a hypothetical spill in a river. A barge containing the material in question (i.e., chloroform) is involved in an accident resulting in the release of a quantity of m kilograms of material. The following sequence of events describes the spill process. Assume a fairly large hole ($\geqslant$ 10cm) is formed in a river barge type carrier

Table 5.1-1. Water-Soluble, High Density ($\rho > 1$), Immiscible Chemicals

Species	Density in air (g/cm^3)	Solubility in water (mg/L)	Interfacial tension (dynes/cm)[a]		
			Air	Water	Vapor
1. Acetic acid	1.06	50,000	$68.0_{30°}$	–	$27.8_{20°}$
2. Acetic anhydride	1.087	500,000	–	–	$32.7_{20°}$
3. Acetophenone	1.03	5,550	–	–	$39.8_{20°}$
4. Aniline	1.022	34,000	44.0	–	$42.9_{20°}$
5. Benzaldehyde	1.04	1,000	40.04	$15.51_{20°}$	–
6. Benzyl alcohol	1.043	46,000	$39.0_{20°}$	$4.75_{22.5°}$	$39.0_{20°}$
7. Bromine	2.93	41,700	$41.5_{20°}$	–	$41.5_{20°}$
8. Carbon disulfide	1.26	2,200	–	$48.36_{20°}$	–
9. Carbon tetrachloride	1.595	500	–	$45_{20°}$	$26.95_{20°}$
10. Chlorine (liquid)	3.2	50,000	–	–	$18.4_{20°}$
11. Chloroform	1.5	5,000	$27.14_{20°}$	$32.8_{20°}$	–
12. Chlorpthalene			–	$40.74_{20°}$	–
13. Dichloroethane	1.256	9,000	$23.4_{35°}$	–	–
14. Ethyl bromide	1.431	10,600	–	$31.2_{20°}$	$24.15_{20°}$
15. Ethylene bromide	2.18	4,300	–	$36.54_{20°}$	$38.37_{20°}$
16. Furfural	1.159	83,100	$43.5_{20°}$	–	$43.5_{20°}$
17. Glycerol	1.26		–	$63.4_{18°}$	–
18. Hydrogen peroxide	1.46	50,000	–	–	$76.1_{18.2°}$
19. Mercury[b]	13.54	.0005	470	$375_{20°}$	–
20. Naphthalene	1.15	30	$28.8_{127°}$	–	$28.8_{127°}$
21. Nitrobenzene	1.205	1900	$43.9_{20°}$	–	$43.9_{20°}$
22. Phenol	1.071	67,000	$40.9_{20°}$	–	$40.0_{20°}$
23. Phenylhydrazine	1.097		–	–	$46.1_{20°}$
24. Phosphorus trichloride	1.5	50,000	–	–	$29.1_{20°}$
25. Trichloroethane	1.325	10	$22_{114°}$	–	–
26. N-Propylbromide	1.353	2,500	–	–	$19.65_{20°}$
27. Quinoline	1.095	60,000	$45.0_{20°}$	–	–
28. Tetrachloroethane	1.60	3,000	$36.3_{22.5°}$	–	–
29. Water[b]	1.00	N.A.	$73.05_{18°}$	N.A.	72

[a]In air, water, and its own vapor. Temperature is °C.
[b]Mercury and water data included for reference.

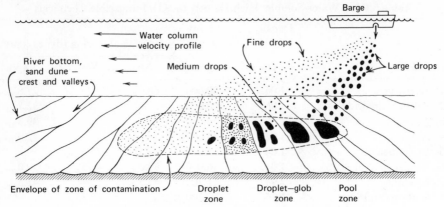

Figure 5.1-1.   Illustration of hypothetical spill incident.

and that a dense liquid is spilled into a deep ($\geq$4m), slow-moving body of water.

1. A jet of liquid emerges from the hole. The jet diameter is the hole diameter.

2. Large globules of liquid are created as the jet breaks up. These globules are roughly the size of the hole.

3. Globules continue to fall through the water. Large globules, settling through water, are unstable and start to break up after falling a short distance.

4. Breakup of droplets and globules continues until a cloud of small, stable droplets of various diameters is formed.

5. Horizontal classification by drop size occurs in the direction of stream flow. The largest drops have a high velocity of fall and contact the bottom at a point downstream but near the spill site. Smaller-diameter drops move further downstream. Fine droplets ($d \leq 1$ mm) remain suspended.

6. As the drops of liquid arrive at the bottom, they tend to accumulate at specific locations downstream from the point of release.

7. As accumulation increases, coalescence commences. Most stream bottoms consist of a somewhat uniform series of triangularly shaped parallel sand waves (Fig. 5.1-3). Droplets coalesce in the valleys of the waves.

8. Immediately below the spill point, the bottom wave structures can become filled with liquid (i.e., saturated). Late-arriving droplets

coalesce at the upper interface. Liquid spills over into the unfilled sand waves in the downstream direction.

9. Drops arriving somewhat further downstream also accumulate in the sand wave valleys but because of their limited numbers they do not fill the structures.

10. Still smaller drops touch bottom at sites well removed from the spill point. These drops remain as isolated particles splattered on the bottom and do not coalesce.

11. The liquid chemical achieves its final bottom form, and the on-bottom dissolution process commences. The liquid is in three basic geometric shapes: spherical drops, globs, and "sand wave valleys." *Globs* are defined as pancake-shaped pools several centimeters in diameter and a few millimeters thick.

There can and will be many variations on this idealized spill mechanism. If a spill occurs in shallow water, a droplet cloud may never form. The liquid may ooze down in large globules and cover the bottom. The spill of a small quantity of liquid in deep water may result in the bottom being splattered with individual drops only. Stream turbulence in a fast-moving body of water can produce significant changes. The high flow velocity may keep the liquid moving along the bottom. Movement is not unlike the sediment bed-load phenomenon that causes sand and silt to move downstream. In this extreme case the spillage moves out as a slug and behaves more or less like the spill of a miscible material.

Dissolution commences immediately on water contact and occurs from globules and droplets in transit to the bottom. Normally, the duration of the in-transit time is short, an hour at the most, but typically minutes. With slightly soluble materials, the major part of the dissolution occurs while the material resides on the bottom. Of all the on-bottom processes that can occur to a spilled liquid chemical, dissolution is possibly the most rapid and produces acute chemical stresses on the water ecosystem. Most of the occurrences noted above have been observed in the laboratory.[1] We now consider the on-bottom interphase mass transfer aspects of the problem.[2]

As shown in Figure 5.1-2, the chemical occupies a portion of the stream bottom, length L and width W. In this case the area of the zone of contamination is simply LW. Since the chemicals of interest are slightly soluble, Eq. 3.1-11b is sufficient to describe the dissolution flux rate in the water:

$$W_{A0} = {}^4k_{A2}A(x_A^* - x_{Ab}) \qquad (5.1-1)$$

W_{A0} is the molar flux rate, ${}^4k_{A2}$ is the water phase mass transfer coefficient

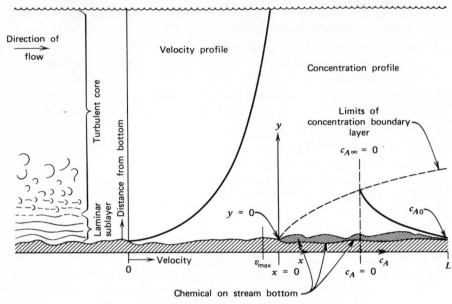

Figure 5.1-2. Chemical dissolution in a flowing stream.

above the liquid (2≡water phase, 4≡chemical phase), A is the interfacial area between the liquid chemical and water, x_A^* is the solubility of the chemical in mole fraction, and x_{Ab} is the "cup-mixing" background mole fraction of the chemical in water.

Consider a quantity of pure liquid chemical A of mass m_A in place on the bottom of a moving stream of flow rate Q. The chemical displays an interfacial area $A(m_A)$ that is a function of the mass remaining. Some simplifying assumptions can be made as follows: $x_{Ab}=0$ is valid if the water approaching the spill site contains no species A in solution and $^4k_{A2}$ is a constant independent of geometric shape. The following simple differential equation describes the dissolution process:

$$\frac{dm_A}{d\theta} = -A(m_A)^4 k_{A2} x_A^* M_A \tag{5.1-2}$$

The downstream cup-mixing concentration of species A, c_{A2}, is the quotient of the molar dissolution rate and the stream volumetric flow rate:

$$c_{A2} = \frac{A(m_A)^4 k_{A2} x_A^*}{Q} \tag{5.1-3}$$

The spillage lifetime on bottom t_A due to the dissolution process is obtained by separating the t and the m_A variables in Eq. 5.1-2 and integrating from $t=0$, $m_A = m_{Ai}$ to $t = t_A$, $m_A = 0$:

$$t_A = \frac{1}{4k_{A2}x_A^* M_A} \int_0^{m_{Ai}} \frac{dm_A}{A(m_A)} \tag{5.1-4}$$

These final two expressions can be employed to obtain important water quality predictions associated with the spill. The importance of $A(m_A)$ in predicting c_{A2} and t_A is readily apparent at this point. Specifying the three bottom geometric forms yields the function $A(m_A)$.

THE BOTTOM-RESIDING GEOMETRIC FORMS

The liquid on the stream bottom is assumed to be present in three geometric forms: spherical drops, globs, and sand wave valleys. The interfacial area displayed is a function of the geometric shape.

DROPS

Liquid drops arriving at the bottom of a water body do not coalesce into pools or globs, but remain as isolated spheres positioned on the bottom. The bottom will undoubtedly be splattered with drops of various diameters. An equation is available for estimating the maximum stable diameter d of a drop falling through water.[3] If these drops are assumed not to break up when they arrive on the bottom, the diameter can be calculated by

$$d = 3.79 \sqrt{\frac{\sigma_{A2}}{(\rho_A - \rho_2)}} \tag{5.1-5}$$

where d is diameter in cm, σ_{A2} is the interfacial tension of the liquid in water in N/m, and $\rho_A - \rho_2$ is the density difference in g/cm^3. The interfacial area of a mass m_{Ad} of uniform drops of diameter d is

$$A_d = \pi \left(\frac{6m_{Ad}}{\pi \rho_A} \right)^{2/3} \tag{5.1-6}$$

If dissolution is assumed to proceed such that the drops remain spherical with decreasing diameter, then Eq. 5.1-6 substituted into Eq. 5.1-4 yields a mass–time relationship from which lifetime can be obtained:

$$t_d = \frac{c_A d}{2^4 k_{A2} x_A^*}$$ (5.1-7)

The downstream concentration of the chemical resulting from the drops is obtained from Eq. 5.1-3:

$$c_{Ad} = \frac{6^4 k_{A2} m_{Ad} x_A^*}{\rho_A d Q} \left(1 - \frac{2^4 k_{A2} x_A^* t}{c_A d}\right)^2$$ (5.1-8)

GLOBS

Liquids residing on flat surfaces do so provided interfacial forces result in "nonwetting" of the bottom material by the liquid chemical. The height of a glob h_g is controlled by the water–chemical interfacial tension σ_{A2} and the density difference $\rho_A - \rho_2$:

$$h_g = \sqrt{\frac{2\sigma_{A2}}{g(\rho_A - \rho_2)}}$$ (5.1-9)

This equation is similar in form and derivation to a relationship for an oil film on water.[4] The interfacial area of globs is

$$A_g = \frac{m_{Ag}}{\rho_A h_g}$$ (5.1-10)

where m_{Ag} is the mass of chemical in the shape of globs. As dissolution proceeds, the interfacial tension forces cause A_g to decrease proportionately to the mass remaining, while h_g remains constant. The globs therefore shrink laterally.

It is assumed that globs are somewhat large so that the perimeter interfacial area is small compared to the top interfacial area. This assumption is invalid as glob diameter approaches h_g. At this point the glob becomes a drop. Dissolution is assumed to occur from the top only so that interfacial area is proportional to mass remaining. Substituting Eq. 5.1-10 into Eq. 5.1-4, we get a mass–time relationship from which the lifetime can be established:

$$t_g = \frac{c_A h_g}{4 k_{A2} x_A^*} \ln\left(\frac{1}{f}\right)$$ (5.1-11)

where f is some small fraction of the liquid remaining, but for all practical purposes the dissolution process is completed. A reasonable f is 0.05 (i.e., 5% undissolved), and $\ln(1/f) \simeq 3$. The downstream concentration resulting from shrinking globs is

$$c_{Ag} = \frac{{}^4k_{A2}m_{Ag}x_A^*}{\rho_A h_g Q} \exp\left(\frac{-{}^4k_{A2}x_A^* t}{c_A h_g} \right) \qquad (5.1\text{-}12)$$

POOLS

The geometric makeup of the bottom of most large actively flowing streams, such as rivers, consists of sand waves as ripples and dunes. The bottom material is mainly sand, and spilled liquids accumulate in the valleys of sand waves.

Yalin[5] maintained that if flow is tranquil (Fr < 1), then two kinds of sand waves can be present: ripples and dunes. Ripples and dunes are similar in their shapes; they both have an upstream surface with a gentle, gradual varying slope and an abrupt downstream face with a constant slope (which is approximately equal to the tangent of the "angle of repose"). The nonsymmetrical shape of ripples or dunes is shown schematically in Fig. 5.1-3. Ripples and dunes are distinguished from each other by the difference in their sizes.

Simons, Richardson, and Nordin[6] have reported on the sedimentary structures generated by flow in alluvial channels. In the low flow regime the bed form is either ripples or dunes or some combination of ripples and dunes, all of which are "triangular shape elements of irregular shape." They also observed that in natural streams and rivers, dunes with ripples superimposed on dunes are the dominant bed forms in the low flow regime. Ripples have a length Λ of about 300 cm or less from crest to crest

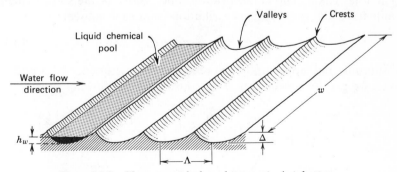

Figure 5.1-3. Nonsymmetrical sand waves on river bottom.

and an amplitude Δ of 0.6 to 6.0 cm in height, and have rather small width normal to the direction of flow. If the boundary shear stress is increased, a magnitude of velocity and a degree of turbulence are soon achieved that causes large sand waves called dunes to form. Viewed in elevation, the dunes are large triangular-shaped elements similar to ripples. Their lengths range from 60 cm to several meters, and their height from 6 cm to a few meters, depending on the scale of flow. In experiments with a large flume the dunes range from 60 cm to 3.0 m in length and from 6 to 30 cm in height. Lengths of dunes in the Mississippi River of several hundred feet and heights as large as 40 feet have been reported!

A sand wave bottom filled to a height h_w with liquid chemical, as shown in Fig. 5.1-3, is considered to be unsaturated. If $h_w = \Delta$, wave depth, then the bottom is saturated with the chemical. The interfacial area for mass transfer for a mass of liquid m_{Aw} in the wave valleys shown in Fig. 5.1-3 is

$$A_w = 2m_{Aw}/h_w \rho_A \tag{5.1-13}$$

The width can extend completely across the stream. Equation 5.1-14 is used to obtain the lifetime for the chemical in the waves, which is

$$t_w = \frac{c_A h_w}{^4 k_{A2} x_A^*} \tag{5.1-14}$$

If the wave structure is saturated, Δ should replace h_w in Eq. 5.1-14. The downstream concentration is obtained from Eq. 5.1-13:

$$c_{Aw} = \frac{2^4 k_{A2} m_{Aw} x_A^*}{\rho_A h_w Q} \left(1 - \frac{^4 k_{A2} x_A^* t}{c_A h_w} \right) \tag{5.1-15}$$

where t is the time that commences when the spillage arrives in place on the stream bottom.

The total mixing-cup concentration time history of the spilled chemical at mile 0.0 is the sum of the contribution from each source:

$$c_{A2} = c_{Ad} + c_{Ag} + c_{Aw} \tag{5.1-16}$$

Substituting the appropriate expressions from Eqs. 5.1-8, 5.1-12, and 5.1-15, we get the concentration-time history in the flowing stream:

$$c_{A2} = \frac{^4 k_{A2} x_A^*}{\rho_A Q} \left\{ \frac{6 m_{Ad}}{d} \left(1 - \frac{2^4 k_{A2} x_A^* t}{c_A d} \right)^2 + \frac{m_{Ag}}{h_g} \exp\left(\frac{-^4 k_{A2} x_A^* t}{c_A h_g} \right) \right.$$
$$\left. + \frac{2 m_{Aw}}{h_w} \left(1 - \frac{^4 k_{A2} x_A^* t}{c_A h_w} \right) \right\} \tag{5.1-17}$$

The $(1 - 2^4k_{A2}x_A^*t/c_Ah_w)$ and $(1 - {}^4k_{A2}x_A^*t/c_Ad)$ terms must be positive or zero; otherwise they are not included. Also

$$m_A = m_d + m_g + m_w \qquad (5.1\text{-}18)$$

must be satisfied.

Equation 5.1-17 is useful for estimating in-stream concentration of spilled liquid chemicals. Many terms in this equation can be estimated a priori. Stream flow (Q) and the total quantity spilled (m_A) can be obtained from the spill site, actual or projected. The mass transfer coefficient (${}^4k_{A2}$), glob height (h_g), and drop diameter (d) can be estimated from equations given in the book. Solubility (x_A^*) and density (ρ_A) are available in handbooks. The remaining four variables (i.e., m_d, m_g, m_w, and h_w) can be reduced to three unknowns by use of Eq. 5.1-18. The three remaining variables or unknowns must be specified from a special knowledge of the stream bottom, water velocity, depth of water, and so on, and their effect in order to obtain a realistic concentration-time prediction. In the absence of this information the individual models taken separately can yield reasonable estimates of maximum concentration and minimum lifetime due to a projected spill. (See Problem 5.1A.)

STREAM BOTTOM MASS TRANSFER COEFFICIENTS

A necessary piece of information for estimating on-bottom lifetimes and in-stream concentrations is the bottom mass transfer coefficient, ${}^4k_{A2}$. Although there are several "standard" mass transfer correlations available, these may not be altogether applicable. Laminar boundary layer theory for tangential flow along a sharp-edged, semi-infinite flat plate with mass transfer is well developed. The general correlation is given by Eq. 3.1-19. Since a laminar-flowing, natural stream is the exception rather than the rule, this equation is of little use. Equation 3.1-21 is a similar correlation for turbulent flow parallel to flat plates. For this equation as well as Eq. 3.1-19, the hydrodynamic boundary layer commences at the same point (i.e., at the sharp edge) as the concentration boundary layer. For the case of a spilled liquid in a flowing stream, the hydrodynamic boundary layer is developed before the flow enters the zone of contamination.

Kramers and Kreyger[7] obtained experimental measurements of the rate of solution on rather short surfaces of benzoic acid in water in laminar and turbulent flow. The analysis is treated as the diffusion of a solute from a plane surface into a laminar flow with a constant velocity gradient. The final correlation is

$$
{}^4k_{A2} = 0.449 \left\{ \frac{(g_x \Gamma_v)^{2/3} \mathfrak{D}_{A2}^2}{\nu L} \right\}^{1/3} c_2 \qquad (5.1\text{-}19)
$$

where g_x is the acceleration of gravity in the direction of flow, and $Re \equiv 4\Gamma_v/\nu > 2360$. Here Γ_v is the volumetric flow rate per unit channel bottom width, $L^3/t \cdot L$. This correlation gave a reasonable fit to the experimental data for the Re range 1500 to 7000. The soluble section was located 330 mm downstream from the water film inlet. A hydrodynamic boundary layer developed before the water encountered the soluble section (5 to 80 mm in length).

The gradient of the water surface of a flowing stream s is related to g_x by

$$g_x = g \sin \alpha \qquad (5.1\text{-}20)$$

and is very nearly equal to the slope of the bottom. Here $s = \sin \alpha$, where α is the angle of the stream bottom from the horizontal. In open channel flow it is possible to estimate the gradient from Manning's formula:

$$\bar{v}_x = \frac{1.486 r_H^{2/3} s^{1/2}}{n} \qquad (5.1\text{-}21)$$

This equation is dimensional, and $\bar{v}_x$ is the mean flow velocity in ft/s, r_H is the mean hydraulic radius of the wetted surface in feet, s is the slope of the water surface, and n is a coefficient of roughness. Values of n may be found in Appendix E.

There are other correlations in this book that may be used for estimating $^4k_{A2}$. Equation 5.1-19 more nearly mimics the streamflow dissolution process; however, the experimental apparatus was small, and the turbulence level low. The major drawback of Eq. 5.1-19 is that it is for a flat geometry, and most streams have a wave-type bottom structure. A wavy bottom reduces the mass transfer coefficient from that observed on a flat geometry because of decreased turbulence in the valleys.[8]

Chang[8] performed a series of dissolution experiments in a laboratory-scale model of a flowing stream. Furfural (C_4H_3OCHO, $\rho_A = 1.159$ g/cm^3) was placed in shallow circular pans of diameter 5.2 to 7.0 cm imbedded into the sand bottom so that only the liquid surface was in contact with the water. The dissolution mass transfer coefficient was computed by

$$^3k_{A2} = \frac{\Delta m_A}{A M_A t x_A^*} \qquad (5.1\text{-}22)$$

where Δm_A is the mass of furfural dissolved in water, A is the interfacial area, M_A is the molecular weight of furfural, t is dissolution time, and x_A^* is the mole fraction solubility of furfural in water. Observations were performed with a flat sand bed and with the sand bed formed into a repeated wave structure.

As was noted earlier, natural stream bottoms consist of sand waves known as ripples and dunes. Various wave amplitudes and periods were observed in the laboratory and the field. A fixed wave structure, amplitude $\Delta = 5.1$ cm and length $\Lambda = 25.4$ cm, was used to measure dissolution rates in the presence of sand waves. The circular pans were placed in the valleys of the waves, this being the likely resting place of the liquid chemical.

Careful observation revealed a water circulation pattern in the valley. The water layers directly above the pan were moving in a direction opposite to that of the average stream. In general, higher coefficients were achieved with the flat bed bottom structure.

Several independent variables were tried in order to find a suitable correlation for the observed mass transfer coefficient. The friction velocity, $v_* = \sqrt{\tau_0/\rho_2}$, proved to be a good choice for unifying both the flat bed and the sand wave dissolution data. The final forced convection correlation for both sets of data was

$$\mathrm{Nu}_{A2} = b\mathrm{Re}_*\mathrm{Sc}_{A2}^{1/3} \qquad (5.1\text{-}23)$$

For flat beds $b = 0.0786$ and for the $\Delta = 5.1$ cm waves $b = 0.0543$. Sc_{A2} is the Schmidt number; however, Re_* is a Reynolds number made of the friction velocity,

$$\mathrm{Re}_* \equiv \frac{v_*L}{\nu_2} \qquad (5.1\text{-}24)$$

where L is pool length or zone of contamination. Nu_{A2} is the Nusselt number for mass transfer. The bottom shear stress τ_0 may be obtained from

$$\tau_0 = \rho_2 shg \qquad (5.1\text{-}25)$$

where s is the slope of the water surface, h is water depth, and g is gravitational acceleration. Equations 5.1-23 to 5.1-25 should give a more realistic estimate of the liquid phase mass transfer coefficient at the bottom of a flowing stream.

Recognizing that life-sustaining mass transfer processes take place between flowing water and the bottom and bank organisms of natural streams, Novotny[9] made a study to explain the process and develop an expression for the mass transfer coefficient. The bottom is a collection point for settleable waste organic matter of biogenic or anthropogenic origin. Organisms that reside on the bottom must draw on the oxygen resources of the flowing stream so that the bottom boundary layer resistance is an important aspect of the transfer of dissolved oxygen. Novotny points out that in some cases bottom organisms may be anaerobic even though the oxygen concentration in the flowing water is high and that this can be explained by the fact that the amount of oxygen diffusing through the boundary layer into the benthal layer is not sufficient for aerobic conditions.

Another aspect of the role of the resistance in the boundary layer is regulation of the mineral composition of the water. As a stream flows over various geologic formations, the water dissolves indigenous minerals, and this determines in part the final mineral content of the stream. An acidic stream can become neutralized as it flows over a section of limestone or dolomite.

The flow in a stream can be divided into two zones: (1) the free turbulent flow zone and (2) the laminar sublayer. When dealing with the diffusion phenomenon, the diffusive boundary layer, a third zone, should be considered. Figure 3.3C shows the three zones.

The thickness of the diffusive sublayer is naturally less than that of the laminar sublayer, and the two are related by Eq. 3.3C-1, which is

$$\delta_{A2} = \frac{\delta_v}{Sc_{A2}^{1/3}} \tag{3.3C-1}$$

In the laminar sublayer the momentum transfer is domained by viscosity, and in the diffusive boundary layer mass transfer is governed by molecular diffusion. In the turbulent medium mass transfer is determined by the turbulence of the liquid.

The velocity profile in a stream conforms to the universal velocity profile given by Eq. 3.2-42, which is

$$v_x = \frac{v_*}{\kappa_1} \ln\left(\frac{y + y_0}{y_0}\right) \tag{3.2-42}$$

where y_0 is a characteristic height of the bottom roughness projections. This roughness length y_0 is greater than the thickness of the laminar sublayer δ_v.

In the region above the bottom $y \sim y_0$ the Prandtl mixing length expression (Eq. 3.2-32) applies. Since turbulent flow prevails, the same expression can be used to represent the turbulent diffusivity $\mathcal{D}_{A2}^{(t)}$ and the turbulent kinematic viscosity $v_2^{(t)}$,

$$\mathcal{D}_{A2}^{(t)} = v_2^{(t)} = l^2 \frac{dv_x}{dy} \tag{5.1-26}$$

Now if one assumes that the mixing length is of the order of the characteristic roughness length, $l \sim y_0$, the right side of Eq. 5.1-26 can be used to create the tautology

$$y_0^2 \frac{dv_x}{dy} = y_0^2 \left(\frac{dv_x}{dy} \right)$$

Equation 3.2-37 is used to replace dv_x/dy, and Eq. 3.2-34 is used for y_0. Both of these equations are results of the mixing length theory. The tautology becomes

$$y_0^2 \frac{dv_x}{dy} = y_0 \kappa_1 y \left(\frac{v_*}{\kappa_1 y} \right) \tag{5.1-27}$$

After simplification this equation can be integrated from $v_x = 0$ at $y = 0$ to yield

$$v_x = \frac{v_* y}{y_0} \tag{5.1-28}$$

This final expression relates that within the region $0 \leqslant y \leqslant y_0$ where the laminar sublayer exists, the velocity of flow is proportional to v_*, the friction velocity. If it is now assumed that the Reynolds number in terms of the laminar sublayer thickness, δ_v, converges to 1 (i.e., the forces of viscosity start to predominate). Using Eq. 5.1-28 in the Reynolds number yields

$$\frac{v_x \delta_v}{\nu_2} = \frac{v_* \delta_v^2}{y_0 \nu_2} = 1$$

from which we obtain

$$\delta_v = \left(\frac{\nu_2 y_0}{v_*} \right)^{1/2} \tag{5.1-29}$$

Since the diffusive sublayer is smaller than the laminar sublayer, molecular diffusion may be used to obtain the flux rate of chemical A near the solid boundary

$$j_A = \frac{\mathcal{D}_{A2}}{\delta_{A2}} \Delta \rho_{A2} \tag{5.1-30}$$

This equation assumes that all the resistance to mass transfer occurs in the diffusive sublayer of thickness δ_{A2}. Employing Eq. 3.3C-1 and 5.1-29 yields

$$j_A = \mathfrak{D}_{A2} Sc_{A2}^{1/3} \sqrt{\frac{v_*}{v_2 y_0}}\; \Delta\rho_{A2} \qquad (5.1\text{-}31)$$

This final equation contains Novotny's result for predicting the mass transfer coefficient at the water–sediment interface of streams. If the mass transfer coefficient is put in terms of the Nusselt number for mass transfer, it appears as

$$Nu_{A2} = \left(\frac{h}{y_0}\right)^{1/2} (Re_*)^{1/2} Sc_{A2}^{1/3} \qquad (5.1\text{-}32)$$

where $Nu_{A2} \equiv ({}^3k'_{A2}h)/\mathfrak{D}_{A2}$ and $Re_* \equiv v_* h / v_2$. From this equation one only needs to obtain a velocity profile, extract v_* and y_0, and use these to obtain a stream bottom mass transfer coefficient. Because the field data employed by Novotny to support his theory are inadequate to the task, Eq. 5.1-32 remains a theoretical development that has not been verified by experimental observation.

CLOSURE

Equations for estimating lifetime and "cup-mixing" in-stream concentrations of pure, soluble liquids that sink have been presented. These liquids are immiscible and remain in place on arriving at the bottom. Some estimate of the mass of the chemical in droplets, globs, and waves along with sand wave amplitude is necessary for a precise prediction.

Results reflecting flat plate and sand wave structures are available for estimating stream bottom mass transfer coefficients. Although the focus has been aimed at liquid-pool interface type microenviroments, the results are applicable to related dissolution problems including the dissolution of trace contaminants from sediments. These problems include the dissolution of trace contaminants from the sediment-water interface of streams. Novotny developed an asymptotic ($L \to \infty$) mass transfer coefficient relation for stream bottoms. However, the theory has not been sufficiently verified with field data.

Natural Convection Dissolution

It frequently occurs that liquid chemicals come to rest on the bottom of water bodies in which the water flow is low or stagnant. Because of the

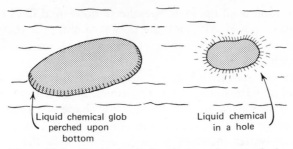

Liquid chemical glob
perched upon
bottom

Liquid chemical
in a hole

Figure 5.1-4. Liquid chemical on bottom of a water body.

almost complete absence of shear between the bottom material and the fluid, turbulence is absent. In the absence of flow and/or mechanical turbulence generated by flow, there can remain a sizable mass transfer rate.

Three important cases of natural convection dissolution are considered. These cases correspond to the microenvironment in which the liquid exists on the bottom. First, quantities of liquid chemicals can come to rest on the bottom and occupy perched positions, such as the globs mentioned previously. Interfacial tension in the range of 0.03 to 0.04 N/m for the liquids of interest will cause the liquid glob surface to rise 1 to 2 cm above the sand bottom. In the second case the liquid may fill a shallow depression, pothole, or sand wave. In this case the liquid chemical–water interface plane is the same as the bottom plane. The first two cases are illustrated in Fig. 5.1-4. Third, the chemical may partially fill a shallow pothole or sand wave. In the extreme case the heavy liquid may ooze down and come to rest at the bottom of a deep pothole filled with rocks. These microenvironments may occur on the bottom of streams, lakes, estuaries, and oceans.

GLOBS OR PERCHED MATERIAL

Material in the form of globs and drops resting quietly on the bottom dissolves at fairly rapid rates. A thin layer of dense fluid mixture, generated by molecular diffusion processes, forms on top of the liquid chemical surface. Since the chemical is heavier than water, this dense fluid mixture is more dense than the surrounding water and tumbles down the edge of the glob. The movement of the dense fluid near the interface induces local water movement, which speeds up the process. This process is one aspect of natural convection dissolution. At steady state a dense layer of liquid is moving radially from the center of the glob toward the edge. Fresh, lower density water is pulled down from above, and the process continues in this fairly rapid fashion until the glob "weathers" down and does not protrude above the bottom plane. Although the flow is laminar for the most part, a small amount of density-induced turbulence may be observed around the edge of the glob.

The author is unaware of any experimental measurements or mass transfer correlations resulting from experimental measurements made on the dissolution of pure liquid chemicals in the shape of globs. The correlations presented in the next subsection, concerning the case in which the liquid chemical–water interface plane coincides with the bottom plane, may be used in the absence of this specific information.

BRIMFUL SAND WAVES AND SHALLOW POTHOLES

If the liquid chemical fills the bottom depressions so that the chemical–water interface coincides with the bottom plane, there is no edge effect, and the process of dissolution may be slower. There have been experimental observations made in the areas of heat transfer and mass transfer that are directly applicable to this microenvironment.

Fujii and Imura[10] observed and quantified a similar phenomena involving natural convection heat transfer from plates in water. In the case of a heated horizontal plate facing downward, they observed that the boundary layer was entirely laminar. This experiment corresponds very closely to one involving a heavy liquid in water. One major difference is that the liquid chemical can and will respond to the laminar flow stress at the interface. Although the liquid can flow, the solid heated plate cannot; however, this may be negligible. The heat transfer data were well correlated for both a 5 cm and a 30 cm heated plate by the equation

$$Nu = 0.58(Gr\ Pr)^{1/5} \tag{5.1-33}$$

for $10^6 \leqslant Gr\ Pr \leqslant 10^{11}$. One may use the analogy theories to obtain a Nusselt (or Sherwood) number for mass transfer. (See Eqs. E3.3-1 and E3.3-2. Equation 5.1H-1 may be used to estimate the Grashoff number.)

Chang, Vliet, and Saberian[11] performed a series of natural convection dissolution experiments with salt samples approximately 1 m long in water and brine. The salt block was tested facing upward, vertically, and downward. The upward-facing position corresponds exactly to the dissolution process of a dense chemical on the bottom of a water body. Unfortunately no upward-facing data beyond a 60° angle from the vertical were obtained. For this angle the correlating equation was

$$Sh_{A2} = a(Gr_{A2}\ Sc_{A2})^{1/3} \tag{5.1-34}$$

and *a* ranged from 0.05 to 0.11. This equation will likely give high estimates of the natural convection mass transfer coefficient associated with a bottom-residing chemical, the interface of which coincides with the bottom plane.

DEEP POTHOLES AND SAND WAVES

For the majority of the lifetime, the liquid within potholes and sand waves lies with the interface below the bottom plane. In this case the density gradients develop inside the depressions, providing a stable mixture. The upward flux must occur by molecular diffusion, and as a consequence the dissolution rate is significantly slower. Once quantities of the dense mixture arrive at the level of the bottom plane or slightly above, a density imbalance occurs, and the mixture moves radially outward from the depression. Figure 5.1-5 illustrates the path molecules must take in such a microenvironment diffusion process. Figure 5.1-5*a* illustrates an unobstructed diffusion path, and Fig. 5.1-5*b* illustrates an obstructed diffusion path.

DEPRESSIONS WITH AN UNOBSTRUCTED DIFFUSION PATH

The pothole illustrated on the left in Fig. 5.1-5 can be characterized by several realistic dimensions. The bottom where the chemical is located can be assumed to have an a bottom area, A_B, with equivalent radius $R = (A_B/\pi)^{1/2}$. The area at the top of the pothole, A_T, is normally greater than the area at the bottom. The rate of movement of chemical A from pothole bottom is expressed by the now familiar equation

$$W_{A_0} = {}^4k_{A2}A_B(x_A^* - x_{Ab})$$ (5.1-35)

where ${}^4k_{A2}$ is the mass transfer coefficient, x_A^* is the mole fraction solubility

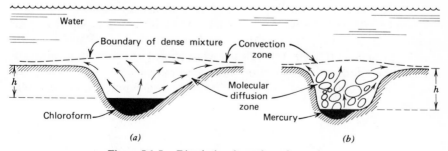

Figure 5.1-5. Dissolution from deep depressions.

of A in water, and x_{Ab} is the background mole fraction concentration in the water well away from the pothole.

The mass transfer coefficient in Eq. 5.1-35 is actually made up of two coefficients, each reflecting a different movement mechanism. Above the chemical–water interface up to a height h, the chemical movement mechanism is molecular diffusion. As the chemical emerges from the hole, natural convection processes driven by density differences dominate the chemical movement mechanism. Ideally these two zones and the associated resistances can be combined to yield an equation containing a convective and a diffusion component:

$$\frac{1}{\mathrm{Nu}_{A2}} = \frac{A_B/A_T}{a(\mathrm{Gr}_{A2}\,\mathrm{Sc}_{A2})^b} + \frac{h}{Rf} \qquad (5.1\text{-}36)$$

The first term on the right-hand side accounts for the resistance in the convective zone. The constants a and b can be obtained from the multiplier and exponent shown in either Eq. 5.1-33 or 5.1-34. The second term on the right-hand side accounts for a resistance in the molecular diffusion zone. The factor f accounts for the physical characteristics that $A_T \geqslant A_B$ and can be obtained from

$$f = \frac{\left[\left(\dfrac{A_T}{A_B}\right) - 1\right]}{\ln(A_T/A_B)} \qquad (5.1\text{-}37)$$

This area correction factor assumes that the pothole area increases linearly from the bottom. The Nusselt number for mass transfer is defined:

$$\mathrm{Nu}_{A2} \equiv \frac{4k_{A2}R}{\mathscr{D}_{A2}c_2}$$

Patterson and Woody[1] simulated the molecular diffusion–natural convection dissolution process occurring in the microenvironment of potholes. To avoid experimental difficulties encountered with heavy liquid chemicals in water, an analogous evaporation process was studied. Wafers of pure solid naphthalene ($C_{10}H_8, M = 128.16$) were placed at the bottom of cylindrical tubes and the evaporation rate determined by weight. For tube depth and radius ratios of $h/R \leqslant 0.8$ the following correlation was reached:

$$\frac{1}{\mathrm{Nu}_{A2}} = \frac{A_B/A_T}{0.788(\mathrm{Gr}_{A2}\,\mathrm{Sc}_{A2})^{1/4}} + 0.323\frac{h}{Rf} \qquad (5.1\text{-}38)$$

The number 0.323 is not unity as directed by the ideal formulation of Eq. 5.1-36. Apparently, laminar convection currents can extend down into the hole and reduce the length of the diffusion path to less than h.

As potholes become deeper and the ratio h/R exceeds approximately 5, the majority of the mass transfer resistance should reside in the molecular diffusion zone so that the Nusselt number for mass transfer becomes simply

$$\text{Nu}_{A2} = \frac{Rf}{h} \tag{5.1-39}$$

and this equation can be used to obtain the necessary mass transfer coefficient.

DEPRESSIONS WITH AN OBSTRUCTED DIFFUSION PATH

Often a pothole is filled in with boulders, rocks, pebbles, and sand as shown in Fig. 5.1-5b. In this case the heavy chemical can ooze downward between the fill, displacing the water, and come to rest in a "pool" a distance h below the stream bottom. Because of the narrowness of the openings between the fill particles and pieces, h/R is likely to be very large and molecular diffusion is the dominant transport mechanism.

The molecular diffusivity alone is not sufficient to describe the diffusion within porous solids that have interconnected voids or pores in the solid. The diffusion is greatly affected by the size and type of the voids. Figure 5.1-6 shows a sketch of a cross-section of such a porous solid. A re-examination of Fick's first law is in order.

Fick's law is replaced with an effective diffusion coefficient, D_{A3}, for the case of diffusion within porous solids and is normally written as follows:

$$n_A = -D_{A3}\frac{d\rho_{A2}}{dy} \tag{5.1-40}$$

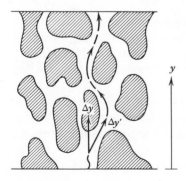

Figure 5.1-6. Pore diffusion in porous media.

Two factors operate to make the effective diffusion coefficient less than the molecular diffusivity. The interfacial area through which the chemical moves is reduced because the free or open cross section for diffusion is but a fraction of the total because of fill particles. The diffusivity is also effectively reduced because the diffusion distance along the tortuous path is greater as shown in Fig. 5.1-6. The effective diffusion coefficient is then defined formally as

$$D_{A3} = \frac{\mathcal{D}_{A2}\epsilon}{\tau_h} \tag{5.1-41}$$

where $\mathcal{D}_{A2}$ is the molecular diffusivity, ϵ is the void fraction of the sand- or sediment-filled hole, and τ_h is the *tortuosity factor*.

The factor τ_h is introduced to allow for the fact that the diffusion path is greater than the distance traveled normal to the face, and for varying cross section of the pores, which are not straight, round tubes. This correction factor must normally be obtained experimentally except for fill of exceedingly uniform structure and pore size. Values obtained from experimental data show that τ_h varies from unity to more than six[13]. If we assume that on the average, the pore makes an angle of 45° with the vertical y direction in a resultant two-dimensional path, then $\tau_h = \sqrt{2}$. If the path is three-dimensional, then $\tau_h = \sqrt{3}$.

Greskovich, Pommersheim, and Kenner[14] performed an experimental study of diffusivities through soils and calculations of hindrance factors to provide a basis for predicting rejuvenation rates of polluted muds. The ultimate goal of the investigation was to estimate the hindrance factor H and hence the rate of diffusion through different types of stream bed media. The hindrance factor is defined as

$$D_{A3} = \frac{\mathcal{D}_{A2}}{H} \tag{5.1-42}$$

This factor relates to particle shape, tortuosity, and bed void fraction. Because the pore size in the bed is usually much larger than the molecular diameter of the diffusing species, a hindrance factor based on one diffusing species is also accurate for other species of similar molecular diameter that exhibit the same effects with the muds.

A Stefan cell was used to measure the hindered diffusivities. Potassium chloride was used as the diffusing species for the determination of the hindrance factors. A solid salt phase was placed under the test diffusion zone containing the appropriate sand, mud, or clay being evaluated. The diffusion of the dissolved salt takes place between the solid salt–mud interface to the top of the diffusion cell at the mud–water interface. By observing the descending salt interface below the mud, the fluxes in the mud phase could be evaluated, and then the values of the hindered diffusivity found. Table 5.1-2 contains the experimental data for potassium chloride through various test media. The results indicate that the diffusion rates differ greatly for various bed types with different packing characteristics and different particle compositions. The hindrance factors shown in Table 5.1-2 vary from 3.04 for fine sand to 1.58 for loosely packed stream bed silt.

Table 5.1-2. Experimental Hindrance Factors and Void Fractions for Test Media

Medium	Particle Diameter (cm)	Void Fraction	Hindrance Factor H
Sand	0.0147–0.0208	0.491	2.82
Sand	0.0208–0.0295	0.507	3.04
Sand	0.0417–0.0589	0.510	2.70
Clayey-silt soil	~50% < 0.0035	–	2.24
Dried stream bed silt	~70% < 0.002	–	2.46
Fresh stream bed silt	~50% < 0.002	–	1.58

Source. Reference 14.

For the case of obstructed potholes it is necessary to modify the Nusselt number for mass transfer to include the effective diffusivity. Equation 5.1-39 should appear as follows:

$$\text{Nu}_{A2} = \frac{\epsilon R f}{h \tau_h} \qquad (5.1\text{-}43)$$

This equation or one containing the appropriate hindrance factor should be used to obtain the mass transfer coefficient in an obstructed pothole or similar bottom depression.

CHEMICAL POOL LIFETIME

The dissolution process takes a finite amount of time. It is of interest to know what the lifetime of a chemical pool at the bottom of a pothole can be from the dissolution process alone. For most organic chemicals, biochemical reaction and adsorption onto bottom solids may be parallel sinks for the chemical. Although other mechanisms are operative in determining the fate of the chemical, the dissolution lifetime is still relevant. Because of the slowness of the molecular diffusion process in liquids, this dissolution time is likely to be the upper bound of its time of existence at this particular position in the environment.

Equation 5.1-4 can be used to obtain specific lifetime expressions when the particular pothole shape is known. For a pothole in the shape of a right circular cone or a right circular cylinder, the expression is

$$t_p = \frac{c_A h}{4 k_{A2} x_A^*} \tag{5.1-44}$$

where h is the depth of chemical at the bottom of the pothole. See Problem 5.1I for a description of an actual situation of a chemical in the bottom of a stream and its environmental impact.

CLOSURE

The student will find that the problems at the end of Chapter 5 contain further useful information. Many problems relate to actual spill incidents, and the problems are constructed around the spill documentation. Problems 5.1A and 5.1B are concerned with a large spill of chloroform in the Mississippi River in 1973. Problem 5.1E concerns the spill of 250 gal of PCB into the Duwamish River in 1974, and Problem 5.1I is concerned with the spill of metallic mercury in the Shenandoah River between 1929 and 1950.

Calculations associated with spills that have already occurred are useful in estimating in-stream concentrations and on-bottom lifetime. Both of these calculations are relevant to exposure level and exposure time of the biota. Calculations are also relevant in deciding whether or not to proceed with cleanup of the chemical. Calculations associated with possible spills can be useful in transportation hazard assessment. Problem 5.1D poses the hypothetical problem of calculating the maximum quantity of a hazardous chemical that can be safely shipped on a certain river.

Problems

5.1A.   ESTIMATING   MAXIMUM   CONCENTRATION   AND MINIMUM LIFETIME OF A SPILL

Approximately 1.75E6 pounds of chloroform were released from a barge that sank near Baton Rouge, La., and the chemical began flowing down the Mississippi River toward the Gulf of Mexico. Although state health officials did not push the panic button, noting that they did not anticipate too much trouble from the accident, the U.S. Coast Guard warned downriver communities to keep a close surveillance on their water supply systems, particularly if intakes were close to the river bottom (chloroform is heavier than water). [*Chem. Eng.*, **80**, (September 3, 1973,) 48.]

Actually the river was at a low flow state, and there is reason to believe that the heavy chemical remained in place on the bottom near the spill site. Estimate the following water quality parameters:

1. Maximum in-stream concentration. Using each shape dissolution model separately, calculate concentrations (ppb).
2. Minimum on-bottom lifetime. Using each shape dissolution model separately calculate lifetimes (hr).

Data: Bottom mass transfer coefficient, $^4k_{A2} = 5.97\text{E}{-5}$ mol/cm^2·s; river flow, $Q = 7590$ m^3/s; ripple (sand wave) amplitude, $\Delta = 7.5$ cm.

5.1B. CONCENTRATION-TIME HISTORY OF A CHLOROFORM SPILL

At the time of the chloroform spill of Problem 5.1A, the Dow Chemical Company obtained river water samples and measured chloroform concentrations.[15] A portion of the data record appears in Table 5.1B. ($A \equiv$ chloroform.) Background concentration in the river is about 5 ppb.

Table 5.1B. Chloroform Spill in Mississippi River[a]

Date—Hour	ρ_{A2} (ppb)	Date—Hour	ρ_{A2} (ppb)	Date—Hour	ρ_{A2} (ppb)
8/19—2330	80	8/20—0630	233	8/24—1330	31
8/20—0030	220	0730	202	8/25—0400	24
0130	264	0830	162	2350	26
0230	352	1230	121	8/26—1600	20
0245	264	1530	81	8/27—1620	21
0330[b]	365	1730	59	8/28—0800	15
0430	326	8/21—0130	70	8/29—0800	6
		1430	70	8/30—0800	13
		8/22—0600	53	8/31—0800	12
		2000	33	9/4—0800	4
		8/23—0400	25	9/6—0800	7
		1330	31		
		8/24—0400	25		

Source. Reference 15.

[a]Sampled at mile 16.3.

[b]Chose 8/20 at 0330 as $t = 0$.

Assume a total of 317,000 kg was accounted for in the river.

1. Using 54,150 kg as the mass of chemical in the shape of spheres, 230,730 kg as the mass of chemical in the valleys of sand waves, and the remainder as globs, calculate the in-stream concentration (ppb) for each half hour for 18 days.
2. Plot the calculated concentrations and the observed concentrations shown in Table 5.1B versus time on a single piece of graph paper.

5.1C.   ESTIMATING   THE   ON-BOTTOM   MASS   TRANSFER COEFFICIENT

Using Eq. 5.1-19 calculate the liquid phase mass transfer coefficient for chloroform in the Mississippi River. Data: Volumetric flow = 7590 m^3/s, width = 1800 m, velocity = 56.3 cm/s, depth = 16.5 m, temperature = 28° C, $\mathcal{D}_{A2} = 1.002E-5$ cm^2/s. (*Hint*: Assume L, the spill length, is approximated by one-tenth of the width of the river.) Report the answer in $mol/cm^2 \cdot s$.

5.1D.   SPECIFYING A SAFE CHEMICAL CONTAINER SIZE

Hazardous chemical A may be accidentally spilled into stream B during shipment by water carrier. The maximum tolerable concentration of A in water is estimated to be 1 ppm based on stream biota and downstream water use. Compute the quantity of chemical A (kg) that can be safely shipped in light of a possible accident that results in a spill. Available data: stream B flow is 50,000 ft^3/s, $\rho_A = 1.45$ g/cm^3, $\sigma_A = 50$ dynes/cm, solubility in water is 1000 ppm, molecular weight is 100, and $^4k_{A2}$ is estimated to be 0.1 lb mol/$ft^2 \cdot$ hr.

5.1E.   SPILL OF PCB IN DUWAMISH RIVER[16]

On September 13, 1974, an electrical transformer fell during loading operations and caused a spill of 250 gal of 100% PCB (Aroclor 1242) into the Duwamish River in Seattle, Washington. (PCB = polychlorinated biphenyl). The lower Duwamish River is affected by tides up to 4 m and regularly flows at approximately 2 m/s. The spill site was a predominantly mud–silt bottom, with fresh water overlaying a saltwater wedge, and approximately 14 m deep and 150 m wide. Environmental Protection Agency divers observed pools of free PCB (specific gravity 1.4) material on the bottom. There was evidence that the river current and tidal action had caused pockets of PCB to move about. Divers observed pools of PCB moving as much as 15 m with the tide from one day to the next.

Using 4 in. hand-held suction dredges, divers picked up pools of PCB from the bottom. The second-stage recovery utilized a special high solids dredge. On March 31, 1976, cleanup operations ceased. It is estimated that 210 to 240 gal of the original 250 gallons of PCB spilled were removed from the river bottom.

1. To investigate the possible fate of the unaccounted-for PCB (i.e., 10 to 40 gal), perform the following dissolution calculations:
   (a) Estimate the dissolution lifetime of the original 250 gal had no recovery operations been attempted.
   (b) Estimate the maximum water concentration in the Duwamish River (ppb).
   (c) Estimate the quantity of material dissolved from September 13, 1974, through March 31, 1976 (gal).
2. In your opinion, where is the unrecovered PCB? List possible fates other than dissolution.

Data needed for calculation: Solubility in water $= 2.4E-4$ mg/L, $\sigma = 40$ dynes/cm, $\rho_A = 1.4$, molecular weight 258, $\mathcal{D}_{A2} \cong 0.8E-5$ cm^2/s, and $^4k_{A2} = 7.0E-5$ mol/cm$^2 \cdot$s.

5.1F. LABORATORY MEASUREMENTS OF THE DISSOLUTION RATE OF HEAVY CHEMICALS IN AQUEOUS ENVIRONMENTS

Experimental measurements of the dissolution rate of heavy chemicals under laboratory conditions are helpful in attempting to predict dissolution rates in real world flowing and nonflowing water environments. Open-top containers filled with model liquid chemicals are placed inside laboratory flowing-stream simulators that reproduce river conditions. Dissolution is allowed to occur for a period of time, after which the quantity of chemical remaining in the container is redetermined. Similar experiments can be performed in quiescent tanks to simulate conditions in deep lakes and/or very slow-moving rivers.

Laboratory flowing-stream simulator experimental results: Average velocity $= 0.28$ ft/s, diameter of container $= 7.05$ cm, temperature $= 18°C$, initial weight $= 685.2$ g, final weight $= 671.2$ g, start time $= 10:25$ A.M., and finish time $= 11:45$ A.M.

Quiescent water body simulator experimental results: Velocity $= 0.0$ ft/s, diameter of container $= 7.05$ cm, temperature $= 27°C$, initial weight $= 321.6$ g, final weight $= 303.1$ g, start time $= 9:14$ A.M., and finish time $= 4:25$ P.M.

The chemical used in both experiments was furfural (C$_4$H$_3$OCHO). $\mathcal{D}_{A2} = 1.31E-5$ cm^2/s.

1. Compute the observed mass flux rates (N_A) and the liquid phase mass transfer coefficient $(^4k_{A2})$ from the data given for both experiments. Report the answer in the SI units of grams, centimeters, seconds, and kelvin.

2. Based on the coefficient in part (1), estimate the mass flux rate of chloroform for the flow condition only. Give your answer in SI units. The diffusivity of chloroform in water is $1.002E-5$ cm^2/s.

5.1G.   ESTIMATING   ON-BOTTOM   MASS   TRANSFER COEFFICIENT, SAND WAVES

Repeat Problem 5.1C by using Eq. 5.1-23. Assume a flat bed.

5.1H.   NATURAL   CONVECTION   DISSOLUTION   MASS TRANSFER COEFFICIENTS FOR GLOBS

Natural convection dissolution experiments like those described in Problem 5.1F were performed with globs consisting of pure furfural and chloroform. Using Eq. 5.1-33 and 5.1-34, calculate the mass transfer coefficients and compare the calculated values with the experimental values.

1. Furfural: Glob radius $=6.5$ cm, temperature $=27.5°C$, $^4k_{A2}$ (experimental) $=1.17E-5$ to $1.71E-5$ mol/cm^2·s.
2. Chloroform: Glob radius $=7.05$ cm, temperature $=24.5°C$, $^4k_{A2}$ (experimental) $=2.44E-5$ mol/cm^2·s.

Use

$$\mathrm{Gr}_{A2} = \frac{R^3(\rho_{A2}^* - \rho_2)g}{\nu_2^2\bar{\rho}_2} \qquad (5.1H\text{-}1)$$

where $\bar{\rho}_2 = (\rho_{A2}^* + \rho_2)/2$ for calculating the Grashoff number. Here ρ_{A2}^* is the density of water saturated with the chemical of interest.

5.1I.   DISSOLUTION   OF   METALLIC   MERCURY   FROM   THE SOUTH FORK OF THE SHENANDOAH RIVER[17]

Because of its great weight $(\rho_A = 13.6$ g/cm$^3)$ and its liquid form, metallic mercury seems to find its way into sheltered nooks and crevices—of which the South River, with its irregular limestone bottom, has plenty—and is not easily dislodged. Major floods, such as the one that followed tropical storm Agnes in 1972, have repeatedly scoured the river bottom, but, although some of the mercury in the bottom sediments undoubtedly has been moved about, it has not been swept away.

Over a period of years and decades, metallic mercury in bottom sediments can be converted by varying rates into ionic inorganic mercury and to organic or methyl mercury, both of which can be transported downstream by river currents. Although relatively insoluble in water, some metallic mercury becomes attached to suspended soil particles and part remains in solution.

It has been estimated that 35 L of liquid metallic mercury escaped to the river from DuPont's Waynesboro "old chemical building," where mercuric sulfate was used as a catalyst in the manufacture of acetate fiber between 1929 and 1950. High concentrations of mercury are still found in sediment samples taken just below the DuPont plant from a natural trap formed by the remnants of a small dam. Estimate the dissolution lifetime for the metallic mercury in this stream.

Assume potholes, nooks, and crevices account for 30% of the stream bottom, and are filled with pebbles 1 cm in diameter. The zone of contamination is assumed to be a bottom area of 100 m^2. Assume the potholes are 30 cm deep and 30 cm wide on the average. Assume the mercury occupies the very bottom of the potholes. Report the lifetime in years.

5.2. THE UPSURGE OF CHEMICALS FROM THE SEDIMENT–WATER INTERFACE OF LAKES

Lakes that have been employed as waste dump sites can contain bottom sediments with a high content of specific chemicals. These chemical-laden bottom sediments can result directly from settlable solids placed in the lake or can result indirectly from algal-stimulating nutrients contained in the inflowing water.

Methoxychlor [2,2-bis (P-methoxyphenyl)-1,1,1-trichloroethane] is a popular substitute for DDT [2,2-bis (P-chlorophenyl)-1,1,1 trichloroethane] used in the control of many insects. Methoxychlor reaches the aquatic environment by direct application such as in the control of black fly larvae in streams or indirectly through terrestial runoff. In a recent study of the Illinois waters of Lake Michigan and its tributaries, concentrations of methoxychlor in water averaged 15.1 ng/L in 1971 and 48.0 ng/L in 1972. Open water sediments tested in 1970 and 1971 averaged 1.24 μg/L. Average DDT concentrations were somewhat higher than those for methoxychlor in both water and open water sediments.

Organic and inorganic nitrogen and phosphorus compounds frequently find their way into lakes, both in particulate and soluble form. Nitrogen and phosphorus are thought to be primarily responsible for the excessive growth of algae and other aquatic plants in lakes. These two chemicals are components of fertilizer and stimulate growth, in the spring, summer, and fall, of the lake vegetative matter. At the end of the growing season, the algal cells settle to the bottom and become part of the sediment matter. On the bottom, chemical and biochemical processes occur to release the nitrogen and phosphorus chemicals from the cellular structure. Available as soluble compounds, these chemicals can work their way upward to the surface waters where the photosynthesis process is active. Once again, these nutrient chemicals become part of the cellular mass of aquatic vegetation, and the cycle has been completed. This is one aspect of the recycling of nutrient chemicals in lakes. Dissolution and upsurge from the sediment–water interface play an important role in the nutrient-cycling processes of freshwater lakes. Lakes characterized by excessive yearly algal growths are termed *eutrophic*.

If the dumping of nitrogen and phosphorus compounds is stopped, the rich bottom sediments retain and recycle these chemicals for many years, and the lake continues to display an eutrophic character. The rate at which these nutrient chemicals emerge from the bottom sediments determines in large part how soon a "damaged" lake will return to a more normal state with respect to algae and aquatic plant productivity. Figure 5.2-1 shows the lake conditions of temperature and concentration during a period of upsurge of chemicals on a typical summer day. The model developed in this section uses the nutrient upsurge process in lakes; however, the results and the concepts presented apply to any chemical emerging from or being transported to the sediment–water interface of a lake or similar body of water.

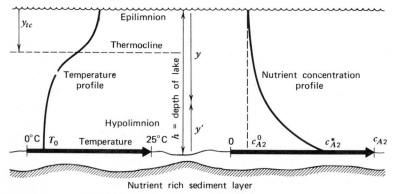

Figure 5.2-1.   Typical lake-water column conditions.

A Fickian Analysis of Upsurge of Chemical Nutrients[18]

The annual thermal history of a particular lake plays a dominant role in the movement of chemicals at the sediment–water interface. The thermal history of lakes as related to the seasons of the year is presented in some detail in Chapter 7. Also presented is a simple thermal model that describes the main features of the lake water column temperature and the thermocline position during the stratified period. Those not familiar with the thermal behavior of lakes are urged to review Section 7.1 prior to proceeding with the following treatment of chemical movement in the lake water column.

The period of time between the vernal equinox (March 21) and the autumnal equinox (September 23) is the dominant productive period for lakes in the temperate zone. During this period, lakes that contain nitrogen and phosphorus witness plentiful growth of algae due to high sunlight levels and warm water. This productive period is roughly from March 21 until September 23, Julian day 80 and Julian day 266, respectively, for a total of 186 days. This is also the heating period for the lake. As is pointed out in Chapter 7, during this period the net heat transfer into the lake is positive, and thermal stratification occurs. Increased air temperature and incoming radiation coupled with a favorable water density temperature relationship cause a stable water column consisting of warm water on top of colder water.

The calendar time of the vernal equinox is convenient for specifying a model time of zero (i.e., $t = 0$). Prior to this time the lake is at nearly uniform concentration with respect to temperature and chemical species concentration. Ice may cover the surface but it commences to melt rapidly. A uniform temperature of 4 to 8°C throughout is not uncommon. The time of the autumnal equinox is also important to the model structure.

The autumnal equinox marks the approximate end of the stratified thermal structure of the lake water column and marks the beginning of a more mixed state with respect to temperature and dissolved chemicals. At this time the net heat flux rate changes from positive to negative. Plunging cold water triggered by heat loss is the primary reason for the rapidly occurring mixed state and the subsequent fall "turnover."

Lake "turnover" is a layman's term for the relatively rapid mixing process that can occur once the lake temperature becomes uniform. High wind drag at the air–water interface causes water motion and forces bottom water to the surface. If the bottom water was anaerobic, H_2S emission through the air–water interface makes "turnover" a noticeable undertaking. Although the water column is at a uniform temperature, many "turnovers" can occur as the water mixing is rapid. This uniform temperature water column condition can remain until the vernal equinox;

however, if the water surface temperature falls below 4°C, reverse stratification occurs.

Reverse stratification occurs during the coldest part of winter. Water having a temperature of less than 4°C or ice is less dense and rides on the surface, and again a state of water column stability prevails. Another mixed period, sometimes called "spring turnover," occurs during the period between the end of the reverse stratified period and the start of the normal stratified period, which occurs about March 21.

SEMI-INFINITE MEDIUM CHEMICAL UPSURGE MODEL

Lakes are extremely complex "living systems." The chemical upsurge model presented here is a very important aspect of the life processes in a lake but is only a small part of the chemical and biochemical processes occurring in a given lake. Even if one considers only the transport of chemicals from the bottom sediments up the water column, it is necessary to consider at least two simple models. Arbitrarily, lakes are classified here as stratified and unstratified.

The stratified lake model accounts for the process of thermal stratification that occurs in most lakes to a greater or smaller degree. Stratification tends to retard the movement rate of chemicals, originating at the sediment–water interface, into the layers next to the air–water interface. The unstratified lake model is applicable to those lakes in which thermal stratification for one reason or another is not a dominant factor in chemical movement. Although both models are simple, they do reflect the major factors that influence chemical movement and serve as a starting point for the construction of more detailed transport models. The major factors include season of the year, time during the season, turbulent diffusivity, eddy thermal diffusivity, depth of the thermocline, initial chemical concentration in the lake water column, and chemical concentration at the sediment–water interface.

The mathematics of diffusion in a semi-infinite medium are employed as the quantitative tool to describe the chemical upsurge phenomena from the sediment–water interface. The medium in this case is water. The lake water is visualized as a medium occupying the space from $y' = 0$ (the bottom) to $y' = \infty$ (far removed from the bottom) with an initial concentration of ρ_{A2}^0. Nutrient A is present in the sediments at concentration ρ_{A3}. Interstitial water of nutrient concentration ρ_{A2}^* is in equilibrium with ρ_{A3}. (See Problem 2.1D for an example of the distribution of phosphorus between soil and water.) At time $t = 0$ it is assumed that the bottom surface of the water is suddenly raised to concentration ρ_{A2}^* and maintained at that concentration for an extended period of time. See Fig. 5.2.-1.

Both models commence with a simplified form of the turbulent multicomponent equation of continuity. The equation for dilute solutions (Eq. 3.2-11) can be simplified to

$$\frac{\partial \rho_{A2}}{\partial t} = \mathcal{D}_{A2}^{(t)} \frac{\partial^2 \rho_{A2}}{\partial y'^2} \tag{5.2-1}$$

where $\mathcal{D}_{A2}^{(t)}$ is the turbulent diffusivity and y' is the distance from the sediment–water interface. The concentration ρ_{A2} is time smoothed. The initial and boundary conditions for Eq. 5.2-1 are

I.C.:   at $t < 0$, $\rho_{A2} = \rho_{A2}^0$ for all y' $\qquad$ (5.2-1a)

B.C.:   at $y' = 0$, $\rho_{A2} = \rho_{A2}^*$ for all $t > 0$ $\qquad$ (5.2-1b)

B.C.:   at $y' = \infty$, $\rho_{A2} = \rho_{A2}^0$ for all t $\qquad$ (5.2-1c)

The solution to Eq. 5.2-1 is

$$\frac{\rho_{A2} - \rho_{A2}^0}{\rho_{A2}^* - \rho_{A2}^0} = 1 - \mathrm{erf}\left(\frac{y'}{\sqrt{4\mathcal{D}_{A2}^{(t)}t}}\right) \tag{5.2-2}$$

where the *error function*, erf, is defined as follows:

$$\mathrm{erf}\left(\frac{y'}{\sqrt{4\mathcal{D}_{A2}^{(t)}t}}\right) \equiv \frac{2}{\sqrt{\pi}} \int_0^{y'/\sqrt{4\mathcal{D}_{A2}^{(t)}t}} e^{-n^2}dn \tag{5.2-3}$$

Numerical values of the error function are presented in Appendix B.

CASE I: STRATIFIED LAKES

This case is applicable to lakes with a well-developed thermal profile. Nutrients entering the epilimnion from the sediment bed are capable of stimulating and supporting primary production. Nutrients enter the mixed epilimnion region because of the gradual deepening of the region and the turbulent diffusion through the thermocline. The flux rate of nutrient A through a horizontal plane located at the thermocline depth y_{tc} is

$$n_{Ay} = \left[(\rho_{A2} - \rho_{A2}^0)v_y - \mathcal{D}_{A2}^{(t)}\frac{\partial \rho_{A2}}{\partial y'}\right]_{y_{tc}} \tag{5.2-4}$$

where v_y is the mass velocity of the thermocline downward. The distance from the bottom y' is related to the distance from the air–water interface y

by $h = y' + y$, where h is the depth of the lake water. From Eq. 7.1-6 the thermocline velocity is

$$v_y \equiv \frac{dy_{tc}}{dt} = \sqrt{\frac{\alpha_2^{(t)}}{2t}} \tag{5.2-5}$$

and from Eq. 5.2-2,

$$\frac{\partial \rho_{A2}}{\partial y'} = \left(\frac{-1}{\sqrt{\mathscr{D}_{A2}^{(t)} t \pi}} \right) \exp\left(\frac{-y'^2}{4 \mathscr{D}_{A2}^{(t)} t} \right) (\rho_{A2}^* - \rho_{A2}^0) \tag{5.2-6}$$

Substituting Eqs. 5.2-2, 5.2-5, and 5.2-6 into 5.2-4 results in

$$n_{Ay} = (\rho_{A2}^* - \rho_{A2}^0) \left\{ \left[1 - \mathrm{erf}\left[\frac{h - \sqrt{2\alpha_2^{(t)}t}}{\sqrt{4\mathscr{D}_{A2}^{(t)}t}} \right] \right] \sqrt{\frac{\alpha_2^{(t)}}{2t}} \right.$$
$$\left. + \sqrt{\frac{\mathscr{D}_{A2}^{(t)}}{\pi t}} \exp\left[\frac{-\left(h - \sqrt{2\alpha_2^{(t)}t}\right)^2}{4\mathscr{D}_{A2}^{(t)}t} \right] \right\} \tag{5.2-7}$$

This equation gives the flux rate of nutrient A into the epilimnion due to the upsurging components released from the sediment bed. In lake ecosystem modeling, this nutrient source term should be included along with the advective entering term, the advective exiting term, the bioconsumption term, the bio-excreta term, settling of particulates, and so on, to yield a complete material balance for the influx of nutrient A into the epilimnion.

CASE II: UNSTRATIFIED LAKES

This case is applicable to lakes with a poorly developed or nonexistent thermal profile. Sunlight penetrates and effectively stimulates primary production down to a depth h_e. This depth depends on the extinction coefficient of the water, self-shading by the algae, and so on. The flux rate of nutrient A into the sunlit surface region of the lake is

$$n_{Ay} = -\mathscr{D}_{A2}^{(t)} \left(\frac{\partial \rho_{A2}}{\partial y'} \right) \bigg|_{h_e} \tag{5.2-8}$$

Employing Eq. 5.2-6 for the derivative expression, we obtain

$$n_{Ay} = \sqrt{\frac{\mathscr{D}_{A2}^{(t)}}{\pi t}} \exp\left[\frac{-(h - h_e)^2}{4\mathscr{D}_{A2}^{(t)}t} \right] (\rho_{A2}^* - \rho_{A2}^0) \tag{5.2-9}$$

Both lake nutrient upsurge rate equations, Eqs. 5.2-7 and 5.2-9, require chemical concentrations and physical characteristics of the particular lakes. These parameters can be obtained from field observations. Equation 7.1-7 can be used to obtain $\alpha_2^{(l)}$ values from the thermal profile, and Eq. 5.2-9 is used to obtain $\mathscr{D}_{A2}^{(l)}$ values from chemical concentration profiles. Model time $t = 0$ can be obtained from inspection of graphs of the thermal and/or chemical concentration profiles for various times of the year. (See Problem 5.2A.) The sediment bed equilibrium concentration of chemical A, ρ_{A2}^{*} and the initial concentration ρ_{A2}^{0} can also be obtained from nutrient profile observations.

GENERALIZATION OF RESULTS OF THE FICKIAN ANALYSIS

The flux rate of nutrient A from the lake sediment bed is

$$n_{A0} = - \mathscr{D}_{A2}^{(l)} \left(\frac{\partial \rho_{A2}}{\partial y'} \right) \bigg|_{y'=0} \tag{5.2-10}$$

The concentration gradient at the bottom of the lake is obtained from Eq. 5.2-6 at $y' = 0$ and used in Eq. 5.2-10 to give

$$n_{A0} = \sqrt{\frac{\mathscr{D}_{A2}^{(l)}}{\pi t}} \; (\rho_{A2}^{*} - \rho_{A2}^{0}) \tag{5.2-11}$$

This equation gives the instantaneous flux rate of nutrient A into the water phase. A more useful result is the average flux rate for the time period t, obtained by integrating Eq. 5.2-11 to obtain

$$\bar{n}_{A0} = 2 \sqrt{\frac{\mathscr{D}_{A2}^{(l)}}{\pi t}} \; (\rho_{A2}^{*} - \rho_{A2}^{0}) \tag{5.2-12}$$

This equation is an expression for the gross rate of nutrient A entering the water through the sediment–water interface for the period t. The rate at which nutrient A arrives in the epilimnion and the photoactive zone of the lake is less. The latter rates are given by Eqs. 5.2-7 and 5.2-9, respectively.

The fraction of nutrient A leaving the sediments that then enter the epilimnion is significant because only this amount can stimulate algae growth. This fraction for Case I is defined

$$\phi_{\mathrm{I}} \equiv \frac{\bar{n}_{Ay}}{\bar{n}_{A0}} \tag{5.2-13}$$

Assuming $\mathcal{D}_{A2}^{(t)} = \alpha_2^{(t)}$, Eq. 5.2-7 is time averaged to obtain $\bar{n}_{Ay}$. The integration must be performed by numerical methods. It is necessary to define time $\theta \equiv t/t_c$, where $t_c \equiv h^2/2\mathcal{D}_{A2}^{(t)}$ is a *characteristic time constant* of the particular lake. Figure 5.2-2 is a plot of ϕ_I versus θ. The upper curve, labeled ϕ_I, gives the fraction of upsurging chemical A that passes to the epilimnion as a function of dimensionless time θ.

In a similar manner the fraction of nutrient A leaving the sediments that enter the photoactive zone of depth h_e for Case II can be defined

$$\phi_{II} \equiv \frac{\bar{n}_{Ay}}{\bar{n}_{A0}} \qquad (5.2\text{-}14)$$

Equation 5.2-9 is time averaged to obtained $\bar{n}_{Ay}$. Numerical values of ϕ_{II} appear in Fig. 5.2-3. For this case it is necessary to define a dimensionless distance $\delta \equiv (h - h_e)/h$, which is the fractional distance from the photoactive zone to the lake bottom. A value of $\delta = 1$ is the lake surface and appears in Fig. 5.2-1 also. Obviously the chemical A cannot pass through the air–water interface; thus $\delta = 1$ is a limiting value presented for completeness only. Figure 5.2-3 is useful for estimating the fraction crossing any plane δ where $\delta < 1$.

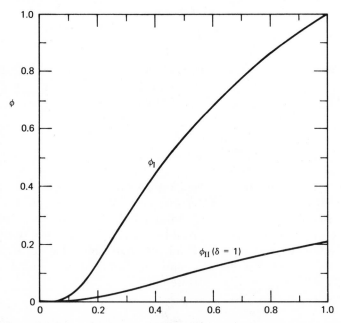

Figure 5.2-2. Fractions of upsurging nutrients in epilimnion (ϕ_I) and at surface (ϕ_{II}). (Reprinted by permission. American Water Resources Assoc., Mpls., MN. **Source.** Reference 18.)

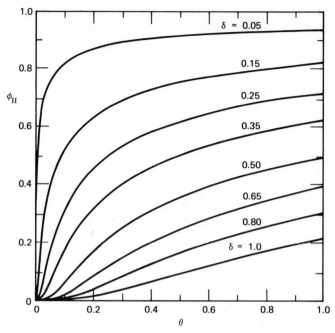

Figure 5.2-3. Fractions of upsurging nutrients in lake-water column (ϕ_{II}). (Reprinted by permission. American Water Resources Assoc., Mpls., MN. **Source.** Reference 18.)

USES AND IMPLICATIONS OF THE MODEL

From the viewpoint of ecosystem modeling, the preceding development provides equations for estimating the rate at which bottom-originating chemical nutrients pass into upper regions of the lake water such as the epilimnion. For this application Eqs. 5.2-7 and 5.2-9 should be used directly. In the interest of average flux rates for a period of time t, Eq. 5.2-12 modified by the fraction ϕ can be used:

$$n_{Ay} = 2\phi \sqrt{\frac{\mathcal{D}_{A2}^{(t)}}{\pi t}} \left(\rho_{A2}^* - \rho_{A2}^0 \right) \tag{5.2-15}$$

Depending on the case, Fig. 5.2-2 or Fig. 5.2-3 is used to obtain ϕ.

The Fickian analysis results can also be used to help determine whether a fall bloom of algae is likely for a given lake based on the availability of nutrients. The spring bloom of algae can deplete the lake surface water of a limiting nutrient, and an additional supply of this nutrient is needed if a fall bloom is to materialize. The depth of the lake h and the turbulent diffusivity $\mathcal{D}_{A2}^{(t)}$ are important lake characteristics regulating the time of travel of chemicals from the bottom sediments into the epilimnion. These two parameters are contained in the lake time constant, $t_c \equiv h^2/2\mathcal{D}_{A2}^{(l)}$.

If one assumes that 10% ($\phi_I = 0.1$) of the limiting nutrient leaving the sediments is sufficient to stimulate increased algal growth in the epilimnion, then from Fig. 5.2-2 $\theta = 0.18$. According to this model, the time required for a significant quantity (i.e., 10%) of the bottom-originating limiting nutrient chemical to arrive at the epilimnion is

$$t_I = \frac{0.18h^2}{2\mathcal{D}_{A2}^{(l)}} \qquad (5.2\text{-}16)$$

In a similar fashion, if 10% ($\phi_{II} = 0.1$) of the botton-originating nutrient A travels 80% of the distance to the surface ($\delta = 0.8$), then from Figure 5.2-3, $\theta = 0.35$. The model for Case II gives the time of travel as

$$t_{II} = \frac{0.35h^2}{2\mathcal{D}_{A2}^{(l)}} \qquad (5.2\text{-}17)$$

Comparison of the nutrient travel times indicates that less time may be required for lakes with a well-developed thermal profile and a downward migrating thermocline if h and $\mathcal{D}_{A2}^{(l)}$ values are similar.

The use of both Eqs. 5.2-16 and 5.2-17 gives a range of times after the vernal equinox in which the photoactive region of a lake commences to receive detectable quantities of nutrients from the sediment bed that may stimulate a fall bloom. If the time computed is less than or equal to 186 days, a fall bloom is likely. If the time is greater than 186 days, then the declining length of the photoperiod and low water temperatures make a fall bloom unlikely.

SHAGAWA LAKE—A COMPREHENSIVE EXAMPLE

Shagawa Lake is located in St. Louis County, Minn., near Ely. The lake is relatively small with a surface area of approximately 9.3 km^2, a mean depth of about 5.7 m, and a volume of about 53E 6 m^3. There are three deep holes of about 13.7 m each, a gradually sloping bottom, and steep sides. A survey conducted in 1937 indicated that the bottom was 70% muck, 29% sand, and the remainder coarser soils.

The lake has probably been eutrophic for at least several decades with local citizens commenting on excessive algae growths for over 40 years. The dominant source of phosphorus was from the wastewater discharged by the city of Ely. An ideal opportunity existed to test the hypothesis that wastewater treated for phosphorus removal and allowed to enter a lake system would no longer exert its previous fertilizing influence, thus allowing the affected system to recover. The experiment commenced when the advanced wastewater plant went on stream in the spring of 1973. The plant was to remove 99% of the phosphorus entering. Only 19% of the phosphorus is entering the lake, as compared to the period before the plant was built.

While the treatment plant was in operation, extensive field data were obtained to monitor the chemical, physical, and biota changes. With regard to phosphorus, the primary nutrient for algae growth, the bottom muds constitute the main source. The upsurge of this nutrient and others from the bottom muds may be observed by studying the field data.[19] (See problem 5.2A.) The lake thermal structure has an important influence on the nutrient upsurge concentration profiles. The following comments summarize a detailed study of the field data for Shagawa Lake:

1. Summary of the temperature profile data:

   December, January, February, March. The lake is reverse stratified with cold water above and warmer water below. Mixing should be mild because of the stability of the water column.

   April, May. The lake is warming up. The density structure has been destroyed, and the lake is fairly isothermal. High winds in the spring keep the lake water fairly well mixed.

   June, July, August. The lake continues to heat up. The lake is normally stratified with a warm water layer above colder water. Because of the stability of the water column, mixing below the thermocline should be at a low level.

   September, October, November. The lake is cooling off. Plunging cold water created near the air–water interface increases mixing up and down the water column.

2. Summary of the total phosphorus concentration profile data:

   December, January, February, March. While the lake is reverse stratified, mixing of the water column is low. A "wave" of high concentration of phosphorus can be seen "upsurging" from the mud–water interface.

   April, May. Mixing destroyed the phosphorus concentration profile of the previous period. Concentration at all points along the water column is fairly constant.

   June, July, August. The lake is normally stratified, and mixing has subsided. A "wave" of high concentration of phosphorus grows upward from the bottom.

September, October, November.   Plunging cold water from the surface causes mixing all along the water column. The total phosphorus concentration is once again fairly uniform.

Field data on orthophosphate, ammonia, nitrate, nitrite ions, and conductivity display the same development as total phosphorus. The preceding summary was for the year 1973. The data were almost identical for the year 1972. Not all lakes stratify twice during a year as Shagawa does. Lakes in southern regions of the United States may stratify only once. For these lakes the upsurge structure is simpler, usually consisting of two periods—one mixed and one unmixed.

Using Larsen's data, Thibodeaux[18] applied the Fickian upsurge model to Shagawa Lake. Equation 5.2-2 was used to determine $\mathcal{D}_{A2}^{(t)}$, turbulent diffusivity values. A total of 1463 observations of concentrations of ammonia, total phosphorus, and orthophosphate was used to calculate $\mathcal{D}_{A2}^{(t)}$

Table 5.2-1.   General Lake Information

	Lake Shagawa	Lake Fayetteville	DeGray Reservoir
Location	northern Minnesota	northwest Arkansas	sourthern Arkansas
Area, km^2	9.3	4.2	54.3
Mean depth, m	5.7	4.3	8.9
Maximum depth, m	14.0	10.5	58.0
Volume, m^3	53×10^6	3×10^6	794×10^6
Drainage basin, km^2	110	–	1173
Eutrophic state	Eutrophic	moderately eutrophic	mesotrophic
$\mathcal{D}_{A2}^{(t)}$, m^2/d^a	0.085	0.10	1.3
$\mathcal{D}_{A2}^{(t)}$ range, m$^2/d^a$	0.022–0.17	0.014–0.453	$\sigma = 0.216$
$\alpha_2^{(t)}$, m$^2/d$	—	0.21	0.743
$\alpha_2^{(t)}$ range, m$^2/d$	—	0.052–0.58	0.18–1.8
ρ_{A2}^{*}, mg/L	0.52 total phororus (P)		12.1 nitrate (N)
ρ_{A2}^{*}, mg/L	0.36 ortho- phosphate (P)		14.6 cal- cium (Ca)
ρ_{A2}^{*}, mg/L	1.0 ammonia (N)	6.63 to 10.3 ammonia (N)	
Julian day for $t = 0$	142 spring 335 fall	91	80

Source. Reference 18.

^{a}Turbulent diffusion coefficients are for the hypolimnion regions of the respective lakes.

values. These data and other results are summarized in Table 5.2-1. Also included are similar results from two other lakes. Figure 5.2-4a contains typical profiles for ammonia concentration in Shagawa Lake. Figure 5.2-4b shows the profiles that result from application of the Fickian model to upsurging ammonia. See Problem 5.2A for an exercise in estimating $\mathcal{D}_{A2}^{(t)}$ from field data.

The Fickian model with its constant $\mathcal{D}_{A2}^{(t)}$ is a simplistic interpretation of a complex diffusion process. The mathematics are tractable, and the model serves as an easy introduction to and an instructional steppingstone to developing and understanding more sophisticated models. The next step in model development may be to use Eq. 3.2H-5 to replace the constant $\mathcal{D}_{A2}^{(t)}$:

$$\mathcal{D}_{A2}^{(t)} = k_1(y + y_0)v_* \tag{3.2H-5}$$

If surface winds are assumed to be a primary reason for water movement, the preceding equation may be of little use. The surface winds change direction constantly; therefore the bottom friction velocity v_* is also changing constantly. The change in direction creates turbulence itself. For this reason and other complexities that render lake turbulence a random process, a constant $\mathcal{D}_{A2}^{(t)}$ may be adequate for many purposes.

Average Annual Upsurge Rate at the Sediment–Water Interface

The average chemical flux rate for a period of time t from a surface of concentration ρ_{A2}^* to a water of background concentration ρ_{A2}° is given by Eq. 5.2-12. This equation is essentially the penetration theory result presented in Section 3.1. The turbulent diffusivity $\mathcal{D}_{A2}^{(t)}$ replaces the molecular diffusivity appearing in Eq. 3.1-16b.

Commencing with the vernal equinox, a year in the life of a lake can be broken up into four time periods. Each period is either a stratified or an unstratified period of varying length. The sum of the four periods must be 365 days ($t_t = 365$ d):

$$t_s + t_f + t_w + t_v = t_t \tag{5.2-18}$$

where t_s is the summer stratified period, t_f is the fall unstratified period, t_w is the winter stratified period, and t_v is the spring unstratified period. During the two unstratified periods the lake is assumed to "turn over" fairly rapidly, characterized by short time periods Δt of the order of days

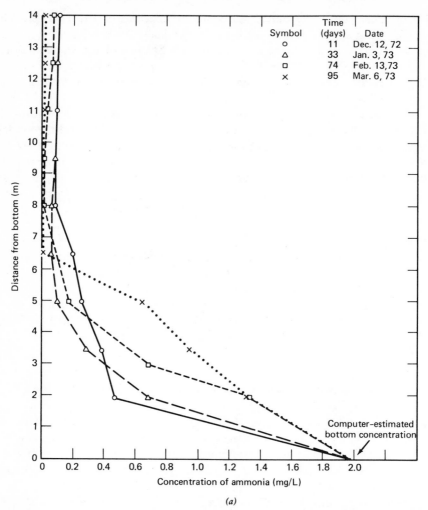

(a)

Figure 5.2-4. Shagawa Lake ammonia concentration station LBS winter 1972–1973. (a) Field data. (Reprinted by permission. American Water Resources Assoc. Mpls., MN. **Source.** Reference 18.)

or weeks. Let Δt_f be the average time between turnovers for the fall period and Δt_v be the average time between turnovers for the spring period.

The average annual flux rate is determined by proportioning the flux rate of each period to the fraction of the year it occupies:

$$\bar{n}_{A0} = \sum_{i=1}^{4} n_{A0,i} \frac{\Delta t_i}{t_t} \qquad (5.2\text{-}19)$$

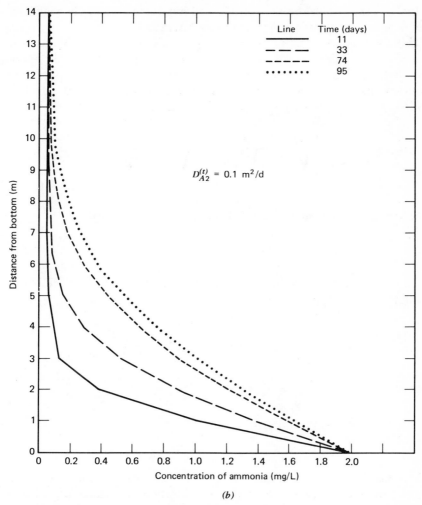

Line	Time (days)
——	11
– – –	33
- - - -	74
••••••	95

$$D^{(t)}_{A2} = 0.1 \ \text{m}^2/\text{d}$$

Figure 5.2-4. (*Continued*) (*b*) Calculated profiles. (Reprinted by permission. American Water Resources Assoc., Mpls., MN. **Source.** Reference 18.)

Applying Eq. 5.2-12 to each time period, we obtain

$$\bar{n}_{A0} = \frac{2}{t_t \sqrt{\pi}} \left[\sqrt{\mathcal{D}^{(t)}_{A2,s} t_s} + \sqrt{\frac{\mathcal{D}^{(t)}_{A2,f} t_f^2}{\Delta t_f}} + \sqrt{\mathcal{D}^{(t)}_{A2,w} t_w} + \sqrt{\frac{\mathcal{D}^{(t)}_{A2,v} t_v^2}{\Delta t_v}} \right] \cdot (\rho^*_{A2} - \rho^\circ_{A2})$$

$$(5.2\text{-}20)$$

Information about lake time periods and hypolimnion turbulent diffusivities is necessary to arrive at an average flux rate.

If it is assumed that a value of $\mathcal{D}_{A2}^{(t)}$ measured during the stratified periods is a fundamental lake parameter applicable to the mixed periods and that the high frequency of turnovers characterized by Δt is paramount in differentiating the mixed periods from the stratified periods, then Eq. 5.2-20 can be simplified greatly and appear as

$$\bar{n}_{A0} = 2\sqrt{\frac{\mathcal{D}_{A2}^{(t)}}{\pi t_t}} \left[\sqrt{\frac{t_s}{t_t}} + \sqrt{\frac{t_f^2}{t_t \Delta t_f}} + \sqrt{\frac{t_w}{t_t}} + \sqrt{\frac{t_v^2}{t_t \Delta t_v}}\right](\rho_{A2}^* - \rho_{A2}^\circ)$$

$$(5.2\text{-}21)$$

The average annual mass transfer coefficient is defined by

$$\bar{n}_{A0} \equiv \overline{^3k_{A2}'}(\rho_{A2}^* - \rho_{A2}^\circ) \tag{5.2-22}$$

A further refinement of Eq. 5.2-21 occurs with the creation of the Nusselt number for mass transfer:

$$\mathrm{Nu}_{A2} = 2\left[\sqrt{\frac{t_s}{t_t}} + \sqrt{\frac{t_f^2}{t_t \Delta t_f}} + \sqrt{\frac{t_w}{t_t}} + \sqrt{\frac{t_v^2}{t_t \Delta t_v}}\right] \tag{5.2-23}$$

where

$$\mathrm{Nu}_{A2} \equiv \frac{\overline{^3k_{A2}'}}{\sqrt{\mathcal{D}_{A2}^{(t)}/\pi t_t}}$$

This equation shows that the average annual lake bottom, water phase mass transfer coefficient is related to the duration of each stratified period, each unstratifed time period, and the respective Δt's within each unstratified period.

The relative importance of the various time periods may be studied by classifying lakes into two categories. The first category includes those lakes that have two stratified periods and two unstratified periods. For simplicity, the stratified periods are assumed to be of the same duration, $t_w = t_s$, and the unstratified periods are also assumed to be of the same duration, $t_v = t_f$, as are Δt_f and Δt_v, so that Eq. 5.2-18 becomes

$$2t_s + 2t_f = t_t \tag{5.2-24}$$

For category I lakes Eq. 5.2-23 becomes

$$Nu_{A2,I} = 4\left[\sqrt{\frac{t_s}{t_t}} + \sqrt{\frac{(t_t - 2t_s)^2}{4t_t\Delta t}}\right] \tag{5.2-25}$$

The second category includes those lakes that have a single stratified period of duration t_s and a single unstratified period of duration t_w for a total of

$$t_s + t_w = t_t \tag{5.2-26}$$

For category II lakes Eq. 5.2-23 becomes

$$Nu_{A2,II} = 2\left[\sqrt{\frac{t_s}{t_t}} + \sqrt{\frac{(t_t - t_s)^2}{t_t\Delta t}}\right] \tag{5.2-27}$$

The Nusselt numbers contain only time ratios. It is now a simple task to choose reasonable values of the ratio t_s/t_t in order to study the effect of period duration and category variation on the annual average coefficient. Table 5.2-2 contains these calculations for Δt periods of 1 and 2 weeks.

Although Nusselt numbers are shown for a range $0 \leqslant t_s/t_t \leqslant 1$, it is unlikely that many lakes in the United States fall beyond the t_s/t_t range of (0.2, 0.5). Within this range the numbers are relatively constant, with a mean of 9.8 and a span of 13.34 to 6.52.

Table 5.2-2. Lake Bottom Nusselt Numbers

t_s/t_t	$Nu_{A2,I}$		$Nu_{A2,II}$	
	$\Delta t=7$ days	$\Delta t=14$ days	$\Delta t=7$ days	$\Delta t=14$ days
0	14.44	10.21	14.44	10.21
0.1	14.26	10.46	13.63	9.82
0.2	13.34	9.96	12.45	9.06
0.3	12.30	9.34	11.20	8.24
0.4	11.20	8.66	9.93	7.39
0.5	10.05	7.93	8.63	6.52
0.6	–	–	7.33	5.63
0.7	–	–	6.01	4.74
0.8	–	–	4.78	3.83
0.9	–	–	3.34	2.92
1.0	–	–	2.00	2.00

Example 5.2-1.   Obtaining an Estimate of the Lake Bottom Water Coefficient

Canale estimated the liquid phase mass transfer coefficient for phosphorus $(A \equiv P)$ at the bottom of Lake Washington to be $^3k'_{A2} = 36$ m/yr. (See Example 3.3-1.) Without any knowledge of the exact diffusivity or the seasonal periods for the lake, estimate the coefficient.

SOLUTION

Without seasonal stratified and unstratified time periods, Eqs. 5.2-25 and 5.2-27 cannot be used, so the average Nusselt number of 9.8 is invoked. A turbulent diffusivity of 0.1 m^2/d is assumed. This value is reasonable and falls within the range of values reported in Table 3.1-4 for deep regions of lakes (0.01 to 1.0 cm^2/s).

$$\text{Nu}_{A2} \equiv \frac{\overline{^3k'_{A2}}}{\sqrt{\mathcal{D}_{A2}^{(t)}/\pi t_t}} = 9.8$$

and for solving for $\overline{^3k'_{A2}}$ yields

$$\overline{^3k'_{A2}} = 9.8 \left(\frac{0.1 \text{ m}^2}{\text{d}} \left| \frac{1}{\pi} \right| \frac{1}{365 \text{ d}} \right)^{1/2} \frac{365 \text{ d}}{\text{yr}} = 33.4 \text{ m/yr}$$

Mass Transfer Coefficients at the Sediment–Water Interface

In the previous section the transport processes within the sediment phase were ignored. It was assumed that the concentration of chemical A at the sediment–water interface, ρ^*_{A2}, was constant and in equilibrium with A in the sediments ρ_{A3}. This assumption is valid when the resistance to mass transfer resides within the water phase. In the latter part of the section on Fickian analysis the liquid phase mass transfer coefficient at the sediment–water interface was developed. In this section a sediment phase mass transfer coefficient is defined, and the two-resistance theory is developed for the transfer of chemical A across the sediment–water interface.

    The sediment on the bottom of a lake, estuary, or ocean consists of solid material that is a combination of both organic and inorganic matter. The organic matter may consist of decayed plant or animal bodies whereas the inorganic is mostly silica in the form of sand and silt. Figure 5.2-5

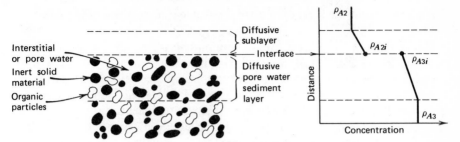

Figure 5.2-5. The sediment–water interface.

illustrates what the sediment–water interface may look like on close inspection of a thin section.

The mechanism for movement of chemical A from the sediment into the overlying water involves the following six individual processes:

1. Release of the molecule from the cell matrix or the solid matrix.
2. Diffusion through the cellular residue or the pores within the solid matrix.
3. Desorption from the residue or solid surface into the interstitial water.
4. Diffusion through the interstitial water.
5. Diffusion through the sediment–water interface.
6. Movement through the diffusive sublayer into the overlying turbulent water.

For the most part the individual rates of each step are unknown. The release of the molecule from the cell matrix involves a biochemical reaction whereas the release of the molecule from a solid particle matrix involves a dissolution surface reaction or an ion exchange reaction. Diffusion through the cellular residue or the pores within the solid matrix may be hindered or enhanced molecular diffusion. Desorption from the residue or solid particle interface into the interstitial water is likely to be a rapid process. Movement through the interstitial water is by molecular diffusion. Movement through the sediment–water interface plane is likely to be a rapid process also. There is a resistance within the diffusive sublayer.

The production of chemical A or its release from the substrate by biochemical processes can be related by a first-order reaction as

$$r_A = k_A''' \rho_{BS} \tag{5.2-28}$$

where ρ_{BS} is the concentration of some characteristic precursor molecule.

Dissolution from the surface of a solid particle is also a first-order reaction. (See Eq. 5.3-8.)

If chemical A is present in the interstitial water or the reactions to produce it are rapid and intercellular diffusion is negligible, then it is possible to characterize mass transfer by a single coefficient. It is further assumed that the sediment bed is "semifluid" such that it is mixed periodically through depth h and has a constant concentration of chemical A below this depth, ρ_{A3}. It is necessary to define a sediment phase mass transfer coefficient to account for the steps 1 through 4 given earlier:

$$n_{A0} = {}^2k'_{A3}(\rho_{A3i} - \rho_{A3}) \tag{5.2-29}$$

where ${}^3k'_{A3}$ is the sediment phase coefficient with dimensions of L/t and ρ_{A3i} is the concentration of A in the sediment phase side of the interface M/L^3. The form of the two-resistance theory applicable to this case is

$$\frac{1}{{}^3K'_{A2}} = \frac{1}{{}^3k'_{A2}} + \frac{\mathcal{K}^*_{A23}}{{}^2k'_{A3}} \tag{5.2-30}$$

where $\mathcal{K}^*_{A23}$ is the partition coefficient of chemical A between the interstitial water associated with the sediment solid phase and the water phase as defined by Eq. 2.1-14.

There is some evidence to indicate that the activities of benthic organisms, bottom currents, and gases evolving from decaying organic matter do mix the upper sediment layer to a certain degree. This activity below the sediment–water interface is slow compared to the water side transport processes above the interface, thus ${}^2k'_{A3}$ is likely an order of magnitude or more less than ${}^3k'_{A2}$. Assuming that a sediment bed is "semifluid" in order to invoke the two-resistance theory may be a case of model simplification to the point of absurdity. Realizing that Eqs. 5.2-29 and 5.2-30 are only rough order-of-magnitude models of the actual sediment layer processes, they do allow estimates of recovery times and water concentrations. A more realistic approach to the problem of the movement of chemicals in a sediment layer toward the interface is taken in Section 5.3.

Minimum Recovery Time for a Lake with Chemicals "Locked" into the Bottom Sediments

Attempts have been made at restoring eutrophic lakes toward an oligotrophic state by reducing the inputs of nitrogen and phosphorus

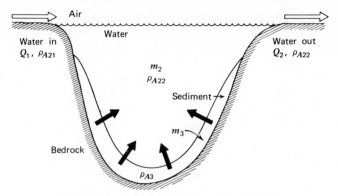

Figure 5.2-6. Lake–water and sediment system.

compounds. Such attempts may prove futile because of the recycling process that occurs with these particular chemicals. Even without the recycling process, the "cleansing" of lake sediments of refractory chemicals by transfer to the water phase may involve hundreds of years. Phosphorus, in particular, is strongly partitioned or "locked" into the soil phase. Partition coefficients (i.e., $K_{A32}^* \equiv \rho_{A3}/\rho_{A2}$) from the data presented in Problems 2.1D range from approximately 23 to 72, indicating a strong preference for the soil phase. The chemical transfer mechanism within the sediment bed that moves the chemical to the interface is slow compared to that on the water side. Figure 5.2-6 contains a schematic of a lake showing the interaction between water and sediments that results in sediment cleansing.

UNSTEADY-STATE MODEL

The development and use of some simplistic lake and sediment models will help focus our attention on the chemical transport process. As is shown, the minimum recovery time takes years, so model time is large, and the annual cycle of events can be ignored. Assume that the lake water is of a uniform concentration ρ_{A22} and nutrient cycling does not occur. Assuming no nutrient recycling within the lake implies that when a chemical crosses the sediment–water interface, it eventually leaves the system in the over-flow water. A component balance of chemical A on the lake water yields

$$Q_1 \rho_{A21} - Q_2 \rho_{A22} + n_{A0} A = \frac{d}{dt} (V \rho_{A22}) \qquad (5.2\text{-}31)$$

where n_{A0} is the input rate from the sediment bed:

$$n_{A0} = {}^3K'_{A2}(\rho^*_{A2} - \rho_{A22}) \tag{5.2-32}$$

Equation 5.2-31 shows that the depletion of chemical A from the lake water is dependent on the input from the sediments as well as the input by hydraulic flow. (See Problem 5.2D.)

A component balance of chemical A on the mass of bottom sediments yields

$$0 - n_{A0}A = \frac{m_3}{\rho_3}\frac{d\rho_{A3}}{dt} \tag{5.2-33}$$

The equilibrium relationship between the sediments and the water can be expressed by

$$\rho^*_{A2} = \mathcal{K}^*_{A23}\rho_{A3} \tag{5.2-34}$$

Assuming ρ_{A22} is negligible, Eqs. 5.2-32, 5.2-33, and 5.2-34 are combined to give a differential equation, which can then be integrated to yield

$$t = \frac{m_3}{{}^3K'_{A2}A\,\mathcal{K}^*_{A23}\rho_3}\ln\left(\frac{\rho^o_{A3}}{\rho_{A3}}\right) \tag{5.2-35}$$

If sediment cleansing of chemical A is assured at $\rho_{A3} = 0.05\rho^o_{A3}$, 95% removal, Eq. 5.2-35 becomes

$$t = \frac{3m_3}{{}^3K'_{A2}A\,\mathcal{K}^*_{A23}\rho_3} \tag{5.2-36}$$

Equation 5.2-36 contains the important parameters that regulate sediment cleansing time and hence lake recovery time. It is seen that recovery time is proportional to the mass of contaminated sediments m_3, inversely proportional to the overall mass transfer coefficient ${}^3K'_{A2}$, the bottom interfacial area A, and the chemical partition coefficient between water and sediment $\mathcal{K}^*_{A23}$. See Example 5.2-2 for a calculation of the minimum recovery time of a lake sediment bed from contamination by a chemical.

STEADY-STATE MODEL

Since recovery time involves years or decades, in-lake water concentration of chemical A changes very slowly. It is now necessary to reduce the time scale to a few years and consider some results of a steady-state lake model.

This model is also simplistic. Features include constant inflow, outflow, and sediment concentration. The only processes occurring are dissolution of a chemical A at the sediment–water interface and hydraulic dilution. The lake water is assumed to be completely mixed. A component balance on chemical A yields

$$Q_1 \rho_{A21} + {}^3K'_{A2}(\rho^*_{A2} - \rho_{A22})A = Q_2 \rho_{A22} \qquad (5.2\text{-}37)$$

Substituting Eq. 5.2-34 for ρ^*_{A2} and solving for ρ_{A22} yields

$$\rho_{A22} = \frac{Q_1 \rho_{A21} + {}^3K'_{A2} A \, \mathcal{K}^*_{A23} \rho_{A3}}{Q_2 + {}^3K'_{A2} A} \qquad (5.2\text{-}38)$$

where $Q_1 = Q_2$. The short-term in-lake concentration of chemical A is dependent on the inflow concentration ρ_{A21} and the concentration of A within the sediments. Elimination of the inflow may not significantly reduce ρ_{A22} because of the input from the sediment bed.

Example 5.2-2. Minimum Cleansing Time of a Chemical "Locked" in Lake Sediments

Table E5.2-2 contains data on the chemical nutrient phosphorus (total P), within the sediments of the Upper Klamath Lake.[20] It appears that phosphorus tends to partition strongly and in a sense is locked into the sediment phase. The average partition coefficient at 10°C is 4370 and 1430 at 23°C. Compute the minimum time for the natural cleansing of the lake sediment bed of phosphorus to occur (yr). Assume a sediment phase mass transfer coefficient of 2.5 m/yr and a sediment bed depth of 10 cm.

Table E5.2-2. Sediment–Interstitial Water Equilibrium[a]

Sample No.	Temperature (°C)	ρ^*_{A3} (μg P/g sed.)	ρ^*_{A2} (μg P/mL)	$\mathcal{K}^*_{A32}$
1a	10	413	0.23	4780
1b	23	406	0.64	1690
3a	10	793	1.66	1270
3b	23	723	6.08	316
4a	10	504	0.19	7060
4b	23	499	0.58	2290

Source Reference 20.

[a] $\mathcal{K}^*_{A32} \equiv \rho^*_{A3} \rho_3 / \rho^*_{A2}$, $\rho_3 = 2.66$ g/cm^3.

SOLUTION

The minimum time for natural cleansing can be estimated from Eq. 5.2-36. If the sediment bed is of uniform thickness, $m_3/A = h\rho_3$. Since $\mathcal{K}_{A32}^* = 1/\mathcal{K}_{A23}^*$, Eq. 5.2-36 becomes

$$t = \frac{3h\mathcal{K}_{A32}^*}{{}^3K_{A2}'} \qquad \text{(E5.2-2A)}$$

The overall mass transfer coefficient is obtained from Eq. 5.2-30:

$$\frac{1}{{}^3K_{A2}'} = \frac{1}{{}^3k_{A2}'} + \frac{1}{\mathcal{K}_{A32}^* {}^2k_{A3}'} \qquad \text{(E5.2-2B)}$$

Using Example 5.2-1 as the source of the ${}^3k_{A2}'$ value, 33.4 m/yr, and $\mathcal{K}_{A32}^* = 4370$ since the bottom water is probably cold, we obtain

$$\frac{1}{{}^3K_{A2}'} = 0.0299 + 0.0000915 = 0.02999$$

The value ${}^3K_{A2}' = 33.4$ m/yr and the preceding calculation suggests that the water phase controls the rate of release of phosphorus. Using Eq. E5.2-2A to calculate the minimum time yields

$$t = \frac{3|0.1 \text{ m}|4370}{33.4 \text{ m/yr}} = 39.3 \text{ yr}$$

It appears that the minimum cleansing time for phosphorus in this lake is the order of two generations if nutrient recycling does not occur.

Problems

5.2A. THERMAL AND PHOSPHORUS PROFILES IN SHAGAWA LAKE

Water temperature and phosphorus concentrations for Shagawa Lake at sample station LBS for 1973 appear in Table 5.2A.

1. Make two graphs in order to visualize the seasonal temperature and phosphorus concentration developments. On one graph show temperature on the horizontal axis and depth on the vertical axis for each

Table 5.2A. Shagawa Lake Data[a]

h	Jan. 3 T_2	ρ_{A2}	Feb. 13 T_2	ρ_{A2}	Mar. 6 T_2	ρ_{A2}	Apr. 20 T_2	ρ_{A2}
0.1	1.0	0.053	–	0.048	1.1	0.052	5.1	0.058
1.5	1.1	0.050	1.5	0.048	1.5	0.053	5.0	0.059
3.0	2.5	0.052	2.5	0.052	2.4	0.061	4.9	0.059
4.5	3.0	0.043	–	0.047	2.8	0.049	–	0.059
6.0	3.2	0.043	3.5	0.048	3.2	0.043	4.9	0.063
7.5	4.0	0.047	3.8	0.051	3.9	0.059	4.9	0.063
9.0	4.5	0.049	4.2	0.089	4.0	0.130	4.9	0.064
10.5	4.9	0.085	4.5	0.145	4.0	0.154	4.9	0.059
12.0	5.0	0.162	4.5	0.192	4.1	0.143	4.9	0.061

h	May 15 T_2	ρ_{A2}	June 12 T_2	ρ_{A2}	July 10 T_2	ρ_{A2}	Aug. 7 T_2	ρ_{A2}
0.1	10.2	0.031	18.0	0.037	22.2	0.028	22.5	0.052
1.5	10.0	0.032	18.0	0.034	22.1	0.039	22.2	0.053
3.0	10.0	0.031	18.0	0.037	21.8	0.032	22.0	0.047
4.5	10.0	0.037	18.0	0.037	21.2	0.032	21.3	0.065
6.0	10.0	0.034	17.3	0.023	19.8	0.029	20.8	0.066
7.5	10.0	0.038	17.0	0.025	19.5	0.043	20.2	0.091
9.0	10.0	0.037	16.5	0.033	19.0	0.051	20.0	0.161
10.5	9.8	0.040	16.3	0.038	18.4	0.097	19.9	0.164
12.0	9.0	0.036	15.5	0.080	18.0	0.176	19.0	0.237

h	Sept. 11 T_2	ρ_{A2}	Oct. 10 T_2	ρ_{A2}	Nov. 6 T_2	ρ_{A2}	Dec. 18 T_2	ρ_{A2}
0.1	18.7	0.079	13.8	0.054	4.0	0.033	1.0	0.025
1.5	18.7	0.082	13.8	0.051	4.2	0.034	1.5	0.024
3.0	18.7	0.081	13.7	0.047	4.2	0.034	1.5	0.016
4.5	18.8	0.080	13.7	0.048	4.2	0.036	1.5	0.020
6.0	18.8	0.085	13.7	0.047	4.2	0.036	1.8	0.023
7.5	18.8	0.079	13.7	0.043	4.2	0.042	2.0	0.026
9.0	18.8	0.080	13.7	0.043	4.2	0.037	2.1	0.037
10.5	18.8	0.079	13.7	0.044	4.2	0.033	2.5	0.043
12.0	18.8	0.083	13.6	0.044	4.2	0.032	2.8	0.067

Source. Reference 19.

[a]Depth (h) in meters; temperature (T_2) in °C; P concentration (ρ_{A2}) in mg/L.

month. On another graph show phosphorus concentration on the horizontal axis and depth on the vertical axis for each month. Once the plots are made, confirm the description of the temperature profile and phosphorus concentration profile given on pages 263 and 264.

2. Calculate turbulent diffusivity values representative of the reverse stratified period. Using Eq. 5.2-2, calculate $\mathcal{D}_{A2}^{(t)}$ values for January 3, February 13, and March 6. Choose $t=0$ as December 10. Assume $\rho_{A2}^{0}=0.04$ mg/L and $\rho_{A2}^{*}=0.25$ mg/L. Do not use data from 0 to 4.5 m. Report the answer in m^2/d.

3. Calculate turbulent diffusivity values representative of the normally stratified period. Using Eq. 5.2-2 calculate $\mathcal{D}_{A2}^{(t)}$ values for June 12, July 10, August 7, and September 11. Choose $t=0$ as June 10. Assume $\rho_{A2}^{0}=0.03$ mg/L and $\rho_{A2}^{*}=0.25$ mg/L. Exclude data from 0 to 4.5 m depth because algae may be active in this zone and reduce the phosphorus concentration. Report the answer in m^2/d. Depth to lake bottom is 14 m.

5.2B. PHOSPHORUS UPSURGING FROM SEDIMENTS OF SHAGAWA

The average quantity of phosphorus leaving the sediment–water interface and entering the photoactive zone of the lake for days 170 through 230 (mid-June to mid-August) was calculated to be 40 kg/d.[21]

1. Using Eq. 5.2-12, calculate the quantity of phosphorus leaving the sediment–water interface in kg/d.

2. Using Eq. 5.2-15, calculate the quantity of phosphorus entering the photoactive zone. Assume Shagawa is a Case I–type lake.

Data: Sediment–water interfacial area of 9.3 km^2, $\rho_{A2}^{*}=0.36$ mg/L (laboratory analysis of interstitial water), $\rho_{A2}^{0}=0.046$ mg/L (average lake surface water), $\mathcal{D}_{A2}^{(t)}=0.085$ m^2/d, $h=5.7$ m, $t=230-170=60$ d.

5.2C. LIKELIHOOD OF A FALL BLOOM

It has been observed in the field that both Shagawa Lake and Lake Fayetteville have indications of a fall bloom of algae. Use Eqs. 5.2-16 and 5.2-17 to confirm by calculation that the travel time of nutrients is sufficiently short that a fall algae bloom is likely to occur on both lakes.

5.2D. IMPORTANCE OF SEDIMENT DISSOLUTION AS A CONTRIBUTOR TO LAKE PHOSPHORUS CONCENTRATION

The advisability of phosphorus removal from a wastewater stream should be assessed against its projected impact on the lake. Compute the additional in-lake phosphorus concentration that is contributed by the dissolution from phosphorus-rich bottom sediments (mg soluble PO_4/L).

A certain lake with a bottom area of 4.2 km^2 has bottom sediments rich in phosphorus, $\mathcal{K}^*_{A23} = 0.0435$, and an in-flow of 204 m^3/hr containing 4.3 mg soluble PO_4/L. The sediment bed has a phosphorus concentration of 532 mg/L. Use Example 5.2-2 as a source of $^3K'_{A2}$ and ρ_3. See Table E5.2-2 for a definition of $\mathcal{K}^*_{A23}$ ($= \mathcal{K}^{*-1}_{A32}$).

5.3. FLUX OF CHEMICALS BETWEEN SEDIMENTS AND THE OVERLYING SEAWATER

Processes at the planetary boundaries control the properties of the atmosphere and the hydrosphere. The benthic boundary layer and the adjoining sediment layer are the least well known of these. The flux of various species across the sediment–water interface may be one of the more important processes controlling the chemical composition of seawater. The mechanism of the fluxes of various species is also important to the chemodynamics of anthropogenic substances that arrive for various reasons at this important environmental interface. In this section transport aspects of natural and synthetic chemicals in and above the ocean bottom sediments are presented.

The first section is concerned with the benthic boundary layer, those layers of water at the very bottom of the ocean. Next, chemical exchange and sediment movement processes within the sediment layer are presented. The last section is concerned with the movement of the natural tracer radon across the sediment–water interface of an estuary.

Movement of Chemicals through the Benthic Boundary Layer

A chemical substance diffusing through the sea floor leaves a region where fluxes are controlled by molecular diffusion and passes into a region where fluxes are controlled by turbulent motions. Specifying the nature of the transition zone becomes one of specifying $\mathcal{D}_{A2}$ as a function of height above the bottom. It has been suggested that the nonturbulent boundary layer at the sea floor could substantially reduce chemical fluxes across this zone.

To model the effect of the benthic boundary layer on the diffusion of a dissolved chemical species, the analysis must extend into the water column above the sediments. In the water column the concentration of chemical A is described by

$$\frac{\partial}{\partial y}\left(\mathcal{D}_{A2}\frac{\partial c_{A2}}{\partial y}\right) = 0 \qquad (5.3\text{-}1)$$

where $\mathcal{D}_{A2}$ is the coefficient of diffusivity specified as a function of height above the bottom. Descending into the boundary region, transport mechanisms are transformed from those dominated by turbulent processes to those dominated by molecular motion. The transformation is not a simple one. Turbulence is created by the drag of the sea floor on water driven by deep-sea currents. This turbulence mixes the waters overlying the sea bed, but the turbulent eddies are damped and become progressively less effective as we approach the interface. This system is further complicated by periodic variations in the velocity of deep-sea currents, so the regime is constantly changing its character.

Wimbush[22] distinguishes two regions in the vicinity of the sea floor: (1) just above the interface is the viscous sublayer; (2) transitional to the sea above is the logarithmic layer. These regions are illustrated in Fig. 5.3-1. In the region where momentum is being transferred mostly by viscosity, heat and chemical tracers can still be transferred primarily by the turbulent motions. Only when turbulent transports of heat and mass become less than their associated molecular transports can one say that the "boundary layer" exists, where heat transfer is dominated by thermal diffusivity and mass transfer by molecular diffusivity.

The viscous sublayer is distinguished from the zone above in that turbulent transfer of momentum has become less than molecular transfer of momentum. This does not require that turbulent fluxes be nonexistent, but only that they be much less effective. In seawater (at 2°C) the thermal diffusivity, α_2(1.4E−3 cm^2/s), and chemical molecular diffusivity, $\mathcal{D}_{A2}$ (5E−6 cm^2/s), are much less than the kinematic viscosity, ν_2(1.7E−2 cm^2/s). Although momentum transport in the viscous sublayer is dominated by viscosity, turbulent transport of heat and chemical tracers can still be accomplished by turbulent processes. Viscosity, then, becomes the major transmitter of momentum at a boundary layer thickness far greater than the thickness of the layer in which molecular diffusion

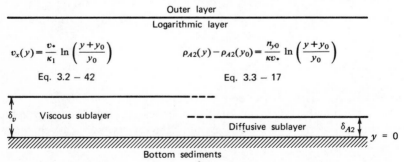

Figure 5.3-1. Dynamical and concentration structure of the benthic boundary layer.

processes dominate turbulent diffusion. The "diffusion sublayer," shown in Fig. 5.3-1, is therefore much thinner than the viscous sublayer. (See Eq. 3.3C-1.)

Deissler [23] compiled a variety of data of momentum transfer in turbulent, isothermal flow in smooth tubes. From this work it is possible to establish that the outer limit at which viscous effects are prominent are at

$$\delta_v = \frac{26\nu}{v_*} \tag{5.3-2}$$

where δ_v is the viscous sublayer thickness. Wimbush[22] suggested a value $\delta_v = 12\nu/v_*$ for the viscous sublayer thickness of a smooth sea bottom. Using this result and Eq. 3.3C-1, the thickness of the diffusion sublayer can be estimated by

$$\delta_{A2} = \frac{26\nu}{v_* \, \mathrm{Sc}_{A2}^{1/3}} \tag{5.3-3}$$

For seawater at 2°C, Sc_{A2} for silica $\simeq 3000$, and the height of the diffusion sublayer is only one-fourteenth of the height of the viscous sublayer.

In our analysis of the benthic boundary layer we assume that transport is by molecular diffusion in the diffusion sublayer and that eddy diffusion obeys a mixing rate law above this layer. Accordingly, $\mathcal{D}_{A2}$ in Eq. 5.3-1 becomes

$$\mathcal{D}_{A2} = \begin{cases} \mathcal{D}_{A2}^{(l)} & 0 < y < \delta_{A2} \\ \kappa_1 v_*(y + y_0) & y > \delta_{A2} \end{cases} \tag{5.3-4}$$

where κ_1 is the von Karman constant (0.4) and y is the height above the bottom.

Measured values of v_* at a location in the deep Pacific Ocean were found to vary with the tidal cycle. Numerical values for v_* varied from about 0.4 to 0.005 cm/s. Any diffusion gradient established when v_* was small was obliterated during the tidal cycle; the viscous sublayer was reduced to its least extent every 6 hr. Transient turbulent effects occasionally eroded even this minimum boundary layer. Accordingly the effective boundary layer for chemical diffusion is a few millimeters at most. (See Example 5.3-1.) A simple calculation shows that concentration gradients disappear a short distance above the sublayer. (See Problem 5.3B.)

The preceding summary of the benthic boundary layer has been abstracted in part from investigations of Schink and Guinasso.[24,25] These investigations have addressed the movement of dissolved silica and calcium carbonate near the sea floor. It was found that the stagnant boundary layers do not play a significant role in regulating the flux of dissolved silica at the sea–sediment interface; however, turbulence in the boundary layer can have a significant influence on calcium carbonate movement patterns.

Example 5.3-1.   Viscous Sublayer and Diffusive Sublayer Thickness at the Sea Floor

(a)   Based on the friction velocity range reported at a location in the deep Pacific Ocean, calculate the range of thickness of the viscous sublayer (mm) at the sea floor.

(b)   Calculate the thickness of the diffusive sublayer for silica (mm) near the sea floor.

SOLUTION

(a)   The viscous sublayer thickness is obtained from Eq. 5.3-2. For $v_* = 0.4$ cm/s

$$\delta_v = \frac{26\nu}{v_*}$$

$$\delta_v = 26 \left| 1.7\text{E} - 2 \, \frac{\text{cm}^2}{\text{s}} \right| \frac{\text{s}}{0.4 \, \text{cm}} \left| \frac{10 \, \text{mm}}{\text{cm}} \right| = 11.05 \, \text{mm}$$

For $v_* = 0.05$ cm/s, $\delta_v = 88.4$ mm.

(b)   The diffusive sublayer thickness is obtained by Eq. 5.3-3 or $\delta_{A2} = \delta_v / \text{Sc}_{A2}^{1/3}$, for $v_* = 0.4$ cm/s and $\text{Sc}_{A2} = 3000$

$$\delta_{A2} = \frac{11.05 \, \text{mm}}{3000^{1/3}} = 0.77 \, \text{mm}$$

for $v_* = 0.05$ cm/s, $\delta_{A2} = 6.1$ mm.

Movement of Chemicals within Bottom Sediments

The distribution of chemicals in the watery portion of the benthic boundary layer below the water–sediment interface is a function of the chemistry of the underlying sediments, factors that disturb the sediment–water interface, and the physics of transport within the bottom water. Discussion here is confined almost entirely to the sediments. As a result of a large ratio of solid surface to interstitial water volume (especially in fine-grained muds), concentrations of species in pore water may change appreciably and give rise to large concentration gradients between sediments and the overlying water. This in turn results in fluxes of dissolved constituents to and from the sediments. In this section some of the more important processes occurring in the upper few centimeters of sediment are discussed, and an attempt is made to show how transport between sediment and overlying water is brought about.

CHEMICALS IN SEDIMENT PORE WATER

Diagenetic chemical reactions, those occurring during and after sediment burial, can be divided into two categories: biogenic and abiogenic. The criterion for classification is whether or not the reactions are mediated by bacteria and other microorganisms. Berner[26] gives an excellent introduction to and a literature review of the chemical reactions and transport processes occurring within abyssal sediments.

Many of the geochemically important bacteria in sediments require the preexistence of organic compounds for their metabolism. Thus biogenic reactions are intimately tied to the deposition of organic matter. High rates of deposition of organic matter are favored by (1) high planktonic productivity in the overlying water and (2) quick settling and burial to avoid decomposition in the water column. Good examples of this situation are provided by many near shore shallow-water muds. Some of the most important diagenetic reactions directly resulting from the bacterial decomposition of organic matter are the removal of dissolved oxygen, the production of carbon dioxide, the reduction of nitrate, the reduction of sulfate, and the production of ammonia, phosphate, hydrogen sulfide, and methane. An example of large differences between pore water and overlying seawater composition is shown in Fig. 5.3-2a.

Reactions that are not biogenically controlled, either directly or indirectly, are less numerous. Examples of these abiogenic reactions include the dissolution of opaline silica, the dissolution of calcium carbonate at the sediment–water interface, the recrystallization of carbonate minerals with the consequent uptake of magnesium and release of strontium, and various documented or suggested silica–seawater reactions. Figure 5.3-2b shows a typical concentration profile of silica in pore water of sediment.

MODELING INTERSTITIAL SILICA (SI0₂) CONCENTRATIONS

Sediments accumulating on the sea floor do not simply reflect what is delivered to the bottom of the ocean; rather they represent a residuum— the net result of deposition, decomposition, dissolution, and reprecipitation. Reactions and transport in pore water play an important role in

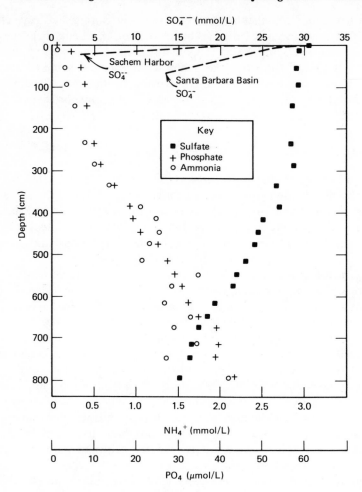

Figure 5.3-2. (a) Sulfate, phosphate, and ammonia versus depth in pore waters of sediment from the West African continental borderland.

determining what stays deposited and what returns to the ocean cycles. Large gradients of dissolved silica are common in the uppermost layers of abyssal marine sediments. Silica has been studied extensively and provides a very good example to study the movement of chemicals within sediments. The silica model presented here is essentially that of Schink and Guinasso.[24]

Apparently the concentrations of dissolved interstitial silica are controlled not by equilibrium processes between pore water and the associated

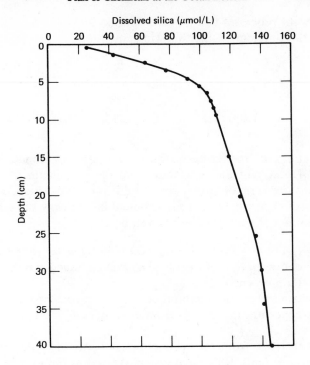

Figure 5.3-2 (*Continued*) (*b*) Dissolved silica versus depth in pore waters of sediment from the Bermuda Rise, North Atlantic. (Reprinted by permission. **Source.** Reference 26.)

solids, but rather by a dynamic balance between dissolution of solids (primarily amorphous biogenic silica, but not necessarily exclusively) and the diffusion of the dissolution products out of the pore waters into the overlying seawater. The dissolved silica comes from particles continuously stirred into the sediments by bioturbation, the principal process that redistributes radioactive particles after they have fallen out of the water column onto the sea floor.

The hypothesis that pore water is not in chemical equilibrium can be tested by calculations using component material balances with parameters chosen as accurately as possible. It is necessary (although not sufficient) that such calculations reproduce concentration profiles found in the sea floor such as that in Fig. 5.3-2*b*. For these calculations and this model, we must include all the effects that have an important influence on pore water composition. These include not only the aqueous phase processes, but also those factors influencing the distribution and reactions of the solid particles that participate.

The relevant processes can be incorporated into a fairly realistic model of the sea floor system by the following equations:

$$0 = D_{A2} \frac{\partial^2 c_{A2}}{\partial y^2} - v_y \epsilon \frac{\partial c_{A2}}{\partial y} + \frac{^3 k''_{A2} c_{A3}(c^*_{A2} - c_{A2})}{c^*_{A2}} \qquad (5.3-5)$$

$$0 = D_{A3}^{(t)} \frac{\partial^2 c_{A3}}{\partial y^2} - v_y \frac{\partial c_{A3}}{\partial y} - \frac{^3 k'''_{A2} c_{A3}(c^*_{A2} - c_{A2})}{c^*_{A2}} \qquad (5.3-6)$$

These equations are the result of steady-state differential material balances on silica ($A \equiv SiO_2$) in the pore water and the particulate matter, respectively. The three terms account for the diffusion, advection, and dissolution processes occurring within the sediment layer below the sea floor. The precise meaning of each symbol is given here:

$\epsilon =$ fraction of the sediments occupied by water (dimensionless)

$c_{A2} =$ concentration of dissolved species in aqueous phase only (μ mol/cm^3)

$c^*_{A2} =$ concentration at saturation in the aqueous phase

$c_{A2i} =$ concentration in the seawater contacting the bottom

$t =$ time (k yr)

$y =$ distance from the sediment–water interface increasing downward (cm). Reference system is fixed to the interface

$D_{A2} =$ diffusion coefficient in porous medium, $\epsilon \mathcal{D}_{A2} / \tau_h^2$ where $\mathcal{D}_{A2}$ is the aqueous diffusivity in homogeneous phase (cm^2/k yr) and τ_h is tortuosity (dimensionless)

$v_y =$ apparent velocity of solid particles and pore water as viewed from the advancing sediment–water interface

$^3 k'''_{A2} =$ first-order dissolution rate constant for substance dissolving to increase c (k yr^{-1})

$c_{A3} =$ concentration of dissolvable particulate matter in bulk phase (solids + fluids) of the sediment (μ mol/cm^3)

$D_{A3}^{(t)} =$ rate of vertical mixing in the solid phase of sediments (cm^2/k yr)

Note that c_{A2} and c_{A3} are concentrations referenced to different volumes so that $-\partial c_{A3}/\partial t = \partial \epsilon c_{A2}/\partial t$ when transport processes are neglected. The boundary conditions for this model are simply (1) that the input flux of particulates be equal to mixing redistribution plus advection (caused by sediment accumulation) at the sea floor boundary:

$$N_{A0} = -D_{A3}^{(t)} \left(\frac{\partial c_{A3}}{\partial y} \right) + v_y c_{A3i} \qquad (5.3-7)$$

(2) that the concentration in the aqueous phase at 50 cm above the bottom be the concentration at the interface, c_{A2i}; and (3) and (4) that there be no gradients in c_{A2} and c_{A3} at the bottom of the model.

Concentration gradients in reactive porous media do produce diffusive fluxes. The pores interconnect, and for abyssal sediments, porosity is 60 to 90%. The presence of nonreactive solid particles slows the diffusive flux by forcing the migrating species into more indirect routes and by reducing the cross-sectional area through which diffusion can proceed. In most marine sediments, with porosities of 0.6 to 0.9, these two effects combine to slow aqueous diffusion by 30 to 40%. (See Eq. 5.1-41.)

Soluble particulate silica can be found at all depths in the sediment. Vertical mixing of solid particles at the sea floor is usually responsible for such burial, and this retards dissolution. Vertical mixing may be accomplished by physical movements of the waters at the sea floor. More important are the myriad activities of the organisms living on or in the sea floor. They eat, bore through, burrow in, spray out, or otherwise disturb and mix the sediments lying near contact with seawater at the floor of the ocean. The combination of all these actions has been termed *bioturbation*, and this process has a significant influence on the chemical interactions between sediment and seawater. See Fig. 3.2-10d for some bioturbation evidence on the sea floor.

Biological activity and ocean bottom currents mix the surface layer of deep-sea sediments to a depth of the order of 10 cm. Guinasso and Schink[27] describe the biological mixing in deep-sea sediments in terms of a time-dependent eddy diffusion model where mixing takes place to a depth h at a constant eddy diffusivity D_{A3}. Microtektite data indicate that abyssal sediments are mixed from the surface to a maximum mixing depth that ranges between 17 and 40 cm below the surface. Estimates of D_{A3} based on dimensional analysis of sediment reworking rates for near shore organisms (10^3 to 10^6 cm^2/k yr) are used to predict abyssal mixing rates between 1 and 10^3 cm^2/k yr by invoking the assumption that mixing is proportional to benthic biomass. It should be noted that bioturbation eddy diffusion coefficients, D_{A3}, are three or four orders of magnitude lower than aqueous molecular diffusivities.

Dissolution rates are, surprisingly, even harder to quantify. The actual processes whereby minerals interact with pore waters via dissolution, precipitation, adsorption, or ion exchange are abiogenic whether or not they are brought about ultimately by biogenic reactions. The kinetics of these reactions, unfortunately, is not at all well understood. Assumptions of first- and null-order mineral reactions make the equations mathematically simpler but may not be correct. The dissolution model presented earlier (Eqs. 5.3-5 and 5.3-6) is a type of first-order dissolution rate expression with $^3k'''_{A2}$ the rate constant. Berner[28] proposes the following general expression for mineral precipitation and dissolution in stagnant (or very-slow-flowing) pore water:

$$N_{A0} = \frac{A\,\mathcal{D}_{A2}}{R}(c^*_{A2} - c_{A2}) \tag{5.3-8}$$

where c_{A2} is the concentration in mass per unit volume of pore water, c^*_{A2} is the concentration in the layer immediately adjacent to the surface of the solid, $\mathcal{D}_{A2}$ is the diffusivity of A in the solution, A is the surface area of solid per unit volume of pore water, and R is the average radius of solid particles. Hurd[29] has described dissolution from biogenic opal in terms of a mass transfer coefficient, $^3k_{A2}$ (mg/cm$^2\cdot$s), which involves consideration of surface area. This is probably the correct approach, but it leaves some uncertainty as to whether the surface area, as measured by nitrogen adsorption, is the appropriate surface area for dissolution.

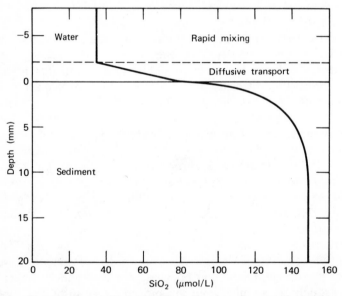

Figure 5.3-3. Concentration profile for dissolved silica in sediment and overlying pore water according to model calculations that include consideration of the laminar-conductive sublayer (the zone labeled "diffusive transport"). (Reprinted by permission. **Source.** Reference 26.)

CLOSURE

The preceding was a semiquantitative description of the silica reactions and transport processes that occur in the sea floor sediments. Figure 5.3-3 is an example of a silica concentration profile generated from such a mathematical model as that presented by Eqs. 5.3-5 and 5.3-6. Although the description involved a "natural chemical" found at the sea floor environment, the same processes are active and a similar model will help describe the fate of an anthropogenic material introduced into this same type of environment.

DISSOLVED CHLORIDE IN THE INTERSTITIAL WATERS OF CHESAPEAKE BAY

The analysis of silica transport in the sediment layer presented earlier involved a steady-state model. If often occurs that conditions at some points in the environment are transient. Because of the seasonal variations in the freshwater input into estuaries, the chemical content of these water bodies is constantly changing with time and provides ideal examples of transient transport. The chloride distribution in the bottom waters is constantly changing, and this produces a continually varying concentration gradient between bottom waters and interstitial waters. By relating the response of the chloride profile in the sediment to changes in the chlorinity of the overlying waters, an estimate of the net rate of transport in the sediment can be made.

The model presented here was developed by Holdren et al.[30] for analyzing the movement of chloride in the sediment of Chesapeake Bay. The bay is very productive, biologically. This is reflected in the organic content of the sediments in the estuary, which is typically 2 to 3% on a dry weight basis. A large infaunal benthic community is supported by these organics. The resulting activity mixes the upper portion of the sediment and enhances the exchange of chemicals between the sediments and the overlying water. To investigate the magnitude of this mixing effect, along with other physical processes such as diffusion, the time-dependent changes that occur in pore water chloride concentration with depth beneath the sediment–water interface were studied.

Chloride is an ideal tracer to study these effects in an estuary such as the bay. It is essentially inert in terms of chemical reactivity in the estuarine environment. Figure 5.3-4*d* shows the results of chloride measurements within the sediment at a single station for over a 1 year period. Easily measurable changes occurred in the chloride profile on a month-to-month

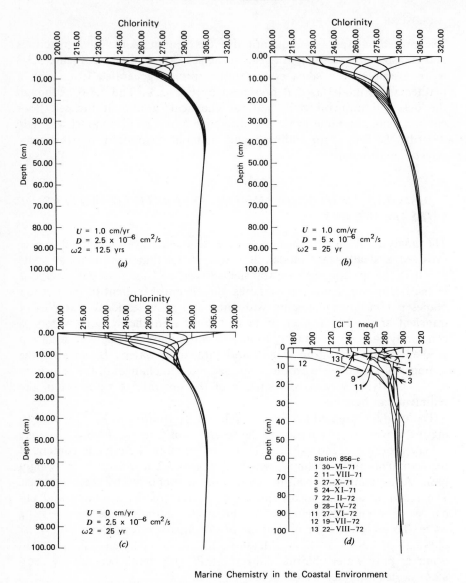

Marine Chemistry in the Coastal Environment

Figure 5.3-4. (*a* to *c*) Vertical profiles of chloride calculated by the diffusion model. These plots show the effect of varying D_{A2}, v_3, and ω_2 on the profiles. (*d*) For comparison to the model profiles, the field data are replotted. (Reprinted by permission. **Source.** Reference 30.)

basis. The surface sediments respond most quickly to the seasonal variations in the chloride concentration of the bottom waters. Seasonal variations result mainly from the discharge of the Susquehanna River, which supplies between 90 and 97% of the fresh water to this portion of the bay.

If the primary mechanism for the transport of chloride is diffusional, the diffusion equation should adequately describe the shapes of the measured profiles. Neglecting the small lateral concentration gradients, the problem is reduced to a one-dimensional diffusion problem. Sediment deposition in the Chesapeake Bay is of the order of 1 cm/yr, and this aspect should be considered in setting up the diffusion model for chloride. The sediment is derived from both shoreline erosion and suspended sediment discharge from the inflowing rivers and streams.

The equation, incorporating a sedimentation term, is

$$\frac{\partial c_{A2}}{\partial t} = D_{A2}\frac{\partial^2 c_{A2}}{\partial y^2} - v_y\frac{\partial c_{A2}}{\partial y} \tag{5.3-9}$$

where c_{A2} = the interstitial chloride concentration
$\quad t$ = time
$\quad D_{A2}$ = the aqueous diffusion coefficient that accounts for porosity and tortuosity
$\quad v_y$ = the sedimentation rate
$\quad y$ = the depth below the sediment–water interface
The boundary conditions are based on the physical observations made of the system.

The first boundary condition describes the chloride concentration in the overlying water as a function of time.

$$c_{A2i}(0,t) = \overline{c_{A2}} + c_{A2s}\cos(\omega_1 t) + c_{A2l}\cos(\omega_2 t) \tag{5.3-10}$$

where $\overline{c_{A2}}$ is the long-term mean chlorinity of the interface water, c_{A2s} and c_{A2l} are the short- and long-term chlorinity variations, respectively, and ω_1, ω_2 are the frequencies of the short- and long-term variations, respectively. The short- and long-term fluctuations account for the seasonal (month-to-month) chlorinity variations and the changes in the mean annual chlorinity, respectively. It is this last term that accounts for the skewing of the upper portion of the profile toward a more dilute concentration relative to the deeper pore waters.

The second boundary condition simply states that there is no net diffusional flux of chloride at great depth in the sediment. This is equivalent to saying that the estuary is "lined" by impermeable bedrock beneath the sediment.

For constant D_{A2} and v_y the solution to this equation is found in Carslaw and Jaeger:[31]

$$c_{A2}(y,t) = \bar{c}_{A2} + c_{A2s}\cos\left[\omega_1 t - ya_1^{1/2}\sin\left(\tfrac{1}{2}\phi_1\right)\right]$$
$$\times \exp\left[\frac{v_y y}{2D_{A2}} - ya_1^{1/2}\cos\left(\tfrac{1}{2}\phi_1\right)\right]$$
$$+ c_{A2l}\cos\left[\omega_2 t - ya_2^{1/2}\sin\left(\tfrac{1}{2}\phi_2\right)\right]$$
$$\times \exp\left[\frac{v_y y}{2D_{A2}} - ya_2^{1/2}\cos\left(\tfrac{1}{2}\phi_2\right)\right]$$

where

$$a_1 = \left[\frac{v_y^4}{16D_{A2}^4} + \frac{\omega_1^2}{D_{A2}^2}\right]^{1/2}, \qquad a_2 = \left[\frac{v_y^4}{16D_{A2}^4} + \frac{\omega_2^2}{D_{A2}^2}\right]$$

$$\phi_1 = \tan^{-1}\left(\frac{-4D_{A2}\omega_1}{v_y^2}\right), \qquad \phi_2 = \tan^{-1}\left(\frac{-4D_{A2}\omega_2}{v_y^2}\right) \qquad (5.3\text{-}11)$$

and all other terms as defined earlier.

By picking values for v_y, D_{A2}, and $\omega_2(\omega_1 = 2\pi/\text{yr})$, time-dependent chloride profiles can be calculated. Values of $\bar{c}_{A2}$, c_{A2s}, c_{A2l} are available from the water analysis data, and the ranges of v_y and ω_2 are available from independent data sources of other investigators. Therefore, an estimate of the diffusion coefficient typical of bay sediments can be made by matching calculated profiles to the field data.

The results of some representative calculations are shown in Fig. 5.3-4a to c. The close resemblance of the model results to the field data lends a degree of credence to the model. All three parameters, D_{A2}, v_y, and ω_2, were varied in the calculations to determine the net effect of each on the profiles. Changes in sedimentation rate had very little effect over the period of one year; however, there are provisions in the model to account for any long-term effects resulting from accumulation. Changes in the diffusion coefficient D_{A2} had the greatest effect of the three parameters. Comparison of the calculated profiles and the field data indicate that the best value for the constant D_{A2} is 5E-6 cm^2/s. This is in good agreement with values that have been reported in other sediment systems, such as the ocean floor.

It is apparent that the numerical model is capable of producing chloride concentrations profiles through time with the selection of reasonable constant parameters. However, the purpose here, as in much modeling work, is not to generate exact replicates of the observed chloride profiles in the bay sediments, but rather to obtain a feeling for the processes and the magnitude of the combined effects of diffusion, bioturbation, and sedimentation on the distribution of any dissolved chemical of the interstitial waters. The simple model described here accomplishes this goal.

RADON-222 AS A TRACER FOR STUDYING MECHANISMS OF EXCHANGE ACROSS THE SEDIMENT–WATER INTERFACE

One problem in formulating material balances for chemicals in aquatic systems is predicting the rates of mass transfer across the sediment–water interface. As shown earlier, the exchange across the sediment–water interface can be calculated by use of reaction–molecular diffusion models. Almost no information exists regarding the accuracy of these models in calculating fluxes. This section seeks to examine such processes in the Hudson Estuary by examining the distribution of radon, a naturally occurring chemical tracer that is generated primarily in sediments and migrates into the overlying water column. The development here follows that of Hammond et al.[32]

The Tappan Zee region of the Hudson Estuary is a broad, shallow channel through quaternary deposits (mean depth 5.3 m) approximately 20 miles in length. The Hudson has been classified as a partially mixed estuary. Bottom salinities are typically 20 to 40% greater than surface salinities. The sediments are fairly homogeneous in grain size, organic carbon, and cation exchange capacity. The dominant sediment type is a clayey silt with an organic content of 3 to 10%. Few tributary streams enter this region, and the groundwater flow should be quite small. The Tappan Zee region provides an ideal natural subsystem upon which to perform a radon material balance in order to study transport processes across the sediment–water interface. Hammond et al. performed such an analysis, and this is a summary of that work.

Radon is generated within the sediment by the decay of dissolved radium:

$$^{226}Ra \rightarrow ^{222}Rn + \alpha$$

Radon near the sediment–water interface escapes and creates a deficiency, as shown in Fig. 5.3-5. Assuming a constant diffusion coefficient for radon in the sediment pore water D_{A2}, a balance can be written for production, decay, and diffusion. At steady state,

$$D_{A2}\frac{d^2 c_{A2}}{dy^2} + k_A'''(c_{A2}^{**} - c_{A2}) = 0 \qquad (5.3\text{-}12)$$

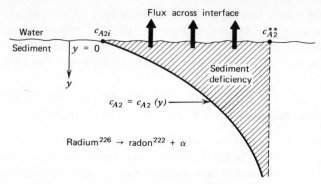

Figure 5.3-5. Radon concentration profile in sediment layer.

where c_{A2} = the pore water concentration of radon

c_{A2}^{**} = the pore water concentration of radon deep within the sediment bed where upward diffusion is insignificant

y = the distance into the sediment from the interface

k_A''' = the fraction of radon decaying per unit time ($k_A''' = 2.1E-6s^{-1}$)

The boundary conditions are

$$c_{A2} = c_{A2}^{**} \quad \text{at } y = \infty \quad \text{and} \quad c_{A2} = c_{A2i} \quad \text{at } y = 0 \quad (5.3\text{-}13a,b)$$

where c_{A2i} is the radon concentration in the overlying water. The solution to Eq. 5.3-12 with boundary conditions 5.3-13a and b is

$$c_{A2} = c_{A2}^{**} - (c_{A2}^{**} - c_{A2i}) \exp\left(-\sqrt{\frac{k_A'''}{D_{A2}}}\, y \right) \qquad (5.3\text{-}14)$$

which is the concentration profile of radon in the sediment pore water. The flux of radon into the overlying water may be obtained by differentiating Eq. 5.3-14 for the interface concentration gradient and applying Fick's first law to yield

$$N_{A0} = \sqrt{k_A''' D_s}\, (c_{A2}^{**} - c_{A2i}) \qquad (5.3\text{-}15)$$

From field measurements a radon material balance can be constructed for the Tappan Zee region. Stream inputs are minor. Input carried by methane gas bubbles escaping from the sediments is negligible. When these two terms are subtracted from the radon losses due to decay in the water column and evaporation to the atmosphere, the remaining excess must be supplied from the sediments. The summertime radon budget for the Tappan Zee region resulted in a flux from the sediments of 200 atoms/m^2·s.

Measurements of c_{A2}^{**} were made by making a slurry from a known volume of wet sediment and distilled water. This slurry was sealed in a glass kettle, purged of radon, and stored, and the regrowth of radon was measured. When radon production equaled radon decay (i.e., steady state in the kettle), a value of 0.33 dpm/cm^3 was measured. The water concentration was 0.0014 dpm/cm^3. Using k_A''', this yielded $c_{A2}^{**} = 2.62$ E3 atoms/cm^3 and $c_{A2i} = 11.1$ atoms/cm^3. Using Eq. 5.3-15, D_{A2} was found to be 2.80E-5 cm^2/s.

This value can be compared to the molecular diffusivity of radon in water, $\mathcal{D}_{A2} = 1.37$E-5 cm^2/s at 25°C. In sediments this value must be reduced for the effect of tortuosity and porosity according to Eq. 5.1-41. In a Pacific red clay it was found that τ_h (tortuosity) was 1.37 when ϵ (porosity) was 50%. Since $\epsilon \approx 0.60$ in Hudson Estuary sediment and grain size is slightly larger, $\tau_h = 1.20$ should be a good estimate. Using these values and the molecular diffusivity $D_{A2} \equiv \epsilon \mathcal{D}_{A2}/\tau_h^2 = 0.95E-5$ cm^2/s. This value is about 35% of that required to produce the observed flux.

Suspecting that some process other than molecular diffusion was at work, a two-layer model[33] was used. It was found that the observed flux could be reproduced by this two-layer model by using a 2 cm upper zone of $D_{A2} = 7.2$E-5 cm^2/s and a lower zone of 0.95E-5 cm^2/s. The upper layer diffusivity is about seven times the molecular diffusivity in the layer below and indicates that chemicals dissolved in the interstitial waters may migrate through the upper few centimeters of sediment at rates above their molecular diffusivity. It is speculated that the source of turbulence for this stirring could be small polychaete worms that populate these sediments, or it could be stirring by tidal currents.

Example 5.3-2.   A Sediment Phase Mass Transfer Coefficient

In Section 5.2 the idea of a sediment phase mass transfer coefficient was proposed. (See Eq. 5.2-29.) In the case of the Hudson Estuary it was found that the "sediments in the estuary were characterized by a surface layer of about 2 cm thickness which is lighter in color and soupier than the material which underlies it." The sediments were also suspected to be stirred by worms and currents.

Using the radon tracer model and the results given earlier, obtain a value for the sediment phase mass transfer coefficient as defined by Eq. 5.2-29.

SOLUTION

Rewriting Eq. 5.2-29 in terms of a molar flux and pore water concentration yields

$$N_{A0} = \frac{^2k'_{A3}(c_{A2} - c_{A2i})}{\mathcal{K}^*_{A23}} \qquad (5.3\text{-}2E)$$

Here c_{A2} is defined as a characteristic concentration throughtout a mixed, semifluid layer of depth h. Choose $h = 2$ cm, based on the above observations of the Hudson Estuary and $D_{A2} = 7.2\text{E}{-5}$ cm^2/s. With $y = 1$ cm, Eq. 5.3-14 can be used to calculate the characteristic concentration c_{A2} representing h:

$$c_{A2} = 2.62\text{E}3 - (2.62\text{E}3 - 11.1)\exp\left[-\sqrt{\frac{2.1\text{E}{-6}}{s}\left|\frac{s}{7.2\text{E}{-5}\ \text{cm}^2}\right|}\ 1\ \text{cm}\right]$$

$$c_{A2} = 4.21\text{E}\,2 \quad \text{atoms/cm}^3$$

Using $N_{A0} = 200$ atoms/m$^2\cdot$s, Eq. 5.3-2E can be solved for $^2k'_{A3}$ assuming $\mathcal{K}^*_{A23} = 1.0$:

$$^2k'_{A3} = \frac{N_{A0}}{(c_{A2} - c_{A2i})} = \frac{200\ \text{atoms}}{\text{m}^2\cdot\text{s}}\left|\frac{\text{m}^3}{(4.21\text{E}2 - 11.1)\text{E}6\ \text{atoms}}\right|\frac{3600\text{s}}{\text{hr}}$$

$$^2k'_{A3} = 0.0018 \quad \text{m/hr (15.4 m/yr)}$$

Problems

5.3A. FLUX RATE OF SILICA THROUGH THE DIFFUSIVE SUBLAYER

1. Assuming that the diffusive sublayer provides all of the resistance to the movement of silica from the sediment–seawater interface, calculate the flux rate of silica in μ mol/cm$^2\cdot$yr for $\rho_{A2i} = 150$ μmol/L, $\rho_{A2} = 0.02$ mg/L, $\mathcal{D}_{A2} = 5\text{E}{-6}$ cm^2/s, and $v_* = 0.40$ cm/s.
2. Estimate the sea bottom liquid phase mass transfer coefficient (m/yr).

5.3B. SOLUTION OF THE DIFFUSION EQUATION IN THE BENTHIC BOUNDARY LAYER

Solve the diffusion equation (Eq. 5.3-1) for the concentration profile of chemical A in the benthic layer. Use Eq. 5.3-4 for the variation of $\mathcal{D}_{A2}$ with y.

1. Obtain a relationship for $c_{A2}(y)$ in the diffusion sublayer. Assume that the flux rate N_{A0}, the concentration at the sediment–water interface c_{A2i}, and v_* are known.

2. Extend the relationship for the concentration profile $c_{A2}(y)$ into the logarithmic layer.

ANSWER

1.

$$c_{A2}(y) = c_{A2i} - \frac{N_{A0}y}{\mathcal{D}_{A2}}, \qquad \delta_{A2} \geqslant y \geqslant 0 \qquad (5.3B\text{-}1)$$

2.

$$c_{A2}(y) = c_{A2}|_{\delta_{A2}} - \frac{N_{A0}}{\kappa_1 v_*} \ln\left[\frac{y + y_0}{\delta_{A2} + y_0}\right], \qquad y > \delta_{A2} \qquad (5.3B\text{-}2)$$

5.3C CONCENTRATION PROFILE OF SILICA IN THE BENTHIC BOUNDARY LAYER

Calculate the concentration of dissolved silica at various heights above the sediment–seawater interface. Assume a silica flux rate of 15 $\mu mol/cm^2 \cdot yr$ and an interface concentration of $c_{A2i} = 10.0$ $\mu mol/L$, $v_* = 0.3$ cm/s, $\mathcal{D}_{A2} = 5E{-}6$ cm^2/s, and $y_0 = 0$. Use the results of Problem 5.3B in your calculations. Obtain silica (SiO_2) concentration ($\mu mol/L$) and make a graph of the concentration for each of the following distances from the sediment–seawater interface: 0.1, 0.2, 0.5, 1.0, 10.0, and 100 cm.

5.3D. RADON PROFILE IN ESTUARY SEDIMENTS

Radon is generated within sediments from the decay of radium. Once generated it either decays or diffuses through the sediments and eventually out through the sediment–water interface. A steady-state material balance is represented by Eq. 5.3-12 with boundary conditions given by Eq. 5.3-13a and b.

1. Solve the differential equation and show the steps to obtaining Eq. 5.3-14.

2. Obtain the flux rate equation of radon across the sediment–water interface (Eq. 5.3-15).

3. Generate and graph the concentration profile of radon in the Hudson

Estuary. Use $c_{A2}^{**} = 2.62E+3$ atoms/cm^3, $c_{A2i} = 1.11E+1$ atoms/cm^3, and $D_{A2} = 2.80E-5$ cm^2/s. Calculate the concentration c_{A2} at each centimeter for a depth of 10 cm.

REFERENCES

1. L. J. Thibodeaux, "Spill of Soluble, High Density, Immiscible Chemicals in Water", U.S. Dept. of Transportation, Report CG-UOA-77-004, Washington D.C., 1978.

2. L. J. Thibodeaux, "Mechanisms and Idealized Dissolution Modes for High Density ($\rho > 1$), Immiscible Chemicals Spilled in Flowing Aqueous Environments," *Am. Inst. Chem. Eng. J.*, **23** (4), 544 (1977).

3. S. Hu and R. C. Kintner, "The Fall of Single Liquid Drops through Water," *Am. Inst. Chem. Eng. J.*, **1** (1), 42 (1955).

4. I. Langmuir, "Oil Lenses on Water and the Nature of Monomolecular Expanded Films," *J. Chem. Phys*, **1**, 756 (1933).

5. M. S. Yalin, *Mechanics of Sediment Transport*, Pergamon, New York, 1972, p. 204.

6. D. E. Simons, E. V. Richardson, and C. F. Nordin, Jr., Report CER 64 DES-EVR-CFN-15, Colorado State University, Ft. Collins, 1964.

7. H. Kramers and P. J. Kreyger, "Mass Transfer Between a Flat Surface and a Falling Liquid Film," *Chem. Eng. Sci.*, **6**, 42 (1956).

8. L-K. Chang, "On-Bottom Aspects of Immiscible, Heavy, Hazardous Chemicals Spilled into Flowing Aqueous Environments", unpublished master's thesis, University of Arkansas, Fayetteville, 1978.

9. V. Novotny, "Boundary Layer Effects on the Course of the Self-Purifcation of Small Streams," in S. H. Jenkins, Ed., *Adv. Water Pollut. Res.*, Pergamon, New York, 1969, p. 39–50.

10. T. Fujii and H. Imura, "Natural Convection Heat Transfer from a Plate at Arbitrary Inclinations," *Int. J. Heat Mass Transfer*, **15**, 755 (1972).

11. C. Chang, G. C. Vliet, and A. Saberian, "Natural Convection Mass Transfer at Salt-Brine Interfaces," Paper 76-HT-33, ASME-AIChE Heat Transfer Conference, St. Louis, Mo, August 1976.

12. M. Patterson and K. Woody, unpublished chemical engineering senior project report, University of Arkansas, Fayetteville, 1978.

13. J. M. Smith, *Chemical Engineering Kinetics*, 2nd ed., McGraw-Hill, New York, 1970, p. 414.

14. E. J. Greskovich, J. M. Pommersheim, and R. C. Kenner, Jr., "Determination of Hindered Diffusivities for Nonadsorbing Pollutants in Mud," *Am. Inst. Chem. Eng. J.*, **21** (5), 1022 (1975).

15. G. W. Daigre, Dow Chemical Company, provided data on the chloroform spill and water sample analysis results, private communication, 1976.

16. J. C. Willmann, "PCB Transformer Spill, Seattle, Washington," *J. Haz. Mater.*, **1**, 361–372 (1975–1977).

17. L. J. Carter, "News and Comment... Chemical Plants Leaves Unexpected Legacy for Two Virginia Rivers," *Science*, **198**, 1015–1020 (December 9, 1977).

18. L. J. Thibodeaux and C. K. Cheng, "A Fickian Analysis of Lake Sediment Upsurge," *Water Resour.* Bul., **12** (1) 1976.

19. P. Larsen, Private communication, U.S. Environmental Protection Agency, Pacific NW Laboratory, Corvallis, Oreg., 1974.

20. R. E. Wildung and R. L. Schmidt, "Phosphorus Release from Lake Sediments," U.S. Environmental Protection Agency, EPA-R3-73-024, Washington, D. C., 1973.

21. D. P. Larsen, H. T. Mercier, and K. W. Malueg, "Modeling Algal Growth Dynamics in Shagawa Lake, Minnesota," in E. Joe Middlebrooks, D. H. Falkenborg, and T. E. Maloney, *Modeling the Eutrophication Process*, Ann Arbor Science Publisher Inc., Ann Arbor, Mich. 1974, p. 15–31.

22. M. Wimbush, "The Physics of the Benthic Boundary Layer", in I. N. McCave, Ed., *The Benthic Boundary Layer*, Plenum, New York, 1976, p. 3–10.

23. R. G. Deissler, National Advisory Committee for Aeronautics Report 1210 (1955).

24. D. R. Schink and N. L. Guinasso, Jr., "Effects of Bioturbation on Sediment–Seawater Interaction," *Mar. Geol.*, **23**, 133–154 (1977).

25. D. R. Schink and N. L. Guinasso, Jr., "Modeling the Influence of Bioturbation and Other Processes of $CaCO_3$ Dissolution at the Sea Floor," in N. R. Andersen and A. Malahoff, Eds., *The Fate of Fossil Fuel CO_2 in the Oceans*, Plenum, New York, 1975, p. 375–399.

26. R. A. Berner, "The Benthic Boundary Layer from the Viewpoint of a Geochemist," in I. N. McCave, Ed., *The Benthic Boundary Layer*, Plenum, New York, 1976, p. 33–55.

27. N. L. Guinasso and D. R. Schink, "Quantitative Estimates of Biological Mixing Rates in Abyssal Sediments," *J. Geophysi. Res.*, **80** (21), 3032–3043 (1975).

28. R. A. Berner, "Kinetic Models for the Early Diagenesis of Nitrogen, Sulfur, Phosphorus and Silicon in Anoxic Marine Sediments," in E. D. Goldberg, Ed., *The Sea*, Wiley-Interscience, New York, 1974, p. 427–450.

29. D. C. Hurd, "Factors Affecting Solution Rate of Biogenic Opal in Seawater," *Earth Planet. Sci. Lett.*, **15**, p. 411–417 (1972).

30. G. R. Holdren, Jr., O. P. Bricker, III, and G. Matisoff, "A Model for the Control of Dissolved Manganese in the Interstitial Waters of Chesapeake Bay," in T. M. Church, Ed., *Marine Chemistry in the Coastal Environment*, Am. Chem. Soc. Symp. Ser. 18, Washington, D.C., 1975, p. 365–381.

31. H. S. Carslaw and J. C. Jaeger, *Conduction of Heat in Solids*, Oxford University Press, London, 1959, p. 43.

32. D. H. Hammond, H. J. Simpson, and G. Mathieu, "Methane and Radon-222 as Tracers for Mechanisms of Exchange Across the Sediment-Water Interface in the Hudson River Estuary," in T. M. Church, Ed., *Marine Chemistry in the Coastal Environment*, Am. Chem. Soc. Symp. Ser. 18, Washington, D.C., 1975, p. 119–132.

33. T. H. Peng, T. Takahashi, and W. S. Broecker, *J. Geophys. Res.*, **79**, 1772–1780 (1974).

CHAPTER 6

CHEMICAL
EXCHANGE RATES
BETWEEN AIR AND SOIL

This chapter deals with transfer processes occurring in the lower atmosphere, across the interface and in the uppermost layers of natural earthen solid materials. Aspects of micrometeorology important to the subject of the movement and residence of chemicals in the air layers immediately above the soil interface are presented. The subject of micrometeorology, according to Sutton,[1] is worthy of serious study because such aspects as the dispersion of pollution from a factory chimney or the draining of cold air into a valley are significant and may profoundly affect human welfare and economy. He continues:

...the physics of the lower strata of the atmosphere is both interesting and important, chiefly because of the large variation in conditions which are found in the layers of air nearest the ground. Such variations are significant, not only for meteorology, but for other sciences. The climate into which a plant first emerges is quite unlike that experienced by man and the larger animals a few feet higher up, for the layers of air within a fraction of an inch of the ground may experience both tropical heat and icy cold in the course of a single day.

The topic of thermal turbulence is presented and related to the physics of the lower air layers. The mixing length theory is then extended to develop the Thornthwaite-Holzman equations, which are useful in calculating chemical flux rates from soil surfaces. The movement of pesticides and ammonia from agricultural land provides classic examples of the movement of chemicals through the air boundary layer. Once again radon is used as a tracer to aid in quantifying movement of chemicals in the upper soil layers. The chapter ends with a presentation of heat transfer processes at the air–soil interface.

6.1. THERMAL TURBULENCE ABOVE THE AIR–SOIL INTERFACE

As wind moves over natural surfaces, the friction with the surface generates turbulence. This is called *mechanical* turbulence and was the subject of Section 3.2. Turbulence is also generated when air is heated at a surface and moves upward because of buoyancy. This is called *thermal* or *convective* turbulence. The size of the eddies produced by these two processes is different, as is shown in Fig. 6.1-1. The fluctuations from mechanical turbulence tend to be smaller and more rapid than the thermal fluctuations. A very good demonstration of these types of turbulence can be seen by watching the plume from a smokestack on a hot day. The plume seems to loop up and down. The plume is called a "looping plume" because, in addition to the small-scale mechanical turbulence that tears the plume apart and spreads it with distance, the thermal updrafts and downdrafts cause the entire plume to be transported upward or downward.

Chapter 4 was concerned with the movement of air over water. Velocity profiles over water are similar to those over land. Mechanical turbulence is present over water just as it is present over land, but thermal turbulence effects are not as great over water because the temperature gradients are more moderate. For this reason the subject of thermal turbulence is presented now. The reader is referred to Section 3.2 for a review of mechanical turbulence.

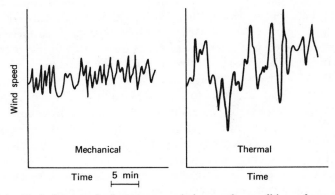

Figure 6.1-1. Typical traces of a fast-response wind sensor for conditions of pure mechanical turbulence and mechanical plus thermal turbulence. (Reprinted by permission. **Source**. Reference 2.)

The Concept of Thermal Stability

Buoyancy plays a major role in the transport and mixing of air, which, when it moves, carries with it the heat, water vapor, carbon dioxide, and other chemicals it contains. When air is stable, little mixing occurs, whereas unstable air results in turbulent conditions. Both the wind speed and temperature gradient determine the stability of air. Here we examine only the temperature gradient effect, as presented by Gifford.[3] For a detailed presentation of stability of the environment, see Eskinazi.[4]

Because of the vertical force exerted on any volume of air by gravity, the pressure of the atmosphere decreases with elevation. This vertical pressure variation implies a certain vertical structure governed by the ideal gas law. Specifically, the temperature of a volume of dry air displaced upward by a process that does not add or remove sensible heat (i.e., an adiabatic process) decreases at the linear rate of 9.66 K per kilometer, called the dry adiabatic lapse rate:

$$\Gamma_A \equiv \frac{dT_1}{dy} = \frac{g}{c_{p1}} \tag{6.1-1}$$

where c_{p1} is the heat capacity of dry air.

Under certain circumstances the lower layers of the atmosphere may possess a dry adiabatic lapse rate; if so, a small isolated air parcel that is undergoing adiabatic vertical motion due to the eddies of mechanical turbulence will at all times adjust so that it will experience no buoyancy force tending to restore it to its original level. It will always possess just the temperature of its environment. Figure 6.1-2 illustrates that the dry adiabatic lapse rate exists in the lower atmosphere for only a fraction of the total time. This figure shows that during a normal clear-day diurnal variation of temperature structure of the lower atmosphere the adiabatic state can be expected just after dawn and at dusk and will last perhaps for a few moments. The reason is that the flow of heat to and from the underlying surface by radiation, conduction, and convection causes the lapse rate in the lower air layers to vary from day to night over wide limits. During the day vertical displaced volumes of air undergoing adiabatic expansion (thermals) must be acted on by positive buoyant forces, and as a result turbulence is enhanced. During the night the converse effects tend ordinarily to suppress turbulence sharply.

The buoyant force on an air parcel is easily calculated, being equal to the weight of the displaced air volume of density ρ_1', minus the weight of the original air parcel ρ_1. Dividing by the mass of the original, the resulting

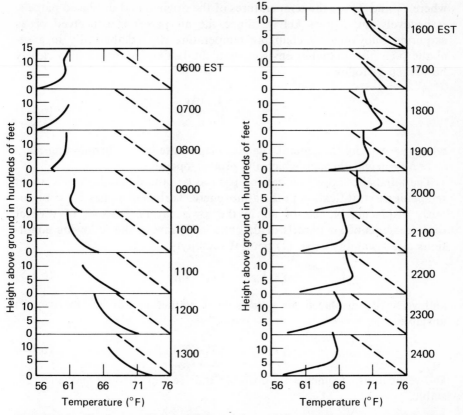

Figure 6.1-2. The average diurnal variation of the vertical temperature structure at the Oak Ridge National Laboratory during the period September to October 1950. The data were obtained from captive-balloon temperature soundings. The dashed line in each panel represents the adiabatic lapse rate. (**Source**. Reference 3.)

acceleration of the displaced parcel is

$$a = g\frac{\rho_1 - \rho_1'}{\rho_1} \tag{6.1-2}$$

From the ideal gas law equation of state, assuming the pressure of each parcel is equal, Eq. 6.1-2 can be written as follows:

$$a = g\frac{T_1 - T_1'}{T_1} \tag{6.1-3}$$

where T_1 and T_1' are the temperatures of the original and displaced parcels, respectively, in degrees kelvin. Since the air parcel is conceived of as acquiring buoyancy by changing temperature dry adiabatically in a diabatic (i.e., nonadiabatic) environment, the last equation can also clearly be written as follows:

$$a = \frac{g(\gamma - \Gamma_A)\Delta y}{T_1} \qquad (6.1\text{-}4)$$

where γ = existing (in general, diabatic) lapse rate in the surrounding air
Δy = height through which the process operates

The adiabatic lapse rate can be used as a natural standard of vertical temperature stratification in the atmosphere. Turbulent eddies are continuously displacing parcels of air in the lower layers. Because of vertical displacements of air parcels of diabatic lapse rate γ, the following conditions of air stability exist: (1) neutral stability,

$$\gamma = \Gamma_A \qquad (T_1 = T_1') \qquad (6.1\text{-}5)$$

and vertically displaced air parcels tend neither to fall nor to rise; (2) unstable,

$$\gamma < \Gamma_A \qquad (T_1 < T_1') \qquad (6.1\text{-}6)$$

and vertical displacements are unstable and are amplified by buoyancy, (3) stable,

$$\gamma > \Gamma_A \qquad (T_1 > T_1') \qquad (6.1\text{-}7)$$

and vertical displacements are strongly damped. In effect, strong heating of the air near the earth's surface causes overturning of the air layers, with resultant increases in turbulence and mixing. Conversely, strong cooling of these layers suppresses mixing and turbulence.

The Richardson Number

As demonstrated earlier, instability results from buoyancy effects, which are a function of temperature differences between air at one level and the air above. However, buoyancy effects can be strongly counteracted by the wind stream, which in generating turbulence breaks up the rising eddies. These counteracting influences, the temperature gradient and the wind speed, are related in the dimensionless gradient Richardson number Ri:

$$\text{Ri} \sim \begin{bmatrix} \text{(rate of consumption of turbulent energy by buoyancy} \\ \text{forces)} \div \text{(rate of production of turbulent energy by} \\ \text{wind shear)} \end{bmatrix}$$

$$(6.1\text{-}8)$$

The ratio is approximated as

$$\text{Ri} \equiv \frac{(g/T_1)(\partial T/\partial y)}{(\partial v_x/\partial y)^2} \tag{6.1-9}$$

where g is the acceleration due to gravity, $\partial T/\partial y$ and $\partial v_x/\partial y$ are the gradients of temperature and wind speed, and T_1 is the absolute temperature at a level usually defined by[5]

$$\bar{y} = \left(\frac{y_1^2 + y_2^2}{2} \right)^{1/2} \tag{6.1-10}$$

or

$$\bar{y} = (y_1 y_2)^{1/2} \tag{6.1-11}$$

For application to the first few meters near the ground, the Richardson number approximation above may be used. The exact Ri employs the potential temperature gradient instead of $\partial T_1/\partial y$. (See Eskinazi.[4]) Expressing the derivatives as finite differences Eq. 6.1-9 becomes

$$\text{Ri} \simeq \frac{g}{T_1} \frac{(T_{12} - T_{11})(y_2 - y_1)}{(v_{x2} - v_{x1})^2} \tag{6.1-12}$$

where T_{12} and T_{11} are temperatures at heights y_2 and y_1 and v_{x2} and v_{x1} are wind speeds at the same heights.

It can be said that when $\text{Ri} = 0$ the dynamic atmosphere is vertically neutral in the absence of buoyant forces; when $\text{Ri} < 0$ turbulence increases because the temperature difference with the adiabatic is negative and thus unstable; and when $\text{Ri} > 0$ turbulence is suppressed because $T_1' < T_1$. A comparison with the adiabatic lapse rate results given by Eq. 6.1-5, 6.1-6, and 6.1-7 indicates that Ri relates the same stability information. The Richardson number has come to be used as a characteristic turbulence parameter rather than an absolute criterion of turbulence. That is, it is regarded as broadly indicating the nature and to some extent the intensity of the turbulence rather than specifying an exact criterion for turbulence to occur.

Problems

6.1A. DIURNAL VARIATION OF THE DIABATIC LAPSE RATE

The average diurnal variation of the vertical temperature structure for a period of 18 hr is shown in Fig. 6.1-2.

1. For each time period shown, compute the existing lapse rate γ in the lower atmosphere at the soil surface (°F/1000 ft). (*Hint:* Estimate dT/dy @ $y=0$ by constructing a tangent line to the temperature structure curve, and determine the slope graphically.)
2. Using the stability criteria given in terms of the lapse rate, determine the stability for each time. Denote the stability by $n=$ neutral, $u=$ unstable, and $s=$ stable.

6.1B. THE DRY ADIABATIC LAPSE RATE

Show that Eq. 6.1-1 does yield the correct numerical value and dimensions for the dry adiabatic lapse rate.

6.2 CHEMICAL FLUX RATES THROUGH THE LOWER LAYER OF THE ATMOSPHERE

The soil surface is both a source and a sink for anthropogenic and biogenic chemicals. In this section we present the mechanisms and rates of chemical movement within the layers of the atmosphere next to the soil interface, $y \lesssim 10$ m. The aerodynamic method of flux estimation based on assumed identities or similarities in exchange coefficients as presented by Brooks and Pruitt[6] is developed.

The Thornthwaite-Holzman Equation

The aerodynamic approach of estimating the flux of chemicals toward and away from a soil or vegetative surface depends on an adequate expression for the transfer of momentum as a function of change of wind speed with height, and a knowledge of the relationship between eddy transfer mechanisms for momentum and for the chemical of interest. The Reynolds analogy, Eq. 3.3-23, is used as the basis for a relationship between the fluxes.

Under conditions of neutral atmospheric stability, wind speed can be described as a function of elevation over relatively smooth surfaces and short crops by Eq. 3.2-41:

$$v_x = \frac{v_*}{\kappa_1} \ln\left(\frac{y}{y_0}\right) \qquad \text{for} \quad y > h > y_0 \qquad (3.2\text{-}41)$$

where $v_* \equiv \sqrt{\tau_0/\rho_1}$, τ_0 is the shear stress (the flux of horizontal momentum transferred vertically and absorbed by the ground and constant in the lower layers of the atmosphere), ρ_1 is the density of air, and h is the crop height. On physical grounds for rough surfaces (e.g., tall crops) the zero plane displacement d is introduced into Eq. 3.2-41 so that it becomes

$$v_x = \frac{v_*}{\kappa_1} \ln\left(\frac{y-d}{y_0}\right), \qquad \text{for} \quad y > h > y_0 + d \qquad (6.2\text{-}1)$$

Means of obtaining d, y_0, and v_* from measured wind speed profiles under neutral or nearly neutral atmospheric stability conditions are given by Rosenberg.[5] He also gives relationships developed from large accumulations of data in the micrometeorology literature for estimating the magnitude of y_0 and d. In very short crops (lawns, for example) y_0 describes the roughness, and little adjustment of the zero plane is necessary. For tall crops y_0 is related to crop height h by

$$\log y_0 = 0.997 \log h - 0.883 \qquad (6.2\text{-}2)$$

In tall crops y_0 is no longer adequate to describe the roughness, and a value of d, the zero plane displacement, is needed. For a wide range of crops and heights, $0.02 \leqslant h \leqslant 25$ m, an equation giving the zero plane displacement has been obtained:

$$\log d = 0.979 \log h - 0.154 \qquad (6.2\text{-}3)$$

Crop height h, roughness coefficient y_0, and displacement d in the preceding equations are in meters.

The aerodynamic approach to quantifying the addition or loss of a chemical from a soil surface requires the quantification of the advection and vertical diffusion components as illustrated in Fig. 6.2-1. Masts are shown at points A, B, and C. At predetermined heights along a mast the

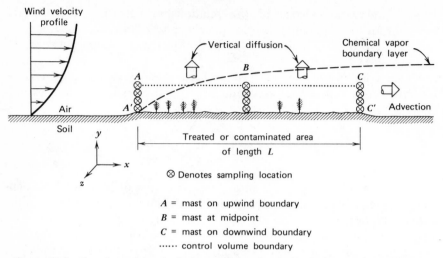

Figure 6.2-1. Aerodynamic method of quantifying chemical emission from a soil surface.

wind speed and chemical concentrations in air are measured. A quantitative assessment of chemical losses is based on a specified soil surface area and a well-defined control volume in the air space above the soil surface. Assume the soil surface area is A (m^2), is rectangular, and is length L (m) and width W (m), and the wind approach is perpendicular to one side. The control volume is then AH (m^3), where H is the mast height in meters above the soil surface.

VERTICAL DIFFUSION

Under adiabatic or near adiabatic conditions, the well-known logarithm law indicates a straight line relationship between wind velocity and the logarithm of height when adjustments are made for the influence of the crop in displacing the effective surface upward. Considering the velocity at two heights above the soil surface, v_{x1} at y_1 and v_{x2} at y_2, Eq. 6.2-1 can be employed to yield

$$\frac{v_*}{\kappa_1} = (v_{x2} - v_{x1})/\ln\left(\frac{y_2 - d}{y_1 - d}\right) \tag{6.2-4}$$

Now employing the definition of the friction velocity, $v_* \equiv \sqrt{\tau_0/\rho_1}$ Eq.

6.2-4 can be used to obtain the shearing stress τ_0:

$$\tau_0 = \frac{\rho_1 \kappa_1^2 (v_{x2} - v_{x1})^2}{\{\ln[(y_2 - d)/(y_1 - d)]\}^2} \tag{6.2-5}$$

The shearing stress is also expressed in the momentum transport equation (Eq. 3.2-33)

$$\tau_0 = -\rho_1 \nu_1^{(t)} \frac{dv_x}{dy} \tag{6.2-6}$$

where $\nu_1^{(t)}$ is the eddy viscosity or eddy transfer coefficient for momentum in L^2/t.

The equivalent transport expression for chemical A is as follows (Eq. 3.1-7 mass rate form):

$$n_{Ay} = +\mathcal{D}_{A1}^{(t)} \frac{d\rho_{A1}}{dy}\bigg|_y \tag{6.2-7}$$

where n_{Ay} = the chemical transfer flux rate in the positive y direction at y, $M/L^2 \cdot t$

$\mathcal{D}_{A1}^{(t)}$ = the turbulent diffusion coefficient for the chemical in air, L^2/t

ρ_{A1} = the concentration of A in air, M/L^3

Dividing Eq. 6.2-7 by Eq. 6.2-6 and converting the gradients to finite differences, we obtain

$$\frac{n_{Ay}}{\tau_0} = -\frac{\mathcal{D}_{A1}^{(t)}}{\rho_1 \nu_1^{(t)}} \frac{\rho_{A12} - \rho_{A11}}{v_{x2} - v_{x1}} \tag{6.2-8}$$

where ρ_{A12} and ρ_{A11} are the concentrations at y_2 and y_1, respectively. Substituting Eq. 6.2-5 for τ_0 and solving for n_{Ay} results in

$$n_{Ay} = \frac{-\rho_1 \kappa_1^2 \mathcal{D}_{A1}^{(t)}}{\nu_1^{(t)}} \frac{(v_{x2} - v_{x1})(\rho_{A12} - \rho_{A11})}{\{\ln[(y_2 - d)/(y_1 - d)]\}^2} \tag{6.2-9}$$

If under adiabatic conditions it is assumed that $\mathcal{D}_{A1}^{(t)}/\nu_1^{(t)} = 1.0$, Eq. 6.2-9 becomes the familiar mass flux equation developed by Thornthwaite-Holzman in 1939.

MODIFIED THORNTHWAITE-HOLZMAN EQUATION

As indicated in the introduction, the preceding approach is reliable only under near adiabatic conditions and then only if $\mathcal{D}_{A1}^{(t)}/\nu_1^{(t)}$ does indeed equal 1.0. When surface heating or cooling produce a nonadiabatic temperature profile, the straight line relationship of v_x versus $\ln(y-d)$ no longer holds, especially under calmer conditions. In the development of Eq. 3.2-41 and 6.2-1 for the neutral conditions the relation

$$\frac{dv_x}{dy} = \frac{v_*}{\kappa_1 y} \qquad (6.2\text{-}10)$$

was used (see Eq. 3.2-37). According to Eskinazi,[4] in the thermally unstable boundary layer Eq. 6.2-10 must be a function of the dimensionless altitude y/L_m, where

$$L_m = \frac{-\rho_1 c_{p1} v_*^{3} T_1}{\kappa_1 g q_0} \qquad (6.2\text{-}11)$$

L_m is the Monin-Obukhov length scale, which is a characteristic vertical length scale of the buoyant effects. The absolute value of L_m is seldom less than 10 m. T_1 should be the adiabatic temperature, and q_0 is the heat flux rate (sensible) from the surface. When q_0 is positive (surface warmer than air), the atmosphere is said to be unstable, and mixing is enhanced. When q_0 is negative, the atmosphere is said to be stable, and mixing is suppressed by thermal stratification.

Under unstable boundary layer conditions it is necessary to modify Eq. 6.2-10 by

$$\frac{dv_x}{dy} = \frac{v_*}{\kappa_1 y}\phi_m \qquad (6.2\text{-}12)$$

where ϕ_m is a function of y/L_m. With the proper value of ϕ_m, Eq. 6.2-12 can be used for all stability conditions. The Richardson number is closely related to y/L_m, and it has been observed that for Ri < 0, a good approximation is given by Ri $= y/L_m$. There have been many efforts to define precisely the functional relationships between Ri and ϕ_m in stable and unstable conditions. Pruitt et al.[7] investigated the functional relationships in field research. In doing so they summarized the results of some previous investigations into Fig. 6.2-2. As can be seen, the general form of the relationship is

$$\phi_m = (1 \pm b\text{Ri})^{\pm 1/n} \qquad (6.2\text{-}13)$$

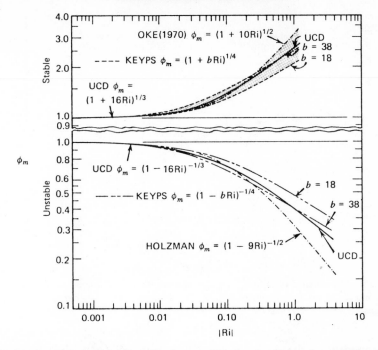

Figure 6.2-2. Functional relationship of ϕ_m on $|Ri|$, showing University of California, Davis (UCD), results and other relationships selected from the literature. KEYPS refers to a relationship developed independently by a number of researchers. (**Source**. Reference 7.)

where b is a constant determined by experiment and n is either 2, 3, or 4, depending on the bias of the individual researcher. The curves plotted agree well in both slightly stable and slightly unstable conditions ($|Ri| <$ 0.01). With increasing value of $|Ri|$ the expressions tend to deviate from one another. Even with the degree of uncertainty of ϕ_m indicated in the graph, adjustments can be made in computation of n_{Ay}:

$$n_{Ay} = \frac{-\kappa_1^2}{\phi_m^2} \frac{\mathcal{D}_{A1}^{(t)}}{\nu_1^{(t)}} \frac{(v_{x2} - v_{x1})(\rho_{A12} - \rho_{A11})}{\left\{\ln\left[(y_2 - d)/(y_1 - d)\right]\right\}^2} \qquad (6.2\text{-}14)$$

The foregoing discussion was concerned with the stability dependence of $\nu_1^{(t)}$. The assumption that the Reynolds analogy is valid must be confronted. In other words, the assumption that the turbulent diffusivities $\nu_1^{(t)}$ and $\mathcal{D}_{A1}^{(t)}$ are similar or identical in neutral or nonneutral conditions must be dealt with.

A summary of field results where $\nu_1^{(t)}$ and $\mathcal{D}_{A1}^{(t)}$ ($A \equiv H_2O$ vapor) were determined independently is given by Pruitt[7] in Fig. 6.2-3. Under near

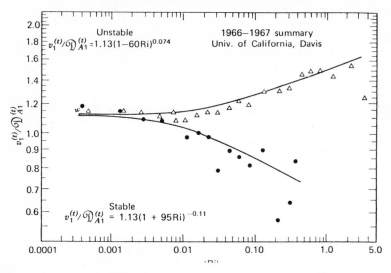

Figure 6.2-3. Dependence of $\nu_1^{(t)}/\mathcal{D}_{A1}^{(t)}$ on atmospheric stability $|Ri|$ conditions. The points represent means of many experimental data. The smooth curves are derived from other calculations. (**Source.** Reference 7.)

neutral conditions the data converge to a ratio $\nu_1^{(t)}/\mathcal{D}_{A1}^{(t)}$ slightly greater than unity. The ratio increases to 1.4 or more in strongly unstable conditions and to 0.7 or more under strong stability. The findings reported by Pruitt et al. compare with those of other researchers who have found that $\nu_1^{(t)}/\alpha_1^{(t)}$ or $\nu_1^{(t)}/\mathcal{D}_{A1}^{(t)}$ usually constant at about 1.0 to 1.3 in nearly neutral conditions, but that the ratio is about 0.7 to 0.8 in strongly stable conditions and may increase to as much as 2.0 to 2.5 in unstable conditions. It is generally agreed that $\alpha_1^{(t)}/\mathcal{D}_{A1}^{(t)}$ ($A \equiv H_2O$ vapor) is near unity and that $\mathcal{D}_{A1}^{(t)}$ ($A \equiv CO_2$) is close in magnitude to $\alpha_1^{(t)}$ and the turbulent diffusivity of water vapor. The preceding summary has been modified from that presented by Rosenberg.[5]

Air stability also affects the profile equations. Equations 3.2-41 and 6.2-1 are applicable only under conditions of neutral atmospheric stability. The diabatic profile equations must now be related to stability and other parameters as discussed earlier. The diabatic profile equations corresponding to the neutral stability profile equations are

$$v_x = \frac{v_*}{\kappa_1}\left[\ln\left(\frac{y}{\delta_v} - \frac{1}{\phi_m}\right)\right] \qquad (6.2\text{-}15)$$

for wind velocity,

$$T_1 = T_1|_{\delta_v} + \frac{q_0}{\kappa_1\rho_1 c_{p1} v_*}\left[\ln\left(\frac{y}{\delta_T} - \frac{1}{\phi_h}\right)\right] \qquad (6.2\text{-}16)$$

for temperature, and

$$\rho_{A1} = \rho_{A1}\big|_{\delta_{A1}} + \frac{n_{A0}}{\kappa_1 v_*}\left[\ln\left(\frac{y}{\delta_{A1}} - \frac{1}{\phi_h}\right)\right] \tag{6.2-17}$$

for the concentration of chemical A. The profile correction factors, ϕ_m and ϕ_h, for momentum and heat, respectively, are shown as functions of the stability parameter y/L_m in Fig. 6.2-4. See Eqs. 3.3-15, 3.3-16, and 3.3-17

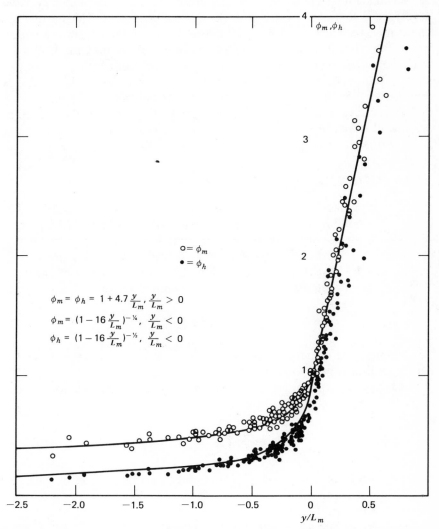

Figure 6.2-4. Dimensionless wind gradient and temperature. (Reprinted by permission. Copyright Hemisphere Publishing Corporation. **Source.** Reference 8.)

for comparison with the neutral atmosphere conditions. See Businger[8] for a review of the aerodynamics of vegetated surfaces and details on the diabatic profiles.

Because of the effect of stability on the wind profile measurements taken to permit accurate estimation of y_0, the roughness parameter, and d, the zero plane displacement, must be done during periods of near neutral stability. There are other precautions that must be taken when using the aerodynamic method of estimating vertical flux of chemicals. Adequate upwind distance (fetch) must be guaranteed so that the boundary layer developed is that equilibrated with the surface of interest. This means that there must be a long expanse of uniform cover upwind from the site at which the wind profile is measured. Parmele[9] relates that the height-to-fetch ratio requirement should be $1:40$ for minor discontinuities and at least $1:100$ or $1:200$ if the discontinuity is severe. Campbell[2]— reports that the wind can usually be assumed to be 90% or more equilibrated with the new surface to heights of 0.01 times fetch. Thus, at a distance 1000 m downwind from the edge of a uniform field of grain, we might expect our wind profile equations to be valid to heights of 10 m.

ADVECTION COMPONENT

Quantification of the vertical diffusion component, illustrated in Fig. 6.2-1, is made tractable by use of the Thornthwaite-Holzman modified equation (Eq. 6.2-14). By use of this equation, it is possible to obtain the vertical flux of chemical A in the vicinity of each mast and obtain an average, $\overline{n_{Ay}}$, representative of the A-B-C plane in Fig. 6.2-1. The mass rate of chemcial A leaving through the top of the control volume by vertical diffusion is then

$$w_{A,diff} = \overline{n_{Ay}} LW \qquad (6.2\text{-}18)$$

To complete the material balance on the control volume the advective component must be obtained.

The advection component is the mass rate of movement due to the bulk air flow. The advection flux rate is

$$n_{Ax} = \frac{1}{H} \int_{y=0}^{y=H} v_x \rho_{A1} \, dy \qquad (6.2\text{-}19)$$

where n_{Ax} = the horizontal component (x-direction) flux rate of chemical A
　　　　at point x
　　v_x = the wind profile
　　ρ_{A1} = the chemical concentration profile
　　H = mast height

Measurements of v_x and ρ_{A1} at specified heights can be used in Eq. 6.2-19 to calculate the advective component of the flux at the $A'-A$ plane and the $C'-C$ plane shown in Fig. 6.2-1. This will yield the respective fluxes $n_{A,x=A}$ and $n_{A,x=C}$. The mass rate due to advection is then

$$w_{A,adv} = (n_{A,x=C} - n_{A,x=A})HW \qquad (6.2\text{-}20)$$

where W is the width of the soil surface.

Quantifying the mass rate of chemical A leaving the air control volume through the A-B-C plane and the difference between that entering the A'-A plane and leaving the C'-C plane determines the amount desorbing from the soil surface of area WL. The total quantity is obtained from the sum of the advective and diffusive components given by Eqs. 6.2-20 and 6.2-18, respectively.

Selected Field Observations of the Movement of Chemicals between Air and Soil Surfaces

VOLATILIZATION OF DIELDRIN AND HEPTACHLOR RESIDUES FROM FIELD VEGETATION[10]

In recent years postapplication losses of pesticides by volatilization have been increasingly recognized as a pathway for general environmental contamination and as a process limiting their effectiveness. The work described here was designed to measure the volatilization of dieldrin and heptachlor, two chemically persistent insecticides, over a period of 3 weeks of warm summer weather after their application to field vegetation. A uniform grass pasture, freshly mowed to 10 cm height, was chosen.

The site was a 3.34 ha rectangular field. Between 0930 and 1030 EDT on July 12, dieldrin and heptachlor were applied together as a single uniform spray containing 5.6 kg/ha of both chemicals. The application was made with a regular farm spray rig equipped with a 21 ft spray boom mounted at about 70 cm in height. The insecticides were applied to a rectangular 2.00 ha area (82 by 244 m) within the total experimental area, leaving untreated strips 27 m wide along the north and south boundaries. These areas were left to ensure a smooth wind fetch over the boundary of the treated field without interference from fences or changes in vegetation height.

The vertical flux intensities were calculated from the concentration gradients by the aerodynamic method, using wind speed profile data obtained from anemometer masts. On each sampling date, insecticide concentrations were measured at five heights (10, 20, 30, 50, and 100 cm) above the grass surface at two locations in the treated area, one in the

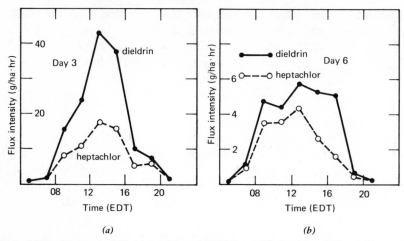

(a)   *(b)*

Figure 6.2-5. (*a*) Vertical flux intensities of dieldrin and heptachlor during 2 hr sampling periods on day 3. (*b*) Vertical flux intensities of dieldrin and heptachlor during 2 hr sampling periods on day 6. (Reprinted by permission. **Source.** Reference 10.)

center and the second on the downwind edge. Differential measurements of the air temperature were taken at 20 and 50 cm heights so that atmosphere stability corrections could be made.

Vertical flux intensities for days 3 and 6 are plotted in Fig. 6.2-5*a* and *b*, respectively. Total quantities of pesticides lost by volatilization on each sampling day are presented in Table 6.2-1. These results were obtained by

Table 6.2-1. Observed Daily Volatilization Losses of Dieldrin and Heptachlor (g/ha. day)

Day	Daily Losses by Volatilization, g/ha. day	
	Dieldrin	Heptachlor
1	654	2554
2	(325)[a]	(335)[a]
3	282	132
6	53.7	33.5
9	40.0	24.1
14	9.2	6.9
23	6.2	7.4

Reprinted by permission. *Source.* Reference 10.

[a]Estimates assuming loss between 0400 and 1200 EDT is 30% of total.

integration of the hourly flux values with the assumption that volatilization was small and could be neglected before 0600 and after 2200–2300 EDT.

Marked diurnal variations were observed in the volatilization of both insecticides during the period of greatest loss early in the experiment, the rates closely following the diurnal variation in solar radiation. Flux intensities were controlled by the rate of evaporation from plant surfaces. Dispersion by turbulent diffusion was never limiting. As overall volatilization decreased because of depletion of the residues remaining on plant surfaces, diurnal variations were less marked. Up to 1300 EDT on the first day about 40% of the dieldrin and 58% of the heptachlor applied can be accounted for as being in or having evaporated from the target area, the remainder being lost directly to the atmosphere as vapor or spray drops that never reached the target area. Estimates of the total postapplication losses were made up to the twenty-third day. The dieldrin loss was 1900 ± 250 g/ha, of which about 35% was lost on the first day and 90% in the first seven days. The heptachlor estimates for the 23 day period were 3200 ± 250 g/ha, with 75% of this on the first day and 95% in the first week.

AMMONIA FLUX INTO THE ATMOSPHERE FROM A GRAZED PASTURE[11]

Ammonia is present in the air in very small concentrations (a few micrograms per cubic meter), but its manner of transfer to the atmosphere and its subsequent dispersion are of interest in many sciences. Direct measurements of the inputs of ammonia to the atmosphere from extensive land surfaces were made. Observations of the flux density of nitrogen as ammonia and related gaseous amino compounds from a grazed pasture are reported.

The method employed is based on the energy balance at the ground surface and is much used in micrometeorology, as is the aerodynamic method. The technique relies on the conservation of energy at the ground surface. From these considerations a heat transfer coefficient can be calculated for the layer of air near the ground. A similar formulation for the transport of other atmospheric constituents can be obtained by using the Reynolds analogy.

With the technique the investigators carried out several calculations of ammonia flux over a grazed alfalfa pasture near Canberra, Australia, during March 1974. The field had an area of approximately 4 ha, and during the measurement period it was evenly grazed by 200 sheep. The surface layers of soils in the area were slightly acidic. Up to half of the nitrogen collected in the traps was in the form of other volatile basic compounds, presumably amines. The proportion of ammonia varied from day to day, and in cool, humid weather virtually pure ammonia was found.

Two representative profiles measured at about the same hour on consecutive days are shown in Fig. 6.2-6. The very large differences in concentration and gradient are due to differences in the turbulence of the air. Profile a, for instance, was measured in light winds, about 0.9 m/s at 1 m above the ground, whereas profile b was measured in strong winds, about 3 m/s at the same height.

The general character of the flux determinations is also shown in Fig. 6.2-6. Those observations were made during a period of 29 hr covering the daylight period and some of the night. A diurnal cycle in the flux is clearly evident with peak flux densities of nitrogen (in excess of 3 mg/m²·hr)

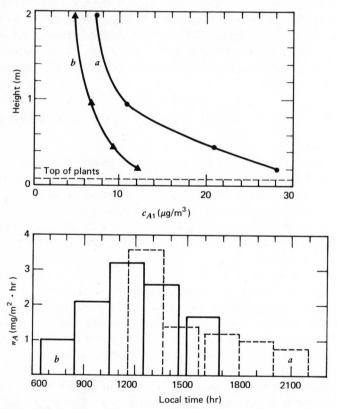

Figure 6.2-6. (*Top*) Profiles of ammonia concentration c_{A1} over an alfalfa pasture: Profile a, 1600–1800 hr on March 14, 1974; profile b, 1500–1700 hr on March 15, 1974. (*Bottom*) Flux densities of nitrogen (as ammonia and related compounds) from pasture: Profile a, March 14, 1974; profile b, March 15, 1974. (Reprinted by permission. Copyright 1977 by the American Association for the Advancement of Science. **Source.** Reference 11.)

occurring around midday and minimum values (about 0.8 mg/m^2·hr) overnight. Twenty-six calculations were made between March 7 and March 29 on seven separate days. These gave an average daily loss of nitrogen of 0.27 kg/ha.

See the problems for additional information on the movement of chemicals between air and soil surfaces.

Evaporation of Liquid Chemicals Spilled or Otherwise Placed on Land

The following discussion is confined to a process that is fairly common in the lower atmosphere, the removal of chemical vapors from free-liquid pools or permanently saturated solid surfaces. A knowledge of the rate of evaporation of spills of liquid hydrocarbons to the atmosphere is important, because not only is this a major method of mass loss from oil spills but also evaporation contributes to the "weathering" or changes in the physical properties of the oil. For example, crude oil on losing the more volatile fractions by evaporation becomes more viscous, and its flow and spreading characteristics change considerably. In spill incidents arising from accidents to road and rail tankers containing gasoline and other volatile hydrocarbon fractions, it is important to know for how long an explosion hazard exists. Evaporation rates estimates are thus invaluable in determining how to deal safely with such incidents. This problem is also of particular significance in the Arctic where spills may occur on ice or from pipelines on land and where evaporation rates may be very much reduced as a result of the low temperatures but possibly enhances by the high wind speeds.

A solution of the general problem requires a determination of the rate of evaporation of a given chemical spill as a function of temperature, wind speed, atmospheric conditions, solar radiation, ground conditions, the dimensions of the chemical spill, and the volatility and diffusion characteristics of the chemical. This estimate involves a description of the total mass transfer process from the bulk of the liquid to the atmosphere. This section is concerned mainly with the vapor phase resistance, for which a theoretical model and an empirical model are presented. A description of the liquid phase resistance may or may not be necessary, depending on conditions, and this and other aspects of the general problem are discussed only qualitatively.

Sutton[1] presents a theoretical study of evaporation by focusing on the diffusional aspects of the removal of water vapor from a free-liquid or permanently saturated soil surface. The problem is concerned with limited

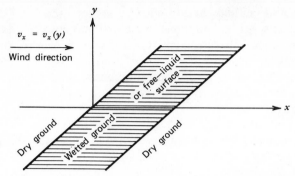

Figure 6.2-7. Two-dimensional evaporation into a steady wind (**Source**. Reference 1.)

areas such as pools of liquid and patches of saturated soil. The two-dimensional evaporation problem is illustrated in Fig. 6.2-7.

A systematic approach to the chemical evaporation problem and similar transport problems commences with the general equation for the fate of chemical A, Eq. 3-1. The expanded version of the equation for the air layers above a soil surface in mass concentration dimensions is

$$\frac{\partial \rho_{A1}}{\partial t} + \rho_{A1}\frac{\partial v_x}{\partial x} + \rho_{A1}\frac{\partial v_y}{\partial y} + \rho_{A1}\frac{\partial v_z}{\partial z} + v_x\frac{\partial \rho_{A1}}{\partial x} + v_y\frac{\partial \rho_{A1}}{\partial y} + v_z\frac{\partial \rho_{A1}}{\partial z}$$

$$= \frac{\partial}{\partial x}\left(\mathcal{D}_{A1}^{(t)}\frac{\partial \rho_{A1}}{\partial x}\right) + \frac{\partial}{\partial y}\left(\mathcal{D}_{A1}^{(t)}\frac{\partial \rho_{A1}}{\partial y}\right) + \frac{\partial}{\partial z}\left(\mathcal{D}_{A1}^{(t)}\frac{\partial \rho_{A1}}{\partial z}\right) + r_{A1}$$

$$(6.2\text{-}21)$$

The coefficient of eddy diffusivity $\mathcal{D}_{A1}^{(t)}$ is used rather than the combined eddy and molecular diffusivity because the applications to be presented are concerned with the turbulent regions of the air boundary layer. If the diffusion process is specified as a constant source and constant density problem, Eq. 6.2-21 may be simplified by use of the continuity equation, Eq. 3.2-2, to give

$$v_x\frac{\partial \rho_{A1}}{\partial x} + v_y\frac{\partial \rho_{A1}}{\partial y} + v_z\frac{\partial \rho_{A1}}{\partial z} = \frac{\partial}{\partial x}\left(\mathcal{D}_{A1}^{(t)}\frac{\partial \rho_{A1}}{\partial x}\right)$$

$$+ \frac{\partial}{\partial y}\left(\mathcal{D}_{A1}^{(t)}\frac{\partial \rho_{A1}}{\partial y}\right) + \frac{\partial}{\partial z}\left(\mathcal{D}_{A1}^{(t)}\frac{\partial \rho_{A1}}{\partial z}\right) + r_{A1}$$

$$(6.2\text{-}21a)$$

By using a single-component wind and neglecting the downwind diffusion term and the crosswind diffusion term and assuming no reaction, Eq.

6.2-21a can be further simplified to

$$v_x \frac{\partial \rho_{A1}}{\partial x} = \frac{\partial}{\partial y}\left(\mathcal{D}_{A1}^{(t)} \frac{\partial \rho_{A1}}{\partial y}\right) \tag{6.2-21b}$$

The final equation contains a horizontal advective component and a vertical diffusive component.

The boundary conditions must be specified for a solution to the two-dimensional problem. It is evident that $\rho_{A1} = \rho_{A1b}$ if $x < 0$, for all y, and that $\rho_{A1} \to \rho_{A1b}$ as $y \to \infty$ for $0 \leqslant x \leqslant L$. For substances that are normally found in the air (such as water) and those that are present in measurable trace quantities, ρ_{A1b} is the background concentration. The value of ρ_{A1} at all points on the evaporating surface is a constant concentration identified with the saturation value ρ_{A1i}, which depends only on the temperature of the surface and the nature of the evaporating liquid.

At this point a mathematical difficulty enters. In conditions of neutral stability the velocity profile is most accurately represented by a logarithmic function, Eq. 3.2-41, and $\mathcal{D}_{A1}^{(t)}$ is proportional to y, Eq. 3.2H-5, but the analytical solution of Eq. 6.2-21b for these two conditions has not yet been found. If a power law relationship for velocity is used, this aids in the solution of Eq. 6.2-21b. The power law profile is of the form

$$v_x = v_x(y_1)\left(\frac{y}{y_1}\right)^m \tag{6.2-22}$$

where $v_x(y_1)$ is the value of v_x at a fixed reference height y_1, which for convenience may be supposed to be unity. When the velocity profile is given by Eq. 6.2-22, m is a constant:

$$m = \frac{n}{2-n} \tag{6.2-23}$$

This relation enables n to be found from observations on wind structure near the ground. If the seventh-root profile is used, the value of n is $\frac{1}{4}$.

The eddy diffusivity is known to be a function of velocity and height. Sutton develops the following expression that enables the eddy viscosity to be evaluated for flow near an aerodynamically smooth surface, provided that the velocity profile can be expressed with a power law:

$$\nu(t) = \left[\frac{\left(\frac{\pi}{2}\kappa_1^2\right)^{1-n}(2-n)n^{1-n}}{(1-n)(2-n)^{2(1-n)}}\right]\nu^n v_x(y_1)^{1-n} y^{2(1-n)/(2-n)}$$

$$\times y_1^{-n(1-n)/(2-n)} \tag{6.2-24}$$

The problem of evaporation from a saturated or free-liquid plane surface, extended indefinitely across wind and of finite length downwind, has been solved by Sutton for the wind profile (Eq. 6.2-22) and the eddy diffusivity expressed by the eddy viscosity (Eq. 6.2-24). The boundary conditions adopted for a chemical of uniform concentration ρ_{A1b} in the air before it reaches the evaporating strip are

$$\lim_{y \to 0} \rho_{A1}(x,y) = \rho_{A1i}, \quad \text{for} \quad 0 \leqslant x \leqslant L \quad (6.2\text{-}25a)$$

$$\lim_{x \to 0} \rho_{A1}(x,y) = \rho_{A1b}, \quad \text{for} \quad y > 0 \quad (6.2\text{-}25b)$$

$$\lim_{y \to \infty} \rho_{A1}(x,y) = \rho_{A1b}, \quad \text{for} \quad 0 \leqslant x \leqslant L \quad (6.2\text{-}25c)$$

The solution of Eq. 6.2-21b, for $\mathcal{D}_{A1}^{(t)}$ given by expression 6.2-24 and subject to the preceding boundary conditions is

$$\rho_{A1}(x,y) = \rho_{A1i} \left\{ 1 - \frac{1}{\pi} \sin\left(\frac{2\pi}{2+n}\right) \Gamma\left(\frac{2}{2+n}\right) \right.$$

$$\left. \times \Gamma\left[\frac{v_x(y_1)^n y^{(2+n)/(2-n)}}{\left(\frac{2+n}{2-n}\right)^2 a y_1^{n/(2-n)} x}, \frac{n}{2+n} \right] \right\} + \rho_{A1b} \quad (6.2\text{-}26a)$$

where

$$a = \frac{\left(\frac{1}{2}\pi\kappa_1^2\right)^{1-n}(2-n)^{1-n}n^{1-n}\nu^n y_1^{(n^2-n)/(2-n)}}{(1-n)(2-2n)^{2-2n}} \quad (6.2\text{-}26b)$$

and $\Gamma(\theta,\rho)$ is the incomplete gamma function defined by

$$\Gamma(\theta,\rho) = \int_0^\theta x^{\rho-1} e^{-x} dx \quad (6.2\text{-}26c)$$

It is important to note that the solution gives the vapor concentration over the wetted area as a function of distance from the leading edge x and height above the surface y. The total rate of evaporation is

$$w_{A0} = \int_0^\infty v_x \rho_{A1}(L,y) \, dy \quad (6.2\text{-}27)$$

since the integral obviously represents the total mass of vapor carried across the plane $x = L$ by the wind. Employing Eqs. 6.2-22 and 6.2-26, the total rate of evaporation per unit crosswind length is

$$W_{A0} = A v_x(y_1)^{(2-n)/(2+n)} L^{2/(2+n)} \qquad (6.2\text{-}28a)$$

where A is a constant:

$$A = (\rho_{Ai} - \rho_{A1b}) \left(\frac{2+n}{2-n} \right)^{(2-n)/(2+n)} \left(\frac{2+n}{2\pi} \right) \sin \left(\frac{2\pi}{2+n} \right)$$

$$\times \Gamma \left(\frac{2}{2+n} \right) a^{2/(2+n)} y_1^{-n^2/(4-n^2)} \qquad (6.2\text{-}28b)$$

These expressions show that the rate of evaporation from a smooth surface can be calculated knowing n, $v_x(y_1)$, and the kinematic viscosity of air, ν_1.

Mackay and Matsugu[12] performed experiments on the evaporation of cumene, water, and gasoline into air. They found it convenient to follow Sutton and assume that the wind velocity profile follows a power law as shown in Eq. 6.2-22 in which v_x is the wind speed (m/hr), $v_x(y_1)$ is the wind speed at a height of $y_1 = 1$m, and y is the height (m). The exponent n is a function of ground roughness and temperature profile in the atmosphere, and typical values are from 0.25 to 1.00. For average atmospheric conditions a value of 0.25 for n was assumed reasonable. Employing Eq. 6.2-28 for obtaining an expression of the gas phase mass transfer coefficient yields

$$^3k'_{A1} = c v_x(y_1)^{(2-n)/(2+n)} L^{-n/(2+n)} \qquad (6.2\text{-}29)$$

where c is a constant. For $n = 0.25$ the equation becomes

$$^3k'_{A1} = c [v_x(10 \text{ m})]^{0.78} L^{-0.11} \qquad (6.2\text{-}30)$$

This final result provides a correlation for the evaporation mass transfer coefficient as a function of wind speed and liquid pool size.

Mackay and Matsugu employed wooden evaporation pans 4 ft by 4 ft and 4 ft by 8 ft by 0.75 in deep painted with epoxy resin. The rate of evaporation was determined for the pure liquids by measuring the rate of inflow to the pan necessary to maintain a constant level. Studies were made of the evaporation of water, cumene (isopropyl benzene), and gasoline. The air temperature varied from $+5$ to $+30°C$, and the wind speed varied from 0 mi/hr to 15 mi/hr at $y = 10$m. The value of 0.78 of the wind velocity exponent has been verified by numerous investigators (see

Sutton). The water evaporation studies suggested a value of -0.12 for the pool equivalent diameter L exponent. This is to be compared with -0.11 suggested by Sutton. From the cumene evaporation data the best value of c, the constant in Eq. 6.2-30, for cumulative mass losses over a 2-day period was 0.0150. It should be noted that this value has dimensions dependent on the units used for the mass transfer coefficient, wind speed, and pool size, here m/hr, m/hr, and m, respectively. For systems other than cumene, c will be given by $0.0150 \, (2.70/Sc)^{0.667}$ or $0.0292 \, Sc^{-0.67}$. The final equation for the gas phase mass transfer coefficient is then

$$^3k'_{A1} = 0.0292 v_x (10\,\text{m})^{0.78} L^{-0.11} Sc^{-0.67} \tag{6.2-31}$$

where $v_x(10\text{m})$ is the wind velocity measured at 10m. If the pool temperature and vapor pressure data are available, the rate of evaporation can be predicted. (See problem 6.2C.)

When a single-component liquid evaporates there is no liquid phase resistance since no concentration gradients exist. For multicomponent liquids, such as gasoline, there is a sizable resistance in the liquid phase. The liquid phase mass transfer resistance that controls the rate of transfer of material from the bulk of the liquid to the interface depends on the eddy and molecular diffusivities and the flow conditions in the liquid. The more volatile components evaporate first, and, as a result, the rate of evaporation falls as evaporation proceeds. As the low volatility fractions remain, liquid mixing rates, resulting from wind or water, become extremely low as a result of high oil viscosities. It has been observed that a type of film forms on the surface of crude oil that essentially stops evaporation of the lighter fractions. The evaporation process for a multicomponent system is obviously a very complex problem, and much work needs to be done before truly realistic quantitative models can be developed.

An evaporating pool of oil generally has a temperature close to that of the atmosphere and ground. Since vapor pressure is a strong function of temperature, accurate evaporation rates require quantifying the diurnal soil or pool surface temperature. At high rates of evaporation the oil may be considerably cooler that the air or the ground because of evaporative enthalpy losses. On the contrary, it may be considerably hotter because of the absorption of solar radiation, an effect commonly observed on black tarred road or roof surfaces on sunny days. To permit accurate predictions of evaporation rates, it is essential to quantify these effects and relate pool temperature to local conditions. The pool temperature is determined by the direct heat transfer from the air above and the ground or water below, the incident radiation, the emitted pool radiation, the rate of evaporation, the enthalpy of evaporation, and the depth and temperature history of the pool. By combining the heat flux terms with the mass of evaporating hydrocarbon, Mackay and Matsugu[12] developed a differential equation describing the temperature variation of a perfectly mixed pool with time. Good agreement between computed pool temperature and evaporation rate and the experimental pool temperature and evaporation was obtained for cumene.

Problems

6.2A PESTICIDE EVAPORATION FROM A FIELD SURFACE

Within the past decade it has become increasingly apparent that atmospheric transport of pesticides contributes substantially to their movement, both by immediate drift of sprays and dusts during application and by long-term postapplication volatilization and contaminated soil erosion. Evaporation can be estimated from field data by accurately measuring pesticide concentration profiles, along with certain micrometeorological variables in the air over a treated surface.

Figure 6.2A-1 shows an air sampling network above a soil surface where postapplication volatilization of trifluralin and heptachlor is occurring. Figure 6.2A-2 shows the field air sampling results at mast B, which is 50 m downwind from the edge of the field. The wind velocity profile above the soil is specified by the logarithm height profile with $v_* = 26$ cm/s, $y_0 = 0.1$ cm, and $d = 0$. Assume a neutral atmospheric stability exists. As part of a project of estimating the volatilization rate of heptachlor at 1 hr and 24 hr

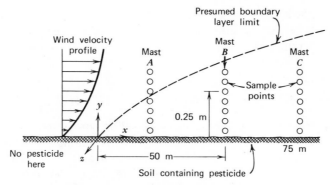

Figure 6.2A-1. Pesticide sampling network.

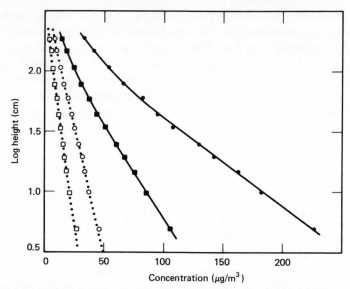

Figure 6.2A-2. Trifluralin (■) and heptachlor (●) concentration profiles in air following soil surface application; 1 hr (—) and 24 hr (···) after application. (Reprinted by permission. Copyright by the American Chemical Society. **Source.** Reference 13.)

after application, the following flux rate calculations must be made in the vicinity of mast B:

1. Compute the advection flux rate ($\mu g/m^2 \cdot s$) of heptachlor from $y = 0$ to $y = 0.25$ m in the vicinity of mast B at 1 hr and 24 hr after application.
2. Compute the vertical diffusion flux rate ($\mu g/m^2 \cdot s$) of heptachlor at a point $y = 0.25$ m in the vicinity of mast B at 1 hr and 24 hr after application.
3. Assume that the treated field area is 50 m wide. Estimate the evaporation loss of heptachlor ($\mu g/s$) from the 50 by 50 m area upwind of mast B for each time period.

6.2B. DEPOSITION VELOCITIES FOR THE RATE OF RE-MOVAL OF GASEOUS CHEMICALS ON EARTHEN SURFACES

Numerical simulations of the behavior of sulfur compounds in the atmosphere as a result of industrial processes require the use of a simple but adequate description of the rate of removal at the surface. Usually the dry surface flux is assumed to be proportional to the concentration at some convenient level near the surface; the constant of proportionality is known as the *deposition velocity* v_d. This property is known to be a function of many meteorological, chemical, and biological factors. Contemporary numerical models appear to favor the use of constant values for the deposition velocities, although there are cases, SO_2 transfer, for example, in which v_d may be as large as 2 cm/s in daytime when stomates are open and winds are strong, or as small as 0.1 cm/s at night in light winds. It is obviously important that deposition velocities be measured in a range of situations, so that means of improving on the assumptions used in numerical simulations can be developed.[14]

1. The deposition velocity v_d is equivalent to the mass transfer coefficient for the transfer of a gaseous chemical to an earthen surface. Develop an equation for ${}^3k'_{A1}$ using Eq. 6.2-14 for a bare surface at neutral stability.
2. Wind speed is usually measured at a convenient level (8 to 10 m). Compute ${}^3k'_{A1}$ values (cm/s) for a range of situations represented by $v_x@10$ m = (2.0, 6.8) m/s and $v_* = (0.16, 0.22)$ m/s.

6.2C. CHEMICAL EVAPORATION FROM AN OPEN PIT

Frequently a used organic chemical is placed in an open holding pit as a means of disposal. Open pits collect rainwater, and immiscible chemicals lighter than water will float. (See Figure 6.2C.)

1. Develop an equation to predict pool lifetime. Perform a transient material balance on a pool of pure chemical A of surface area A and depth h. Assume in the model development that evaporation occurs from the top of the pool and peels away successive layers (so to speak) of constant area A. *Answer:*

$$t = \frac{\rho_A h}{{}^4k'_{A1}\rho^*_{A1}} \tag{6.2C-1}$$

where t is lifetime, ρ_A is the density of chemical A, ${}^4k'_{A1}$ is the gas phase

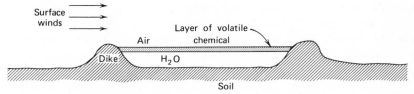

Figure 6.2C. Open pit evaporation of volatile liquids.

mass transfer coefficient, and ρ_{A1}^{*} is the air concentration of chemical A in equilibrium with the liquid pool (4≡volatile liquid phase).

2. Estimate the time (hr) required to evaporate 112,200 gal of styrene (C_8H_8) from an earthen pit 300 by 300 ft surface dimensions. Neglect the dissolution of styrene into the water, and consider only the evaporation into air. Assume this pit is located in an industrial park near Ft. Smith, Arkansas. Use typical winter day conditions of 42°F and 8.2 mi/hr wind speed.

Styrene data: Molecular weight is 104.14, density 0.903, boiling point 145°C, melting point −31°C, colorless liquid, $\mathcal{D}_{A1} \simeq 0.07$ cm^2/s, vapor pressure (mm Hg) versus temperature (°C): $(1, -7.0)$, $5,18.0)$, $(10, 30.8)$, $(20, 44.6)$, $(40, 59.8)$.

This problem was suggested by Albert Hood.

6.2D. POWER LAW VELOCITY PROFILE

Using the wind velocity data of Problem 3.3A, show that the power law formula (Eq. 6.2-22) is a reasonable expression for the wind profile near the soil surface.

1. Make a graph of $\log v_x$ versus $\log y$ (v_x in cm/s and y in cm).
2. Determine m from the graph. Compute n from Eq. 6.2-23. Note that n is normally between 0.25 and 1.0.
3. Using $y = 1.0$ m as the reference point, obtain the final expression for the power law profile.

6.2E. MEASUREMENTS OF α-PINENE FLUXES FROM A LOBLOLLY PINE FOREST[15]

In recent years air pollution regulations have focused on the need to reduce hydrocarbon emissions to reduce photochemical smog. The standard has been criticized as being unattainable since it has been estimated that vegetative hydrocarbon emissions exceed anthropogenic sources. It is desirable to quantify α-pinene emissions from a pine forest without disturbing the vegetation. Estimates of the flux of α-pinene emanating from a stand of loblolly pine trees have been obtained by measuring the net radiation and the vertical gradients of α-pinene, temperature, and water vapor above the forest, using an energy balance–Bowen ratio approach.

The area of study was a uniform 19 year old loblolly pine plantation in rural Alamance County in the central piedmont of North Carolina. Equip-

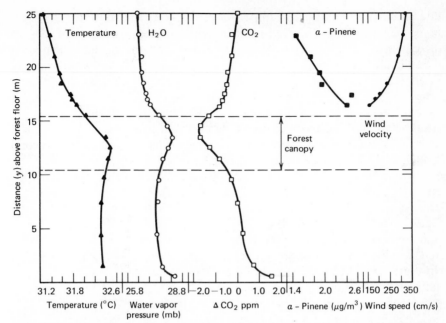

Figure 6.2E. Loblolly pine forest vertical gradients, July 18, 1977, 1035–1100 hr EST. (Reprinted by permission from book, "4th Joint Conference on Sensing of Environmental Pollutants". **Source.** Reference 15.)

ment for micrometeorological measurements included a 25 m scaffold tower that extends 10 m above the forest canopy. The measurements were made around solar noon when the atmosphere was in quasi-steady state. Air samples were pumped into 20 L FEP Teflon bags. The α-pinene analysis was made by chromatograph. Figure 6.2E contains a typical set of profiles.

Using the data shown in Fig. 6.2E, compute the emission rate of α-pinene ($\mu g/m^2 \cdot min$) for the 1035 to 1100 hour period of July 18, 1977. The emission rate as determined by the energy balance–Bowen ratio approach was 46.9 $\mu g/m^2 \cdot min$. Use the aerodynamic method for computing the flux.

6.3. CHEMICAL FLUX RATES THROUGH THE UPPER LAYER OF EARTHEN MATERIAL

This section is related to Section 5.3, which focused on the movement of chemicals within bottom sediments, in that both are concerned with interstitial fluid passages within a solid matrix as the route for chemical movement. Unlike chemicals that lie on the soil surface and desorb directly into the lower layers of the atmosphere, chemicals that originate in the soil column below the interface must find open passages within the soil in order to move upward and eventually escape to the atmosphere. After a brief consideration of the nature of the upper soil layer environment, simple transport models are presented. The first model involves the unsteady-state pore diffusion of a volatile chemical from a contaminated soil layer into the overlying air. Radon transport is studied as the second model of chemical movement within unsaturated earthen material.

Natural earth materials, whether consolidated or unconsolidated, contain varying amounts of internal space not occupied by mineral material. This space is due to the presence of individual pores, and structural features such as joints and bedding planes. Generally, this internal space is interconnected, permitting the movement of water through the material and the associated transport of chemical species dissolved or suspended in the water. The bulk of the soil and rock material near the earth's surface contains interconnected pore spaces that allow circulation of air or other gases and the transport of water and associated materials.

The individual particles composing granular material vary from irregular spheroids to flat plates. The shape and configuration of the intervening pores are dependent on the arrangement and relative size of the particles. Porosity is a measure of the total pore space contained in a given volume of material and is dependent more on the arrangement and size distribution of the particles than on their absolute size. A wide range of particle sizes tends to reduce the porosity, by filling spaces between large particles with smaller particles. In addition to this interstitial porosity, secondary structures such as joints, fractures, or soil aggregates contribute to the total porosity of a material.

The curves in Fig. 6.3-1 illustrate the particle size distribution in soils representative of three textural classes. Note the gradual change in percentage composition in relation to particle size. This figure emphasizes that there is no sharp line of demarcation in the distribution of sand, silt, and clay fractions, which also suggests a gradual change of properties with change in particle size. There is considerable difference in the total pore space of various soils, depending on conditions. Sandy surface soils show a range of from 35 to 50%, whereas medium to fine textured soils vary from 40 to 60% or even more in cases of high organic matter and marked granulation. Pore space also varies with depth; some compact subsoils drop to as low as 25 to 30%. Cultivation and cropping can appreciably reduce the total pore space. Table 6.3-1 shows the effect of cropping on

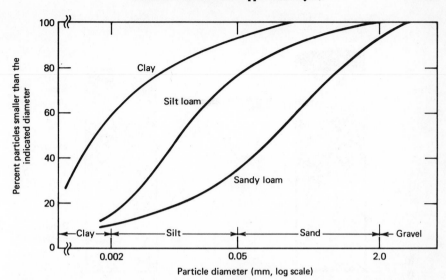

Figure 6.3-1. Particle size distribution in three soils varying widely in their textures. Note that there is a gradual transition in the particle size distribution in each of these soils. (Reprinted by permission. Copyright © 1974 by Macmillan Publishing Co., Inc. **Source**. Reference 16.)

Table 6.3-1. Effect of Continuous Cropping for at Least 40 to 50 Years on Total Pore Space and Macro- and Micropore Spaces in a Houston Black Clay from Texas

Sampling Depth (in.)	Soil Treatment	Organic Matter (%)	Pore Space		
			Total (%)	Macro (%)	Micro (%)
0–6	Virgin	5.6	58.3	32.7	25.6
	Cultivated	2.9	50.2	16.0	34.2
6–12	Virgin	4.2	56.1	27.0	29.1
	Cultivated	2.8	50.7	14.7	36.0

Reprinted by permission. Copyright © 1974 by Macmillan Publishing Co., Inc. *Source*. Reference 16.

pore space, and Fig. 6.3-2 shows typical profiles of virgin and arable soils. This reduction is usually associated with a decrease in organic matter content and a consequent lowering of granulation.[16]

Two types of individual pores in general occur in soils—macro and micro. Although there is no sharp line of demarcation, the macropores characteristically allow the ready movement of air and percolating water.

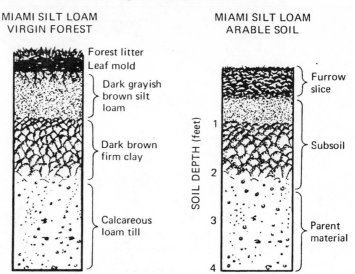

Figure 6.3-2. Generalized profile of the Miami silt loam, one of the alfisols or gray-brown podzolic soils of the eastern United States. A comparison of the profile of the virgin soil with its arable equivalent shows the changes that may occur as the land is plowed and cultivated. The surface layers are mixed by tillage. If erosion occurs, they may disappear, at least in part, and some of the *B* horizon will be included in the furrow slice. (Reprinted by permission. Copyright © 1974 by Macmillan Publishing Co., Inc. **Source.** Reference 16.)

In contrast, in the micropores air movement is greatly impeded, and water movement is restricted primarily to slow capillary movement. Thus, in a sandy soil in spite of the low total porosity, the movement of air and water is surprisingly rapid because of the dominance of the macropore spaces. Typical distributions of total pore space and macro-, and micropore spaces are given in Table 6.3-1.

The occurrence of water in granular earth materials greatly reduces the pore spaces filled with air. The pore spaces in granular material are irregularly shaped, with cusps or necks between adjacent pores, as illustrated in Fig. 6.3-3. Water placed in contact with granular materials tends to displace the air or any other gases present in the pores. The term *saturation* is used to describe a condition where all of the pore spaces are filled with water. When some portion of the pore spaces are only partially water filled, the material is termed *unsaturated* or *partially saturated*. In its natural state, the soil is normally unsaturated, and it contains both water and air. In truth, it is exceedingly difficult to accomplish saturation of any granular material, in part because of the presence of pores that are not interconnected.

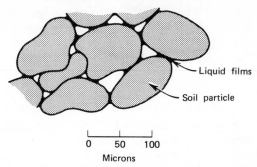

Figure 6.3-3. Soil water and internal pore structure. (Reprinted by permission. Copyright ©
1974 by Macmillan Publishing Co., Inc. **Source**. Reference 16.)

Two forces tend to hold water in contact with granular material:
adhesion between the water and the solid surface, and cohesion between
the water molecules. This situation is identical to water rising in a thin
capillary tube. If water is allowed to drain from an initially saturated
material, it will be removed first from the center of the individual pores,
leaving behind water in the cusps around the edges of the pores and a
relatively thin film of water on the surface of the grains. In Fig. 6.3-3 the
expanded view of the partially filled pores reveals various radii of curvature around the perimeter of the pores, dependent on the grain shape and
the dimensions of the various cusps. As for capillary tubes, the smaller the
radius of curvature, the more tenaciously the water is held. As greater
amounts of water are removed from the material, the water surfaces retreat
further into the cusps, and the films become thinner. Eventually these films
become discontinuous, leaving only isolated pockets of water in the
material. In some minerals (such as clays) water can be incorporated into
their crystal lattice or chemical reactions may occur that incorporate water
into the mineral structure. This "bound" water does not participate in most
intergranular flow and is not of significance to our discussion of pore
space.

Pore Diffusion Transport of a Volatile Liquid
Chemical within a Soil Column

A highly idealized pore diffusion transport model is presented to introduce
the topic of chemical movements within the extremely complex phase
called soil. This model is used as an entree into discussions of the true
complexity of water and chemical movements in unsaturated soil pores
near the surface.

It often happens that when a light liquid is spilled onto the ground, it moves quickly into the soil. Gasoline is an example of a multicomponent chemical that is frequently spilled when the lawnmower is being filled. Because of its low viscosity and interfacial tension, the mixture quickly disappears into the ground.

A well-publicized incident in this category pertains to an industrial solvent reprocessing firm in Maryland that dumped large quantities of volatile organic liquid waste into a sand and gravel quarry. Wide-scale complaints by area residents about nauseating fumes resulted in state action that banned the dumping in August 1974; however, the public health implications of this incident lingered for many years.

For simplicity, consider the spill of a pure chemical such as styrene onto a soil surface. Figure 6.3-4 shows the consequences of such a spill. The soil column becomes partially saturated with the chemical to a depth h. For a light fluid such as styrene the partial saturation process occurs fairly rapidly and is completed in a matter of minutes. Although the actual process of cleansing the soil of this contaminant is extremely complex, we nevertheless invent a simple, naturally occurring process that will decontaminate the zone.

Assume that the soil column is isothermal and of constant porosity ϵ. The soil was dry prior to the spill and therefore contains no water. The natural path for decontamination is upward movement of the chemical to the soil surface and not downward or lateral movement. The actual mechanism is desorption of molecules from a thin layer of liquid that coats each soil particle uniformly, followed by diffusion through the air-filled pore spaces to the soil–air interface, and finally movement into the overlying air phase. Within each pore where liquid is present the pure component vapor pressure is exerted. As the uppermost soil layers become

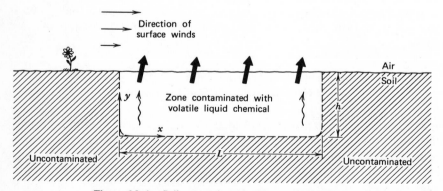

Figure 6.3-4. Soil contaminated with a volatile liquid.

depleted of chemical by the gas phase pore diffusion process concentration gradients develop within the soil column. The bottom layers remain saturated for a longer period of time; however, because of the lengthening diffusion path, the rate of chemical dissipation to the overlying air decreases. If the chemical does not partition irreversibly onto the organic matter within the soil column, then after a considerable length of time, the zone will become relatively free of the contaminant.

DIFFUSION OF A CHEMICAL FROM VAPOR-FILLED PORE SPACES

Assume that the chemical is such that the liquid vaporizes and fills the soil pore spaces with a saturated vapor of chemical A to a depth h. This situation may occur with the spill of a low boiling liquid onto a soil column during a hot summer day. For this diffusion problem the multicomponent continuity equation reduces to

$$\frac{\partial \rho_{A1}}{\partial t} = D_{A3}\frac{\partial \rho_{A1}}{\partial y^2}, \qquad 0 \leqslant y \leqslant h \qquad (6.3\text{-}1)$$

where D_{A3} is a constant diffusion coefficient that characterizes the movement of chemical A as a vapor within the porous solid and ρ_{A1} is the mass concentration of A within the pore spaces. Initially the pore spaces are saturated with the contaminant: $\rho_{A1}=\rho_{A1}^*$ at $t=0$ for all values of y. Boundary conditions: (1) at the soil–air interface, $y=h$; the pore air concentration is maintained at $\rho_{A1}=\rho_{A1i}$; and (2) at the bottom of the contaminated zone, $y=0$; the flux is assumed to be zero to give $\partial \rho_{A1}/\partial y = 0$.

This problem has been solved by Carslaw and Jaeger[17] for the analogous problem of heat conduction in a solid. For the case of linear flow of heat in a solid bounded by two parallel planes, the following solution is offered. The concentration profile of chemical A in the pore space as a function of distance from the bottom, y, and time after contaminant is in place, t, is

$$\rho_{A1} = \rho_{A1i} + (\rho_{A1}^* - \rho_{A1i})\frac{4}{\pi}\sum_{n=0}^{\infty}\frac{(-1)^n}{(2n+1)}\exp\left[-\frac{D_{A3}(2n+1)^2\pi^2 t}{h^2}\right]$$

$$\cdot \cos\frac{(2n+1)\pi y}{2h} \qquad (6.3\text{-}2)$$

This series solution converges slowly for small values of $D_{A3}t/h^2$, say $D_{A3}t/h^2 < 0.01$, and Carslaw and Jaeger present alternative series involving error functions or their integrals. Some numerical results for this problem

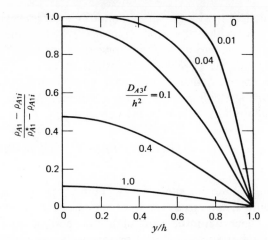

Figure 6.3-5. Concentrations in the contaminated zone. (Reprinted by permission. **Source.** Reference 17.)

are given in Fig. 6.3-5. The average concentration in the pore space of the slab (or contaminated zone) at time t is

$$\bar{\rho}_{A1} = \rho_{A1i} + (\rho_{A1}^* - \rho_{A1i}) \frac{8}{\pi^2} \sum_{n=0}^{\infty} \frac{1}{(2n+1)^2} \exp\left[-\frac{D_{A3}(2n+1)^2 \pi^2 t}{4h^2} \right]$$

(6.3-3)

The quantity of contaminant A remaining in the zone per unit area at time t is just $\epsilon h \bar{\rho}_{A1}$ where ϵ is the fraction of pore spaces per volume of soil. The fraction of A remaining is

$$F_{A1} \equiv \frac{\bar{\rho}_{A1}}{\rho_{A1}^*}$$

(6.3-4)

and Fig. 6.3-6 is a useful graphical interpretation of the solution from which lifetime and half-life can be estimated.

The flux of chemical A at the surface is

$$n_{A0} = \left\{ \frac{2D_{A3}}{h} \sum_{n=0}^{\infty} \exp\left[-\frac{D_{A3}(2n+1)^2 \pi^2 t}{4h^2} \right] \right\} (\rho_{A1}^* - \rho_{A1i}) \epsilon \quad (6.3-5)$$

This solution has been used to determine D_{A3} (and α for the case of heat transfer) of earth materials.

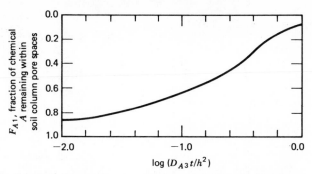

Figure 6.3-6. Fraction of contaminant remaining versus dimensionless time.

As in diffusion within any granular matrix, the effective area is reduced by the effective porosity ϵ and the tortuousness of the path expressed by the tortuosity τ_h. The effective diffusivity within the solid matrix can then be expressed by

$$D_{A3} = \frac{\mathcal{D}_{A1}\epsilon}{\tau_h} \qquad (6.3\text{-}6)$$

where $\mathcal{D}_{A1}$ is the molecular diffusivity of chemical in air. An alternative to using the ratio τ_h/ϵ is to use the hindrance factor to correct the molecular diffusivity for particle shape, tortuosity, and void fraction. (See Eq. 5.1-42 and Table 5.1-2.)

The preceding analysis does give some insight into the transient pore diffusion process that occurs below the soil surface and eventually dispels the chemical to the overlying air. A major limitation of this model is that no liquid is assumed to exist in the soil pore spaces. In all likelihood liquid does exist and vaporizes to replenish the vapor lost by diffusion. It should be noted that the time to dispel a chemical by the preceding process is independent of its vapor pressure. The only property that reflects the spilled chemical is the molecular diffusivity. It is highly unlikely that most chemicals will immediately vaporize into the pore spaces. For this reason the preceding model will only give an estimate of the minimum half-life or lifetime for a spilled volatile liquid to dissipate from a soil section. An analogous gradientless vapor pore diffusion model is presented in Problem 6.3E.

EVAPORATION AND DIFFUSION WITHIN PORE SPACES

Assume that the spilled liquid soaks into the dry soil and contaminates the soil column to a depth h, as shown in Fig. 6.3-7. The liquid coats the pore

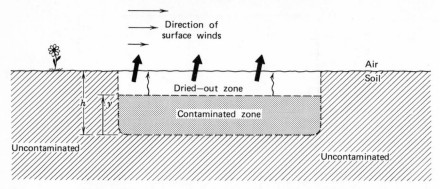

Figure 6.3-7. Evaporation and diffusion within pore spaces.

walls and particle junction sites in the same manner as water. (See Figure 6.3-3.) The chemical evaporates from the interstitial soil surfaces, and the vapor diffuses through the pores upward toward the air–soil interface. In a very short time a "dried-out" zone develops near the surface, and liquid vaporization occurs from the plane formed between this zone and the remaining contaminated zone. It is further assumed that the soil column is isothermal, that no vertical liquid movement occurs by capillary action, no adsorption on soil particles, and so on.

As vaporization occurs the "dried-out" zone increases in depth and the contaminated zone decreases. If the limiting mechanism is vapor diffusion within the pore spaces of the dried-out zone, the diffusion path length increases as the chemical dissipates. The rate of vaporization is

$$n_{A0} = \frac{D_{A3}}{h - y} (\rho_{A1}^* - \rho_{A1i}) \qquad (6.3\text{-}7)$$

where ρ_{A1}^* is the concentration of chemical A within the pore spaces at the evaporating plane y and ρ_{A1i} is the concentration at the air–soil interface. A component balance for chemical A in the contaminated zone yields the time for the dried-out zone to replace the contaminated zone completely. The time for all liquid to vaporize in a vapor pore diffusion limited process is

$$t = \frac{h m_A}{2 D_{A3} (\rho_{A1}^* - \rho_{A1i}) A} \qquad (6.3\text{-}8)$$

where A is the surface area of the contaminated region. After this time the pores still contain vapor of average concentration $\rho_{A1} = (\rho_{A1}^* + \rho_{A1i})/2$ and the previous model concerned only with vapor applies. For the evaporation–pore diffusion period the interface flux rate is

$$n_{A0} = \sqrt{\frac{D_{A3}}{2t}} \; \{\rho_{A3}(\rho_{A1}^* - \rho_{A1i})\}^{1/2} \tag{6.3-9}$$

where $\rho_{A3} \equiv m_A / hA$ is the bulk soil concentration of chemical A.

The results of this model are more pleasing to the senses. Equation 6.3-8 suggests that the time to dispel the contaminant increases with the mass spilled and decreases with chemical vapor pressure. Both these characteristics reflect the chemical. The total time for decontamination is the sum of the vaporization time (Eq. 6.3-8) and the vapor diffusion time (Fig. 6.3-6).

It was pointed out as we began to develop the pore diffusion transport model that it is a highly idealized quantitative description of a complex process. A glimpse of the complexity of the movement of a volatile chemical in the upper layers of the soil may be appreciated in a brief study of water movement in this region. The review presented here is extracted in part from Jackson, et al.[18]

Water is not free in the thermodynamic sense because of capillarity, adsorption, and electrical double layers. Capillarity is dominant in wet coarse-textured media, and adsorption assumes its greatest importance in dry media. Double-layer effects may be significant in fine-textured media exhibiting colloidal properties.

In addition to the diffusion term in Eq. 6.3-1 there should appear an advective term. This term is particularly important at moisture saturations of 40% and greater. The advective term or hydraulic conductivity is usually quantified by a form of Darcy's law:

$$v = - K \nabla \Phi \tag{6.3-10}$$

where v is the vector flow velocity of water, K is the hydraulic conductivity, and Φ is the total potential. The diffusivity employed in Eq. 6.3-1 is not a constant but a strong function of the moisture saturation. (See Fig. 6.3-8.)

The upper layers of soil are not isothermal as assumed in the pore diffusion model. During the course of a typical summer day it is possible for the soil temperature to vary from 17 to 28°C at 1 cm and 18 to 22°C at 20 cm. It is the rule rather than the exception that both heat and moisture are being transferred simultaneously in the upper soil layers. Water movement in soil is influenced by temperature and temperature gradients (*Soret effect*), and heat flow is simultaneously influenced by the movement of soil water (*Dufour effect*). The following expression has been proposed and used to calculate the soil–water flux.[18]

$$j_{A3} = -D_{A32}\overline{\nabla}\rho_{A3} - D_{A31}\overline{\nabla}\rho_{A3} - D_{A32}^T\overline{\nabla}T - D_{A31}^T\overline{\nabla}T - v_y\rho_{A3}$$

$$(6.3\text{-}11)$$

where D_{A32} is the diffusion coefficient for chemical A in soil pore spaces filled with water, D_{A31} is the diffusion coefficient of chemical A in soil pore spaces filled with air, D_{A32}^T is the thermal diffusion coefficient of chemical A in soil pore spaces filled with water, D_{A31}^T is the thermal diffusion coefficient of chemical A in soil pore spaces filled with air, and v_y is the vertical component of the hydraulic conductivity. Figure 6.3-8 contains values for each diffusivity and illustrates how each is a strong and highly nonlinear function of water content.

The soil heat flux expression is

$$q = -\lambda_A D_{A31}\overline{\nabla}\rho_{A3} - k\overline{\nabla}T \qquad (6.3\text{-}12)$$

where λ_A is the latent heat of vaporization, ρ_{A3} is the soil water content, k is the thermal conductivity, and T is temperature. The coupling of heat

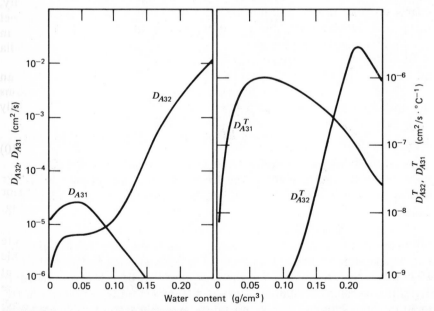

Figure 6.3-8. Mass and thermal diffusion coefficients of water in a soil column. (Reprinted by permission. Copyright Hemisphere Publishing Corporation. **Source.** Reference 18.)

and moisture transfer is imbedded in Eqs. 6.3-11 and 6.3-12.

It is likely that volatile chemicals placed in dry soils will behave very similar to water. Detailed analytical models of the transport can be constructed to aid in determining the fate of such chemicals. However, because of the complexities and interrelationships of the transport mechanisms in soils it cannot be inferred with any degree of certainty what effect water content has on chemical movement.

Radon Exhalations of Emanations
from the Ground

The earth's crust contains the radioactive nuclides uranium-238, uranium-235, and thorium-232, which by decay produce isotopes of the noble gas radon. After formation in the ground, radon diffuses into the atmosphere. The mechanism of radon exhalation is strongly influenced by the varying local conditions of the soil and the atmosphere. Therefore, it is difficult to establish quantitative relations in individual cases, but the basic processes involved are fairly well understood. Israel[19] gave a simple but useful model for these processes, which demonstrates all the essential features and parameters involved and which was expounded on by Junge.[20]

If the soil is sufficiently porous, diffusion proceeds as if the soil were replaced with small tubes filled with air. The solution of the diffusion transport equations for emanations toward the earth's surface can be expressed with an equation similar to 5.3-14:

$$c_{A1} = c_{A1}^{**} - (c_{A1}^{**} - c_{A1i}) \exp\left(-\sqrt{\frac{k_A'''}{D_{A3}}}\, y \right) \tag{6.3-13}$$

where y is depth from the surface, k_A''' is the radon decay rate, c_{A1}^{**} is the concentration of emanation in undisturbed soil air in deep layers, and c_{A1i} is the radon concentration at the air–soil interface. The exhalation rate is similar to Eq. 5.3-15 and is

$$N_{A0} = \sqrt{k_A''' D_{A3}}\ (c_{A1}^{**} - c_{A1i}) \tag{6.3-14}$$

The radon decay constant is $2.1E-6\ s^{-1}$. Israel assumed D_{A3} to be $0.05\ cm^2/s$ for radon. Values of c_{A1}^{**} can be calculated from uranium and thorium content of the soil, if one considers that only a fraction of the equilibrium production of radon escapes into the soil air prior to decay within the soil particles. Assuming this fraction to be 0.10, it is possible to

Table 6.3-2. Radon Exhalation from Soil

	c_{A1}^{**} (curies/cm^3)	N_{A0} (curies/cm$^2\cdot$s)
Calculated values	1.3E−13	4.0E−17
Observed values	0.5E−13	0.1E−17
	to 10,000E−13	to 25E−17
	average ∼3E−13	average ∼4E−17

Radon Concentrations in Air Near the Ground	
Location	c_{A1} (curies/cm^3)
Normal continental areas	70E−18 to 330E−18
Disturbed continental areas	
Innsbruck; Nauheim	400E−18 to 600E−18
Ocean	0.5E−18 to 3E−18
South America	20E−18 to 70E−18
Antarctica	0.2E−18 to 2E−18

Source. Reference 20.

calculate c_{A1}^{**} and N_{A0}. The calculated values and observed values appear in Table 6.3-2.

The large variation of radium concentrations in soil and in soil conditions is reflected in the wide range of radon concentrations in soil. In certain geologically disturbed areas the concentration of radon in soil can be extremely high, as, for example, in Nauheim. The variation around the average is much less if only normal conditions are considered. The exhalation rate N_{A0} and the radon concentration near the ground vary over a range expected from the variation of soil and meteorological conditions. The average values are in satisfactory agreement with the calculated values.

It can be expected that the exhalation rate will depend on soil conditions. Decreases of 70% are observed during rainfall. Table 6.3-2 shows that radon concentrations over land are higher than over the ocean by about two orders of magnitude. In Section 5.3 the Tappan Zee region of the Hudson River sediment displayed a radon diffusion coefficient of $D_{A2} = 2.8E-5$ cm^2/s. The ratio $\sqrt{D_{A3}/D_{A2}}$ is 42, which is two orders of magnitude. The effect of soil temperature and soil moisture is not consistent. Generally, variations in soil moisture result in large fluctuations of N_{A0} but do not much affect the average values. In the eastern Alps there is an indication of a yearly variation with a maximum in late spring, and it is suggested that this is caused by the accumulation of radon during the winter when the soil is frozen and by its escape when the soil thaws.

The preceding discussion on radon transport in the upper soil layers was excerpted from Junge.[20] The information suggests some things to be expected when studying the transport of synthetic chemicals in the soil near the air interface. The agreement between the calculated flux and the observed average flux suggests that radon transport is by molecular diffusion in air-filled soil pore spaces. The calculated flux was based on a numerical value of a diffusion coefficient very near that of the molecular diffusivity of most chemicals in air. The observed low radon concentrations in air above the oceans suggest a higher diffusion resistance in water-filled pores since the molecular diffusivity of radon in water is much lower than in air. And finally, the decrease of exhalation rates after a rainfall is likely due to a temporary saturation or partial saturation of the soil pore spaces with water.

Problems

6.3A. SIMPLIFICATION OF THE MULTICOMPONENT EQUATION OF CONTINUITY FOR THE SOIL CONTAMINANT PROBLEM

Commencing with the multicomponent equation of continuity (Eq. 6.2-21):

1. Simplify it to yield Eq. 6.3-1. State reasons for each simplification made and/or term excluded.
2. Reformulate the problem to account for transport of the chemical by groundwater movement in the vertical direction.

6.3B. CONTAMINANT LIFETIME FOR ACETALDEHYDE SPILLED ON SOIL

A quantity of acetaldehyde (CH_3CH_0) was spilled onto a dry soil. The soil column was contaminated to a depth of 35 cm. This particular soil is estimated to have a porosity of 30% and a tortuosity of 3.5.

1. Estimate the contaminant half-life (hr) for the acetaldehyde spill if the soil temperature is uniform at 25°C.
2. Estimate the time (hr) required for 90% of the acetaldehyde to dissipate into the atmosphere at 25°C.

6.3C. CONCENTRATION PROFILE AND FLUX RATE FOR A BENZENE SPILL ON SOIL

For a benzene (C_6H_6) spill perform the following calculations:

1. Compute the concentration profile of benzene vapor within the soil pore spaces 3.38 hr after the spill. Obtain values of the concentration at $y=0$ cm (bottom of the zone), $y=12$ cm, $y=25$ cm, and $y=35$ cm (the soil–air interface). The term ρ^*_{A1} can be obtained from the pure component vapor pressure of benzene at 25°C. Assume that the interface concentration can be estimated by

$$\rho_{A1i} = \frac{\rho^*_{A1}\,{}^1k'_{A3}}{{}^1k'_{A3} + {}^3k'_{A1}} \qquad (6.3C\text{-}1)$$

Use ${}^1k'_{A3} = 4.2\mathrm{E}-4$ cm/s and ${}^3k'_{A1} = 0.5$ cm/s. Report all benzene vapor concentrations in $\mu g/m^3$.

2. Compute the flux rate of benzene through the soil–air interface 3.38 hr after the spill in $\mu g/m^2 \cdot$ min.

3. Compute the average pore vapor concentration of benzene 3.38 hr after the spill in $\mu g/m^3$.

6.3D. CONTAMINANT LIFETIME FOR BENZENE SPILLED ON SOIL

A quantity (1000 kg) of benzene (C_6H_6) was spilled onto a dry soil. The soil was contaminated over an area of 2.17 m^2 and to a depth of 35 cm. This particular soil is estimated to have a porosity of 30% and a tortuosity of 3.5. Assuming the interface concentration is zero:

1. Compute the pore-liquid vaporization time (hr).
2. Estimate the time (hr) for 90% of the vapor to be dispelled from the pores.
3. Estimate the total contaminant lifetime (hr) within the soil.

6.3E. GRADIENTLESS VAPOR PORE DIFFUSION MODEL

The vapor pore diffusion model presented in Section 6.3 involved a partial differential equation (Eq. 6.3-1). The solution yielded vapor concentration gradients within the soil column (Eq. 6.3-2 and Fig. 6.3-5), average concentration (Eq. 6.3-3), fraction chemical remaining (Fig. 6.3-6), and flux rate at the air–soil interface (Eq. 6.3-5).

A gradientless vapor pore diffusion model may be developed as follows. Assuming that the length of the diffusion path is $h/2$, the flux rate from the pores is

$$n_{A0} = \frac{2D_{A3}}{h}(\rho_{A1} - \rho_{A1i})$$ (6.3E-1)

where ρ_{A1} is a uniform concentration of chemical vapor in pore space, D_{A3} is the diffusion coefficient, h is the depth of the contamination zone, and ρ_{A1i} is the interface concentration of the pore space. The gradientless concentration within the contaminated zone pore spaces is

$$\rho_{A1} = \rho_{A1i} + (\rho_{A1}^* - \rho_{A1i})\exp\left[\frac{-2D_{A3}t}{h^2}\right]$$ (6.3E-2)

The half-life is

$$t_{1/2} = \frac{0.693h^2}{2D_{A3}}$$ (6.3E-3)

and the flux rate at the surface is

$$n_{A0} = \left\{\frac{2D_{A3}}{h}\exp\left(\frac{-2D_{A3}t}{h^2}\right)\right\}(\rho_{A1}^* - \rho_{A1i})$$ (6.3E-4)

The gradientless model equations for concentration and flux rate can be compared to those for the gradient model (Eqs. 6.3-2 and 6.3-5).

1. Verify Eqs. 6.3E-2, 6.3E-3, and 6.3E-4 by first performing a component balance on chemical A to obtain a simple differential equation of ρ_{A1} in t.
2. Using Fig. 6.3-6, obtain an expression similar to Eq. 6.3E-3 for the half-life of chemical A as predicted by the gradient model. Which model will predict the longer vapor half-life?
3. Rework Problem 6.3B using the gradientless model.

6.4 HEAT TRANSFER AT THE AIR–SOIL INTERFACE

Soil surface temperatures and soil column temperatures down to a depth of 50 to 100 cm can have a pronounced effect on chemical movement from and within this region of the air–soil interface. Because of daily and seasonal patterns of atmospheric conditions, the soil temperature changes during the course of a day and during the course of a year. Soil temperature changes can be observed as waves during the course of a day. The amplitude of the temperature wave at the ground surface is great, but it diminishes with depth below the surface.

Fundamental concepts of heat transfer were presented in Section 3.3. The reader should have a firm understanding of these concepts prior to attempting to apply the results of this section to a chemodynamics problem. The first topic to be considered in this section is the basic mechanisms of heat exchange at the air–soil interface. Details for computing radiant, evaporative, and sensible heat flux rates are presented. This material is extended to show how computations of soil surface temperatures and soil column temperatures can be estimated.

The energy balance equation at the air–soil interface may be written as

$$q_{13} = q_{lw} + q_e + q_c + q_s \tag{6.4-1}$$

where q_{lw} is the net longwave radiation flux, q_e is the evaporative (water) heat flux, q_c is the sensible heat flux (conduction) between the surface and air, q_s is the shortwave solar radiation, and q_{13} is the sensible heat flux into the soil column. Here q_{13} is directly proportional to the rate at which the column is warming. The sign convention adopted is that energy additions to the soil column are positive. All energy or heat flux relations are in $J/m^2 \cdot s$ (1 Btu/ft^2·hr = 3.15 J/m^2·s, 1 J/m^2·s = 4.187 cal/m^2·s). With the adopted sign convention, if the right-hand side of Eq. 6.4-1 is positive, the soil column is heating up; if it is negative, the soil column is cooling off.

Equation 6.4-1 for the air–soil interface is similar in form and content to Eq. 4.3-2, which is for the air–water interface. In general, the mechanisms for transfer at the respective interfaces are similar, and it is unnecessary to repeat much of the material in this section. The reader is therefore referred to portions of Section 4.3 for the appropriate material.

Components of the Energy Balance Equation

SHORTWAVE RADIATION

Shortwave radiation originates directly from the sun. The material presented in Section 4.3 on shortwave radiation is directly applicable to obtaining q_s values at the air–soil interface. The reader is urged to review that material.

Although there is some reflectivity at a water surface, much of the incoming solar radiation arriving at a soil surface can be reflected. The total incoming solar radiation needs to be modified by the fraction absorbed at the surface

$$q_s = q_{s0}(1-\rho) \tag{6.4-2}$$

where ρ is the solar reflectivity and q_{s0} is the incoming radiation from Tables 4.3-1 or 4.3-2, depending on the method employed. Typical values of the solar reflectivity of interest to this topic are given in Table 6.4-1. The absorption of opaque surfaces such as soil is related to the reflectivity by

$$a = (1-\rho) \tag{6.4-3}$$

where a is the absorptivity.

Table 6.4-1. Solar Reflectivity (ρ) of Natural Surfaces

Surface	Reflectivity, ρ
Water, summer, mid-lot	0.06
Desert, loamy surface	0.29–0.31
Sand, yellow	0.35
Plowed field, moist	0.14
Newly plowed field	0.17
Grass, high dense	0.18–0.20
Grass, green	0.26
Grass, dried	0.19
Alfalfa, lettuce, beets, potatoes	0.18–0.32

Source. Reference 21.

LONGWAVE RADIATION

The longwave sky radiation emitted downward by the overlying atmospheric constituents (primarily water vapor, carbon dioxide, ozone, and clouds) and that emitted by the soil surface is quantified by an equation similar to Eq. 4.3-4:

$$q_{lw} = a\sigma T_1^4 - e_3\sigma T_3^4 \tag{6.4-4}$$

where e_3 is the surface infrared emissivity and T_3 is the absolute temperature (K) of the soil surface. Most natural surfaces have an infrared emissivity lying between 0.90 and 0.96. (See Section 4.3 for further information on longwave radiation.)

SENSIBLE HEAT EXCHANGE

The sensible heat exchange term accounts for energy exchange due to atmospheric convection of heat to or from the surface. The sensible heat flux is directly proportional to the temperature difference between the surface and the air. Thus,

$$q_c = \rho_1 \bar{c}_{p1} h'_{13}(T_1 - T_3) \tag{6.4-5}$$

where ρ_1 is air density, $\bar{c}_{p1}$ is the specific heat of air at constant pressure (0.057 J/g·K), h'_{13} is the nebulous heat transfer coefficient at the air–soil interface (cm/s), T_1 is air temperature, and T_3 is soil temperature.

The heat transfer coefficient h'_{13} is most difficult of all to approximate. Typically it is assumed that h' is a "general transfer coefficient" for all atmospheric properties found from[22]

$$h'_{13} = h' \equiv \kappa_1^2 v_x \left[\ln\left(\frac{y-d}{y_0}\right) \right]^{-2} \tag{6.4-6}$$

where v_x is the horizontal wind velocity at the height y, d is the zero plane displacement, and y_0 is the roughness height. The reader should recognize the origin of this "general transfer coefficient" from Section 6.2. (See Eq. 6.2-5.)

When the air temperature does not vary greatly with height, as is often the case near sunrise or sunset or under cloudy skies, the transfer coefficient may be estimated from Eq. 6.4-6. When the temperature varies with height different from the adiabatic lapse rate, h'_{13} does not equal h', and Eq. 6.4-6 is no longer valid. The proper relationship to use in these situations is not known. It is known, however, that h'_{13} may be several times larger than h' under very unstable conditions (rapid temperature decrease with height and light winds) and only a fraction of h' under stable conditions.

On the basis of observations made by Kreith[22] over dry bare soil in the Tucson area, a relationship of the form

$$h'_{13} = h' \left(1 + 14 \frac{\Delta T}{v_x^2} \right)^{1/3} \tag{6.4-7}$$

where v_x is the wind speed (m/s) at a height of about 2 m, is not inappropriate, at least under unstable conditions when ΔT in K is positive. Under stable conditions the expression

$$h'_{13} = h'\left(1 - 14\frac{\Delta T}{v_x^2}\right)^{-1/3} \qquad (6.4\text{-}8)$$

seems to fit the data better. Appropriate ϕ_m^2 corrections factors to h' appearing in Fig. 6.2-2 may be substituted for those in Eqs. 6.4-7 and 6.4-8.

LATENT HEAT FLUX

The following presentation of the latent heat flux and the sensible heat flux into the soil column has been adapted from a similar development by Kreith and Sellers[22].

As suggested in Section 3.3 for the case of water evaporation, energy movement is related to water movement. Equation 3.3-5 relates the two and can be put in the form

$$q_e = \lambda_A{}^3 k'_{A1}(\rho_{A1} - \rho_{A13}) \qquad (6.4\text{-}9)$$

By employing the approximation that concentration is proportional to pressure:

$$\rho_{A1} \simeq \frac{M_A p_{A1}}{M_B p_{\text{tot}}}\rho_1 = \frac{0.622 p_{A1}\rho_{A1}}{p_{\text{tot}}}\rho_1 \qquad (6.4\text{-}10)$$

where M_A is the molecular weight of water, M_B is the molecular weight of air, p_{A1} is the partial pressure of water vapor, and p_{tot} is the total pressure. Equation 6.4-9 can be transformed to

$$q_e = \frac{-0.622\lambda_A{}^3 k'_{A1}(p_{A13} - p_{A1})}{p_{\text{tot}}}\rho_1 \qquad (6.4\text{-}11)$$

Equation 4.3-6 may be used to obtain p_{A1} in inches of mercury. If the surface of the soil is wet enough so that the vapor pressure is equal to the saturation value at the temperature of the soil T_3, then the vapor pressure difference may be expressed as a function of temperature difference by using the Clapeyron equation to yield

$$p_{A13} - p_{A1} = \frac{\lambda_A p_{A1}^*}{RT_3^2}(T_3 - T_1) + p_{A1}^* - p_{A1} \qquad (6.4\text{-}12)$$

where p_{A1}^* is the saturation vapor pressure at the air temperature and R is the universal gas content. Equation 6.4-11 can then be transformed to give the potential evapotranspiration, q_{e0}:

$$q_{e0} = -0.622\frac{\lambda_A^2 p_{A1}^* \,{}^3k_{A1}'}{RT_3^2 p_{\text{tot}}}(T_3 - T_1)\rho_1$$

$$-\frac{0.622\lambda_A \,{}^3k_{A1}'(p_{A1}^* - p_{A1})}{p_{\text{tot}}}\rho_1 \qquad (6.4\text{-}13)$$

Equation 6.4-13 defines what is called potential evapotranspiration, that is, the evapotranspiration that occurs in the presence of a nonlimiting water supply. If the soil column is partially dry so that moisture is limiting, it is usually assumed with considerable accuracy that the actual evapotranspiration rate is proportional to the potential rate. The factor of proportionality may be defined as the relative soil moisture content m. It depends on the amount of moisture in the active soil layer of depth h, ρ_{A3}, and on the moisture-holding capacity of the soil, ρ_{A3}^0. Sellers recommends that

$$m = 1.0 \qquad \text{when} \quad \rho_{A3} \geqslant 0.75\rho_{A3}^0$$

and

$$m = \frac{\rho_{A3}}{0.75\rho_A^0} \qquad \text{when} \quad \rho_{A3} < 0.75\rho_{A3}^0 \qquad (6.4\text{-}14)$$

The depth of the active soil layer coincides roughly with the root zone for vegetation-covered surfaces. For bare soil h depends on soil type and ranges from 1 to 5 cm.

Forms similar to Eqs. 6.4-7 and 6.4-8 can be used for ${}^3k_{A1}'$ in Eqs. 6.4-9, 6.4-11, and 6.4-13 if the value of the constant 14 in these equations is reduced, on the basis of theoretical considerations, to 10.5.

Sensible Heat Flux into a Soil Column

There are a number of ways of estimating the sensible heat flux into a soil column. Kreith and Sellers[22] present a simple time-dependent model of the evapotranspiration process applied to a special case of a small field of short grass with variable rooting in an arid environment. The method used is derived from the standard theory of heat transfer in a homogeneous medium. Assuming, as a first approximation, a sinusoidal variation of the surface temperature, the following result is obtained:

$$q_{13} = \frac{\sqrt{2}}{2} \left(\omega \bar{c}_{p3} k_3 \right)^{1/2} \left(1 + \frac{2}{\omega \Delta t} \right) (T_3 - T_1) - \frac{\sqrt{2}}{2} \left(\omega c_{p3} k_3 \right)^{1/2}$$

$$\left[\left(\bar{T}_{S1} - T_1 \right) + \frac{2}{\omega \Delta t} \left(T_{S1} - T_1 \right) \right] \qquad (6.4\text{-}15)$$

where ω is the frequency of the oscillation ($\pi/12$ hr^{-1}), $\bar{c}_{p3}$ and k_3 are the heat capacity and thermal conductivity, respectively, of the soil, Δt is the model time interval involved (typically 1 hr), $\bar{T}_S$ is the mean 24 hour soil surface temperature, T_{S1} is the soil surface temperature at the start of a given hour, and T_1 is the air temperature. Table 6.4-2 gives some thermal properties of soil materials.

This model is based on the energy and water budgets of the particular site. Outputs of the model are the soil surface temperature T_3 and the water evaporation rate n_{A0} in g/cm$^2\cdot$d. As with most environmental

Table 6.4-2. Density and Thermal Properties of Some Soil and Reference Materials[a]

Material	Density, ρ (g/cm^3)	Specific Heat, c_p (cal/g$\cdot$K)	Thermal Conductivity, k (cal/cm$\cdot$K$\cdot$s)	Thermal Diffusivity, α (cm^2/s)
Clay	1.8	0.8	2.88×10^{-3}	2×10^{-3}
Light soil with roots	0.3	0.3	2.70×10^{-4}	3×10^{-3}
Wet sandy soil	1.6	0.4	6.40×10^{-3}	1×10^{-2}
Silver metal	10.5	0.06	1.08	1.72
Dead air	1.3×10^{-3}	0.24	4.99×10^{-5}	1.6×10^{-1}

Reprinted by permission. *Source.* Reference 5.

[a] 1 J = 4.1868 cal.

models, a number of basic meteorological facts must be known for each hour of the day. Included are the incoming solar radiation, the cloud cover, the surface albedo, the air temperature, the relative humidity, and the wind speed at a height of 1 or 2 m above the surface. Given this information and the composition of the surface cover and soil, the model yields hourly values of the net radiation, the energy used in evaporation, the sensible heat transfer to the air, the sensible heat transfer to the soil, and the surface temperature. The authors present details of the working equations, a procedure of computation, and the input and output for a specific computation.

PENETRATION OF HEAT INTO THE GROUND

The thermal diffusivity of soil is small. It is considerably less than that for air at rest as shown in Table 6.4-2. Thermal diffusivity of a soil is a parabolic function of increasing moisture content. A small amount of water reduces the insulating effect of the pore space filled with air, but further increases in water content markedly increase the heat capacity. Soil organic matter lowers thermal diffusivity because of its influence in increasing porosity. Compaction increases the thermal diffusivity by decreasing the volume of the insulating pore space.

The computation of daily and seasonal patterns of soil temperature below the surface is beyond the scope of this book; however, a brief overview of the nature and magnitudes of variation is in order. Soil temperature changes can be observed as waves during the course of a day. The amplitude of the temperature wave at the soil surface is great, but diminishes with depth below the surface. The pattern of soil temperature profiles changes rapidly during a normal day, as illustrated in Fig. 6.4-1 for measurements made at Argonne National Laboratory. The pattern of decreasing amplitude of the soil column temperature wave between summer and winter is well illustrated in Fig. 6.4-2. In general, soil temperature decreases with depth during the daytime in summer. Temperature gradients direct heat into the soil. At night, however, the temperature is highest between 20 and 40 cm; from that level heat is directed both upward and downward. In winter the daily amplitude of the surface temperature is very small. The 81 cm level is warmest, and heat is transferred upward from this level throughout the day and night.

By employing Eq. 6.4-1 it is possible to estimate the range of temperatures the soil surface witnesses during the course of a day. A somewhat crude method of estimating the range of soil surface temperature may be

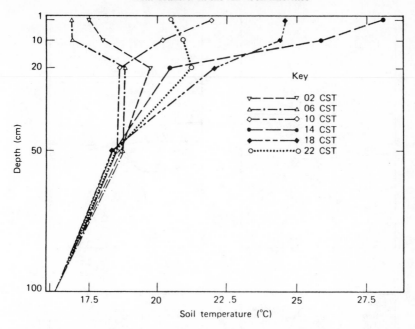

Figure 6.4-1. Vertical temperature profiles in soil during the course of a typical summer day at Argonne, Ill., July 27, 1955. (Reprinted by permission. **Source.** Reference 5.)

developed by assuming that the heat transfer to the soil column is zero. (See Problem 6.4A.) Once the range of soil surface temperature, ΔT_S, has been estimated, it is possible to also estimate the range of subsurface temperature variation. Rosenberg[5] suggested that the range of temperature at any depth in the soil can be obtained by

$$\Delta T_3(y) = \Delta T_{3i} \exp\left[-y\left(\frac{\omega}{\alpha_3}\right)^{1/2} \right] \tag{6.4-16}$$

where $\Delta T_3(y)$ is the temperature range at depth y, ΔT_{3i} is the temperature range at the surface, α_3 is the thermal diffusivity of the soil material, and ω is the period of oscillation ($\pi/12$ hr^{-1}). Equation 6.4-16 assumes that the soil properties, including porosity, water content, and organic matter, are uniform with depth so that a constant value of the thermal diffusivity may be used.

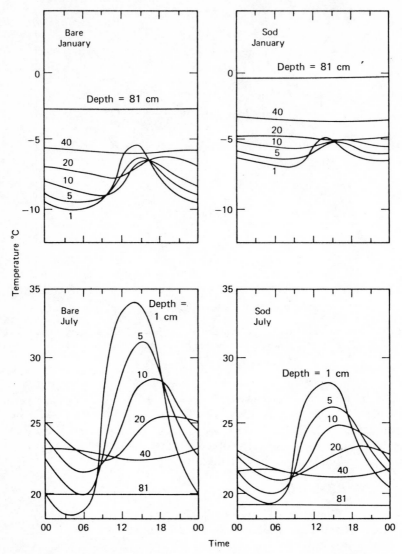

Figure 6.4-2. Average hourly soil temperature under bare and sod-covered soil at St. Paul, Minnesota, in January (*top*) and July (*bottom*) 1961. Soil depth is shown in cm. (Reprinted by permission. **Source.** Reference 5.)

Closure

Soil surface and subsurface temperature is an important parameter with respect to chemodynamics in the soil column near the air interface. Although a general treatment of heat and moisture transfer in soil is beyond the scope of this book, the preceding material does provide a qualitative description of the major process occurring and presents simple relationships to obtain order-of-magnitude estimates of temperature ranges. Additional information on applications of the material is presented in the context of problems.

Problems

6.4A. ESTIMATING SOIL SURFACE TEMPERATURE

It is possible to obtain an estimate of a soil surface temperature by assuming $q_{13} = 0$ in Eq. 6.4-1 and choosing T_3 so that the equality is satisfied. Note that the $q_{13} = 0$ condition implies that the soil is a perfect insulator and that all energy arriving at the interface from the atmosphere is returned to the atmosphere.

It is necessary to estimate the mid-day soil surface temperature of a newly plowed field near Little Rock, Ark. The soil moisture is at 50% of its holding capacity, and the air can be assumed to be in a neutral stability condition.

1. Estimate the temperature (°C) for the month of February.
2. Estimate the temperature (°C) for the month of August.

6.4B. PENETRATION OF HEAT INTO THE GROUND

The range of subsurface temperature variation can be estimated by Eq. 6.4-16 provided the surface temperature variation and the soil thermal diffusivity are known. In general, the subsurface temperature oscillates around some seasonal, slowly changing temperature, $T_3(\infty)$, that is somewhat constant beyond a certain depth. The deviation of the maximum subsurface temperature, $T_{3,MX}(y)$, is related to $T_3(\infty)$ and $T_{3,MX}(0)$ by

$$T_{3,MX}(y) = \exp\left[-y\left[\frac{\omega}{\alpha_3}\right]^{1/2}\right]\left[T_{3,MX}(0) - T_3(\infty)\right] + T_3(\infty)$$

$$(6.4B\text{-}1)$$

The deviation of the minimum subsurface temperature, $T_{3,MN}(y)$, is obtained from

$$T_{3,MN}(y) = T_{3,MX}(y) - \exp\left[-y\left(\frac{\omega}{\alpha_3}\right)^{1/2}\right] \cdot \left[T_{3,MX}(0) - T_{3,MN}(0)\right]$$

(6.4B-2)

From the preceding equations it is possible to calculate the subsurface temperature variations with depth.

1. For a light soil with roots known to have a surface temperature of $T_{3,MX}(0) = 30°C$ and $T_{3,MN}(0) = 17°C$, compute the subsurface temperature variations for $T_3(\infty) = 19°C$. Use $y = 0$, 1, 3, 5, 9, 15, 25, and 30 cm in your calculations. Construct a graphical representation similar to Fig. 6.4-1.

2. Repeat part 1 for $T_3(\infty) = 16°C$.

3. Based on the data shown in Fig. 6.4-1, estimate the thermal diffusivity (cm^2/s) of the Argonne, Ill., soil.

6.4C. SUBSURFACE HEAT FLOW

1. Show that Eq. 3.3-1 may be integrated to obtain

$$q_y = \frac{k}{h}(T_{si} - T_s)$$

(6.4C-1)

where T_{si} is the soil surface temperature and T_s is the subsurface temperature at depth h. Assume that heat flow is constant and thermal conductivity k is independent of depth.

2. Calculate the heat flux rate for Argonne, Ill., soil temperature profile shown in Fig. 6.4-1. Use the data between $y = 1$ cm and $y = 20$ cm (i.e., $h = 19$ cm) and calculate the flux rate $(J/cm^2 \cdot s)$ and the direction of heat transfer for each time shown. Assume $\alpha_3 = 4.4E-2$ cm^2/s, $\rho_3 = 1.0$ g/cm^3, and $\bar{c}_{p3} = 0.5$ $cal/g \cdot °C$.

REFERENCES

1. O. G. Sutton, *Micrometeorology*, McGraw-Hill, New York, 1953.

2. G. S. Campbell, *An Introduction to Environmental Biophysics*, Springer-Verlag, New York, 1977, p. 34.

3. F. A. Gifford, Jr., "An Outline of Theories of Diffusion in the Lower Layers of the Atmosphere," in D. H. Slade, Ed., *Meterology and Atomic Energy*, U.S. Atomic Energy Commission, Oak Ridge, Tennessee 1968, p. 66–116.

4. S. Eskinazi, *Fluid Mechanics and Thermodynamics of our Environment*, Academic, New York, 1975, p. 83–129.

5. N. J. Rosenberg, *Microclimate: The Biological Environment*, Wiley-Interscience, New York, 1974, p. 105.

6. F. A. Brooks and W. O. Pruitt, "Investigations of Energy, Momentum and Mass Transfer Near the Ground," U.S. Army Electronics Command, Atm. Sci. Lab., Res. Div., Fort Huachuca, Ariz., DA Task IV0-14501-B53A-08, 1966, p. 371, Chap. 4.

7. W. O. Pruitt, D. L. Morgan, F. J. Lourence, and F. V. Jones, Jr., "Energy, Momentum and Mass Transfer above Vegetative Surfaces," ECOM Report 68-G10-F, University of California, Davis, Chap. 3.

8. J. A. Businger, "Aerodynamics of Vegetated Surfaces," in D. A. deVries and N. H. Afgan, Eds., *Heat and Mass Transfer in the Biosphere, Part 1—Transfer Process in the Plant Environment*, Scripta, Washington, D.C., 1975, p. 139–165.

9. L. H. Parmele, E. R. Lemon, and A. W. Taylor, "Micrometeorological Measurement of Pesticide Vapor Flux from Bare Soil and Corn under Field Conditions," *Water, Air, Soil Pollut.*, **1**, 433–451 (1972).

10. A. W. Taylor, D. E. Glotfelty, B. C. Turner, R. E. Silver, H. P. Freeman, and A. Weiss, "Volatilization of Dieldrin and Heptachlor Residues from Field Vegetation," *Agric. Food Chem.*, **25** (3), 542 (1977).

11. O. T. Denmead, J. R. Simpson and J. R. Freney, "Ammonia Flux into the Atmosphere from a Grazed Pasture," *Science*, **185**, 609 (1974).

12. D. Mackay and R. S. Matsugu, "Evaporation Rates of Liquid Hydrocarbon Spills on Land and Water," *Can. J. Chem. Eng.*, **51** 434–439 (1973).

13. B. C. Turner and D. E. Glotfelty, "Field Air Sampling of Pesticide Vapors with Polyurethane Foam," *Anal. Chem.*, **49** (1), 7–10 (1977).

14. B. B. Hicks, "Some Micrometeorological Methods for Measuring Dry Deposition Rates," Paper 44f, 70th Annual Am. Inst. Chem. Eng. Meet., New York, November 1977.

15. R. R. Arnts, R. L. Seila, and R. L. Kuntz, "Measurements of α-Pinene Fluxes from a Loblolly Pine Forest," Proceedings Fourth Joint Conference on Sensing of Environmental Pollutants, page 831 (1978), American Chemical Society, Wash. D.C.

16. N. C. Brady, *The Nature and Properties of Soils*, 8th ed., Macmillan, New York, 1974.

17. H. S. Carslaw and J. C. Jaeger, *Conduction of Heat in Solids*, Oxford University Press, London, 1959, p. 96.

18. R. D. Jackson, B. A. Kimball, R. J. Reginato, S. B. Idso, and F. S. Nakayama, "Heat and Water Transfer in a Natural Soil Environment," in D. A. deVries and N. H. Afgan, *Heat and Mass Transfer in the Biosphere, Part 1—Transfer Processes in the Plant Environment*, Scripta, Washington, D.C., 1975, p. 67–76.

19. H. Israël, "Die natürliche Radioaktivitat in Boden, Wasser and Air," *Beitr. Phys. Atmos.*, **30**, 177–188 (1958).

20. C. E. Junge, *Air Chemistry and Radioactivity*, Academic, New York, 1963, p. 209–221.

21. D. R. DeWalle, "An Agro-Power-Waste Complex for Land Disposal of Waste Heat," Res. Pub. 68, University of Pennsylvania, Philadelphia, 1974.

22. F. Kreith and W. D. Sellers, "General Principles of Natural Evaporation," in D. A. deVries and N. H. Afgan, *Heat and Mass Transfer in the Biosphere, Part 1—Transfer Processes in the Plant Environment*, Scripta, Washington, D.C., 1975, p. 207–214.

CHAPTER 7

INTRAPHASE CHEMICAL EXCHANGE RATES

Intraphase transport of chemicals is a topic concerned with the movement of chemicals from point to point within a single phase, whereas interphase transport is concerned with movement across phase boundaries. The most common environmental, chemodynamic, intraphase transport process is the dispersion of air pollutants from an elevated point source. This much-studied topic is briefly presented in this chapter along with other less common but possibly no less important intraphase transport processes relevant to the subject of chemodynamics.

The chapter commences with a section on thermal stratifications of water bodies. Stratification plays an important role in the transport of chemicals within the water column on lakes and the ocean. Next, various models are presented that describe the rate of chemical movement, chemical lifetime, and the time to steady state. Finally, a brief introduction of the classical topic of air pollutant dispersion is presented. As was done in previous chapters, problems are presented in a context that extends the text material and provides realistic exercises for application of the theory.

7.1 TEMPERATURE PROFILES AND STRATIFICATION IN DEEP LAKES AND THE OCEANS

The Annual Thermal Structure of Deep Reservoirs and Lakes

Lakes and reservoirs display seasonal temperature cycles. The energy from the sun causes lakes to display warm surface temperatures in the summer.

Diminished solar energy in the winter results in colder water during that period. The spring and fall seasons are transition periods in the direction of lake heat transfer. In the spring lakes are in a state of receiving heat, and they lose it in the fall. This annual cycle plays an important role in the chemodynamics of lakes. Details of temperature profiles and thermal stratification of lakes follow. Some chemodynamic aspects of lakes appear in Chapter 5, Section 2.

Water bodies such as natural lakes and artificial reservoirs in the temperate zones of the world are in an isothermal condition in the spring. The water temperature at all depths is constant and somewhere between 4 and 10°C. At approximately the vernal equinox (March 21) the net heat transfer becomes positive, and the upper layers of water become warmer than the colder bottom (winter) water. Mixing between hot and cold does not easily occur because the density of water above 4°C decreases with temperature. As spring proceeds into summer, a definite thermal structure develops. (See Fig. 7.1-1.) The upper layers of water contain warm water at a somewhat uniform temperature; this region is called the *epilimnion*. Located below the epilimnion is a smaller layer of water that contains some large temperature gradients. In this region the water temperature decreases rapidly with increasing depth, and it is called the *metalimnion*. The particular point within the metalimnion where the temperature gradient is maximum is defined as the *thermocline*. The region below the metalimnion is the *hypolimnion*; this region contains cold water. This water

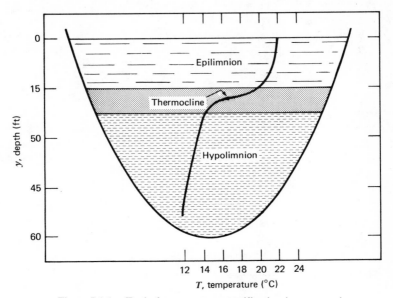

Figure 7.1-1. Typical temperature stratification in a reservoir.

remains cold throughout the year, increasing above 4°C only by a few
degrees. In very deep lakes the bottommost water may not deviate from
4°C. The annual thermal history data for a typical reservoir is presented in
Problem 7.1A.

Around the time of the autumnal equinox (September 23), the net heat
transfer becomes negative, and the lake begins to cool. The upper layers
cool rapidly, especially on clear nights when the sky is a near perfect black
body absorber. The upper layers are cold and more dense so that the water
"plunges" from the surface and causes a high degree of mixing in the

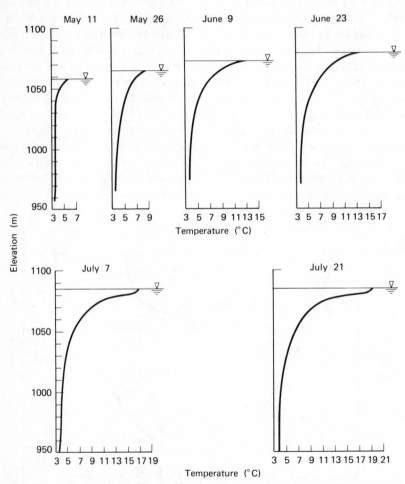

Figure 7.1-2. Observed temperature profiles, Hungry Horse Reservoir, 1965. (**Source.** Reference 1.)

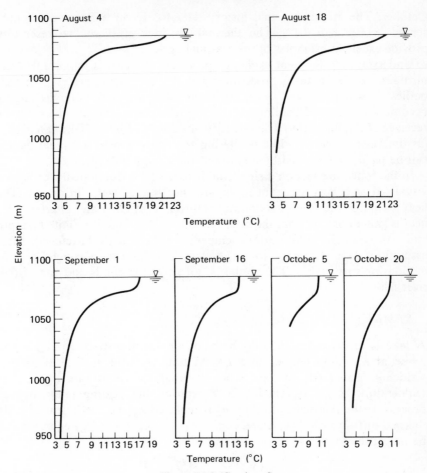

Figure 7.1-2 (Continued)

epilimnion. As fall proceeds into winter, the cooling process continues. In late November the lake is at a near isothermal condition. While at an isothermal condition the water is of uniform density throughout the column and in a neutral state. Surface winds can usually exert enough force to cause complete or near complete mixing of the water. This happening is commonly called "lake turnover" by fishermen. Many turnovers can occur throughout the winter until the spring, when the heating cycle commences again and a stable water column develops. A graphical illustration of the buildup and collapse of the thermal structure of Hungry Horse Reservoir in 1965 is shown in Fig. 7.1-2 for May through

October. The annual thermal history data for Keowee Reservoir is presented in Problem 7.1A. The thermal histories of these two reservoirs provide excellent examples of the seasonal cycles of lakes.

Understanding the heat exchange processes in lakes is important. Man intrudes on the natural processes when he employs lakes and similar bodies of water as heat sinks. Artificial lakes and reservoirs frequently become the source of cooling water for power plants and also become the receiver of the hot water. The quantity of additional heat a lake can accept (assimilative capacity) without altering the ecosystem to a critical degree can be partly determined by a study of the energy balance.

In the following section a simplistic lake model is developed. The lake is visualized as a very stagnant mass with no flow of water in or out. The heat exchange with the surroundings is through the air–water interface. No heat is gained or lost from the sides or bottom. The lake is infinite in depth and has a nonchanging bottom temperature. Seemingly farfetched, these assumptions are very good for large, deep lakes and are a rough approximation for small lakes. The model is not good for ponds and run-of-the-river lakes.

SEMI-INFINITE SOLID MODEL[2]

A lake is visualized as a solid body occupying the space from $y=0$ to $y=\infty$ at an initial temperature $T°$. At time $t=0$ the surface at $y=0$ is suddenly heated (or cooled by an external source at rate $q_{12}(t)$. The temperature at $y=\infty$ remains at $T°$ during the heating (and cooling) process. If the thermal diffusivity α_2 is replaced by the coefficient of eddy thermal diffusivity $\alpha_2^{(t)}$ (assumed constant), the general energy equation can be put in the following familiar form:

$$\frac{\partial T(y,t)}{\partial t} = \alpha^{(t)} \frac{\partial^2 T(y,t)}{\partial y^2} \qquad (7.1\text{-}1)$$

where $\alpha_2^{(t)} \equiv k_2^{(t)}/\rho_2 \hat{c}_{p2}$, $k_2^{(t)}$ is the turbulent coefficient of thermal conductivity and $T(y,t)$ is the water temperature as a function of depth y and time t. The initial and boundary conditions are

initial condition: at $t \leqslant 0$, $T = T°$ for all y (7.1-2a)

boundary condition 1:   at $y=0$, $\rho_2 \hat{c}_{p2} \dfrac{\alpha^{(t)} \partial T}{\partial y} = q_{12}{}^{(t)}$,   for all $t>0$

$$(7.1\text{-}2b)$$

boundary condition 2:   at $t=\infty$, $T=T°$ for all $t>0$

$$(7.1\text{-}2c)$$

Additional assumptions in this simplistic lake model described by Eq. 7.1-1 are enthalpic energy, biochemical reaction, and radiant energy absorption beneath the surface are excluded. The only source or sink term for energy addition is at the surface. The movement of energy within the lake occurs in the vertical direction only (i.e., one-dimensional model) and is characterized by one coefficient, $\alpha_2^{(t)}$, which is assumed constant for purposes of obtaining an analytical solution. Carslaw and Jaeger[3] give the following expression for the solution of Eqs. 7.1-1 through 7.1-2c:

$$T(y,t)=T^\circ + \frac{1}{\left(\alpha_2^{(t)}\pi\right)^{1/2}\rho_2\hat{c}_{p2}} \int_0^t q_{12}(t-\tau)\exp\left(\frac{-y^2}{4\alpha_2^{(t)}t}\right)\frac{d\tau}{\tau^{1/2}} \quad (7.1\text{-}3)$$

where τ is a dummy variable of integration.

This analytical solution may be rearranged to a more interpretable form. The lake surface temperature can be defined and expressed as follows:

$$T(0,t)\equiv T^\circ + \frac{1}{\left(\alpha_2^{(t)}\pi\right)^{1/2}\rho_2\hat{c}_{p2}} \int_0^t q_{12}(t-\tau)\frac{d\tau}{\tau^{1/2}} \quad (7.1\text{-}4)$$

where $T(0,t)$ is the surface temperature. Equation 7.1-3 then expresses the temperature at any depth as a function of the surface temperature and becomes:

$$T(y,t)=T^\circ + \{T(0,t)-T^\circ\}\exp\left(\frac{-y^2}{4\alpha_2^{(t)}t}\right) \quad (7.1\text{-}5)$$

This final expression yields an equation for the time locus of the thermocline position:

$$y_{tc}=\sqrt{2\alpha_2^{(t)}t} \quad (7.1\text{-}6)$$

after the operation $\partial^2 T/\partial x^2=0$ is performed. (See Problem 7.1B for an exact definition of the thermocline.) The term y_{tc} is the depth of the thermocline from the surface of the lake. Equations 7.1-4, 7.1-5, and 7.1-6 are useful working forms of the semi-infinite solid model.

Equation 7.1-5 is a useful interpretative expression for evaluating $\alpha_2^{(t)}$ values from field temperature profile data when rearranged to:

$$\alpha_2^{(t)}=\frac{y^2}{4t\ln\{[T(0,t)-T_0]/[T(y,t)-T_0]\}} \quad (7.1\text{-}7)$$

Model time t used in all the preceding equations is the lapse time since the lake was isothermal. The time of this isothermal condition varies with geographic location but typically occurs during March and April.

The coefficients of eddy thermal diffusivity as computed by Eq. 7.1-7 from field data are not constant. The coefficients vary with time and depth and are closely related to the turbulence in the lake. Flow turbulence in a lake is induced by bulk flow of the lake water, water movement due to the wind and waves (i.e., Langmuir circulations), and water density variations. Coefficients of eddy thermal diffusivity display some general characteristics, as reported by Orlob.[1] The coefficients typically display large values in the epilimnion, small values in the metalimnion, and large values in the hypolimnion. Plunging cold water caused by a net heat loss at the surface results in large epilimnion coefficients in the fall. (See Problem 7.1A.) An average $\alpha_2^{(t)}$ should be used with Eqs. 7.1-3 and 7.1-4.

Figures 7.1-3 and 7.1-4 show the results of a simulation with a single value of $\alpha_2^{(t)}$. Figure 7.1-3 is a simulation of the surface temperature of Beaver Reservoir in northwest Arkansas by Eq. 7.1-4. The flux at the surface is a combination of solar radiation, longwave radiation, natural convection, and evaporation, that is,

$$q_{12}(t) = q_s(t) + q_{lw}(t) + q_c(t) + q_e(t) \qquad (4.3-2)$$

All the heat flux terms in Eq. 4.3-2 are time dependent (i.e., Julian day). The time dependence of the solar radiation was obtained from Weather Bureau insolation data. The time dependence of the other three terms is related mainly to the air temperature, wind speed, and relative humidity.

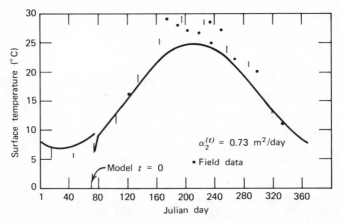

Figure 7.1-3. Semi-infinite slab model, simulation temperature—Beaver Reservoir. (Reprinted by permission. American Water Resources Assoc., Mpls., MN. **Source.** Reference 2.)

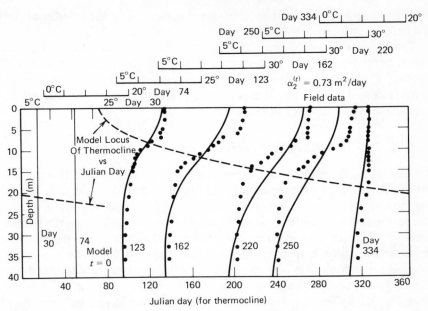

Figure 7.1-4. Semi-infinite slab model, simulation temperature—Beaver Reservoir. (Reprinted by permission. American Water Resources Assoc., Mpls., MN. **Source.** Reference 2.)

These data were compiled from existing local data. (See Chapter 4, Section 3, for details on the heat transfer across the air–water interface and the specific nature of each term in Eq. 4.3-2.) A simple numerical integration (computer) technique can be devised for Eq. 7.1-4. Once the surface temperature history is obtained, Eq. 7.1-5 is used to obtain the temperature at various depths from the surface. Figure 7.1-4 shows the "in-lake" temperature profiles for both the field data and the simulation.

As a realistic lake model the semi-infinite solid approach is very simplistic and incapable of producing a perfect simulation. As with all models, it has some advantages and disadvantages. Concise, explicit representation of the surface and water column temperature is a major advantage of the model. Another advantage of the model is the qualitative simulation of the gross features of the thermal changes with depth and time, within and at the surface of the water body. A major implication of this result is that the general shape of the temperature profile in deep freshwater bodies is structured primarily by the heat input at the air–water interface and the time lapse since an isothermal state existed. The action of surface winds, precipitation, condensation, bulk warm water flow, and plunging cold surface water in the fall on the shape of the profile in the epilimnion and the effects of bulk water movement on reshaping the profile in the hypolimnion are but minor modifications compared to the interfacial heat exchanges, induced thermal regime.

The science of modeling lake thermal behavior is quite advanced. Sophisticated predictive models exist that have the capability of a high degree of precision in simulating lake water temperatures. Parker et al.[4] have reviewed these types of models and present essential features, advantages, and disadvantages of each.

THE THERMOCLINE IN THE OCEAN

Unlike lakes, the oceans have a permanent thermocline. This feature is so named because its character is virtually unchanged seasonally. In the Arctic and Antarctic regions the water is cold from top to bottom. As this dense water flows south and north, respectively, it sinks beneath warmer water that moves outward from the equator. This gives rise to the temperature discontinuity known as the *permanent thermocline*. The top of the permanent thermocline is quite shallow at the equator, reaches maximum depth at midlatitudes, and becomes shallow again at about 50° latitude. (See Fig. 7.1-5) The thermocline disappears between 55 and 60° N or S.

As with lakes, the world's oceans have a seasonal thermocline. This feature is a summer phenomenon found at shallower water depths than the

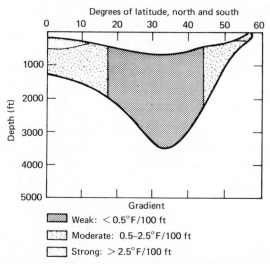

Figure 7.1-5. The permanent thermocline, based on averages for depth, thickness, and gradient within thermocline. ("From ENCYCLOPEDIA OF ENVIRONMENTAL SCIENCE. Copyright © 1974 by McGraw-Hill, Inc. Used with permission of McGraw-Hill Book Company." **Source**. Reference 5.)

permanent thermocline in all the world's oceans except those perennially ice infested. As air temperatures rise above ocean temperatures in the spring season and the sea surface receives more heat than it loses by radiation and convection, the surface water begins to warm so that a negative temperature gradient develops in the first few feet. The surface waters are mixed by transfer of energy from the wind. Although this mixing serves to lower the surface temperature, the net effect is a down-ward transport of heat and formation of an isothermal layer whose temperature is warmer than the underlying water. A strong temperature gradient, or seasonal thermocline, is thus formed between the isothermal surface layer and the water beneath. From July through September a surface layer of mixed water underlain by a strong negative temperature gradient is found in most of the ocean. As air temperatures fall in autumn, the water loses heat to the atmosphere by convective and radiative processes, and the surface layer is cooled to the temperature below. The seasonal thermocline breaks up to form again the following spring. Figure 7.1-6 shows three examples of the oceanic seasonal thermocline. The salinity profile is also shown in the figure. The thermohaline structure of the upper strata in the oceans is governed by the thermocline.

Thermoclines are semipermeable structures to the vertical movement of chemicals in lakes and in the oceans. This physical structure plays a dominant role in the movement of chemicals up and down the water

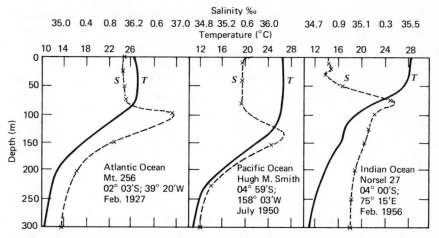

Figure 7.1-6. Examples of temperature and salinity distribution with depth in the upper 300 m layer of the tropical part of the oceans. (Gerhard Neumann, Willard J. Pierson, Jr., PRINCIPLES OF PHYSICAL OCEANOGRAPHY, © 1966, p. 447. Reprinted by permission of Prentice-Hall, Inc., Englewood Cliffs, New Jersey.)

column. Aspects of chemical movement through thermoclines are presented in Chapter 5, Section 2, and Chapter 7, Section 2.

The Vertical Turbulent Diffusivity in Lakes and the Ocean

One result of the semi-infinite solid lake model was a simple analytical expression for obtaining the vertical component of the turbulent thermal diffusivity $\alpha_2^{(t)}$ from field temperature profile data. The analytical expression to use is Eq. 7.1-7. The turbulent thermal diffusivities as computed by this equation are not constant. Problem 7.1A, which contains field data of Keowee Reservoir, demonstrates that $\alpha_2^{(t)}$ is a function of depth and time. The variation of $\alpha_2^{(t)}$ with depth is illustrated in Fig. 7.1-7 for two lakes and a reservoir. The trend of high diffusion coefficients in the epilimnion, low coefficients at the thermocline, and high coefficients in the hypolimnion is typical of most deep lakes and the oceans.

In general, $\alpha_2^{(t)}$ has its maximum value in the surface layer; in the open ocean $\alpha_2^{(t)}$ at the surface varies between 10 and 100 cm^2/s; in coastal areas, 10 to 50 cm^2/s; in lakes, approximately 10 cm^2/s. Below the surface mixed layer (or epilimnion) $\alpha_2^{(t)}$ drops to a minimum in the thermocline, of the order of 1 cm^2/s in the open ocean; in lakes $\alpha_2^{(t)}$ may drop as low as 0.05 cm^2/s. Below the thermocline $\alpha_2^{(t)}$ may increase again.

The presence of density stratification tends to suppress the vertical exchange of thermal energy. It is obvious from a casual study of the data in Fig. 7.1-7 that in the regions of steep density gradients, $\partial \rho_2 / \partial y$, that $\alpha_2^{(t)}$ is lowest. Therefore, one expects the vertical diffusivity to be a decreasing function of density stratification. The presence of shear tends to be destabilizing and increases vertical exchange. It is to be expected that for similar flows the vertical diffusivity should be related to the Richardson number defined by Eq. 6.1-9 but used here in the form

$$Ri \equiv \frac{(g/\rho_2)(\partial \rho_2 / \partial y)}{\left(\dfrac{\partial v}{\partial y}\right)^2} \tag{7.1-8}$$

where ρ_2 is the density of water. Water density is a function of temperature and salinity. An empirical expression for the density of fresh water as a function of temperature is given in Appendix D. Numerous proposed relations between $\alpha_2^{(t)}$ and Ri are of the form

$$\alpha_2^{(t)} = \alpha(1 \pm bRi)^{\pm 1/n} \tag{7.1-9}$$

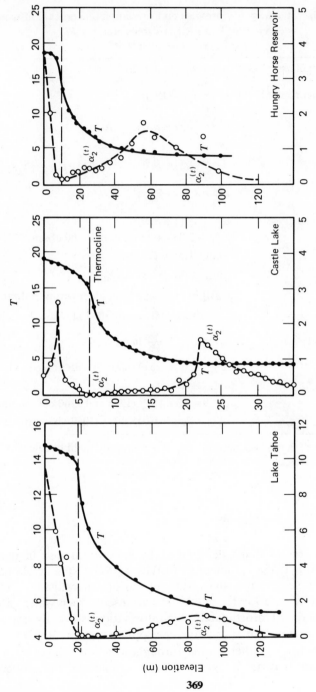

Figure 7.1-7. Effective diffusion and temperature profiles for selected impoundments. (**Source.** Reference 1.)

Table 7.1-1.   Summary of Formulas on Correlation of Vertical Diffusion coefficient $\alpha_2^{(t)}$ with Richardson's Number Ri (or Density Gradient Ω_2)[a]

Rossby and Montgomery (1935)	$\alpha_2^{(t)} = \alpha_{20}{}^{(t)}(1 + \beta Ri)^{-1}$
Rossby and Montgomery (1935)	$\alpha_2^{(t)} = \alpha_{20}{}^{(t)}(1 + \beta Ri)^{-2}$
Holzman (1943)	$\alpha_2^{(t)} = \alpha_{20}^{(t)}(1 - \beta Ri) \qquad Ri \leqslant \dfrac{1}{\beta}$
Yamamoto (1959)	$\alpha_2^{(t)} = \alpha_{20}^{(t)}(1 - \beta Ri)^{1/2}, \qquad Ri \leqslant \dfrac{1}{\beta}$
Mamayev (1958)	$\alpha_2^{(t)} = \alpha_{20}^{(t)} e^{-\beta Ri}$
Munk and Anderson (1948)	$\alpha_2^{(t)} = \alpha_{20}^{(t)}(1 + \beta Ri)^{-3/2}$ $[\beta = 3.33$ based on data by Jacobsen (1913) and Taylor (1931)]
Harremoes (1968)	$\alpha_2^{(t)} = 5 \times 10^{-3} \times \Omega_2^{-2/3}$, cm²/s ($\Omega_2$ in m^{-1}; approximate experimental range $5 \times 10^{-9} < \Omega_2 < 15 \times 10^{-5}$ m^{-1})
Kolesnikov et al. (1961)	$\alpha_2^{(t)} = \alpha_{2\min}^{(t)} + \dfrac{\beta}{\Omega_2}$ in cm²/sec $\alpha_{2\min}^{(t)}$ and β are empirically determined to be $\alpha_{2\min}^{(t)} = 12,$ $\beta = 8.3 \times 10^{-5}$ (1958 and 1960 observations) $\alpha_{2\min}^{(t)} = 2, \; \beta = 10.0 \times 10^{-5}$ (1959 observations)

[a] $\alpha_{20}^{(t)}$ is $\alpha_2^{(t)}$ at $Ri = 0$, that is, the neutral case. β is a proportionalility constant which varies from case to case.

Source. Reference 7.

where b and n are constants determined by experiment and in part by theory. The term α is the turbulent diffusivity at $Ri = 0$, the stable condition. Table 7.1-1 summarizes the results of various investigations of the relation between $\alpha_2^{(t)}$ and Ri. These relations are useful when water velocity profiles are available. Unfortunately this information is usually not available for water columns. An alternative development is desirable.

Since water density gradients exist throughout a water body such as a lake or the ocean, it is not altogether absurd to consider the property of density ρ_2 as a "diffusing species." Commencing with Eq. 3-1 and assuming no reaction, $R_A = 0$, and only diffusion due to turbulence, so that molecular diffusion may be ignored, we obtain

$$\frac{\partial \rho_2}{\partial t} + v_x \frac{\partial \rho_2}{\partial x} + v_y \frac{\partial \rho_2}{\partial y} + v_z \frac{\partial \rho_2}{\partial z} = \frac{\partial}{\partial x}\left(\alpha_{2\,x}^{(t)} \frac{\partial \rho_2}{\partial x}\right) + \frac{\partial}{\partial y}\left(\alpha_{2\,y}^{(t)} \frac{\partial \rho_2}{\partial y}\right)$$

$$+ \frac{\partial}{\partial z}\left(\alpha_{2\,z}^{(t)} \frac{\partial \rho_2}{\partial z}\right) \qquad (7.1\text{-}10)$$

where v_x, v_y, and v_z are the mean currents in the x, y, and z directions, respectively. Because of the strong relationship between density and temperature, eddy turbulent diffusivities are employed, and $\alpha_{2x}^{(t)}$, $\alpha_{2y}^{(t)}$, and $\alpha_{2z}^{(t)}$ are the variables for the x, y, and z directions, respectively. Since the horizontal variations of ρ_2 are usually much smaller than the vertical variations, we assume $\partial \rho_2 / \partial x = \partial \rho_2 / \partial z = 0$; also we assume $v_x = v_y = v_z = 0$. Then Eq. 7.1-10 becomes

$$\frac{\partial \rho_2}{\partial t} = \frac{\partial}{\partial y}\left(\alpha_{2\,y}^{(t)} \frac{\partial \rho_2}{\partial y}\right) \qquad (7.1\text{-}11)$$

The term $\partial \rho_2 / \partial t$ is usually very small except for the near surface waters, which may undergo some diurnal changes. Considering an instantaneous thermal profile we can assume steady state; then

$$\frac{d}{dy}\left(\alpha_{2\,y}^{(t)} \frac{d \rho_2}{dy}\right) = 0 \qquad (7.1\text{-}12)$$

or

$$\alpha_{2\,y}^{(t)} = \frac{a}{d\rho_2 / dy} \qquad (7.1\text{-}13)$$

This final expression indicates that the turbulent diffusivity is related to the density gradient of water.

Rather than relating $\alpha_{2y}^{(t)}$ to Ri, which is physically a more logical approach, it is also possible to attempt a correlation of $\alpha_{2y}^{(t)}$ with

$$\Omega_2 \equiv \left| \frac{1}{\rho_2} \frac{d\rho_2}{dy} \right| \qquad (7.1\text{-}14)$$

the density gradient alone. Intuitive considerations also suggest that the eddy duffusivity should to some extent depend on the steepness of the density gradient. When the density gradient is strong, the degree of turbulence and, hence, the magnitude of the eddy diffusivity in the pycnocline may be expected to be low, or the gradients would be destroyed by turbulent eddies. The steepness of the density gradient within the water layer determines the stability of stratification: The greater the density gradient, the more stable is the stratification. Koh and Fan[7] collected all readily available data on $\alpha_{2y}^{(t)}$, where Ω_2 is simultaneously measured to construct the graphical relationship shown in Fig. 7.1-8. It can be seen that almost all data fall within a factor of 10 of the empirical relation

$$\alpha_{2}^{(t)}{}_y = \frac{10^{-4}}{\Omega_2} \tag{7.1-15}$$

for $4\text{E}-7\,\text{m}^{-1} \leqslant \Omega_2 \leqslant \text{E}-2\,\text{m}^{-1}$, where $\alpha_{2y}^{(t)}$ is in cm^2/s and Ω_2 is in m^{-1}. Unless independent field data are available, Eq. 7.1-15 can be used to estimate $\alpha_{2y}^{(t)}$.

The relationship between density gradient and the degree of stability of stratification can also be expressed by the quantity known as the Brunt-

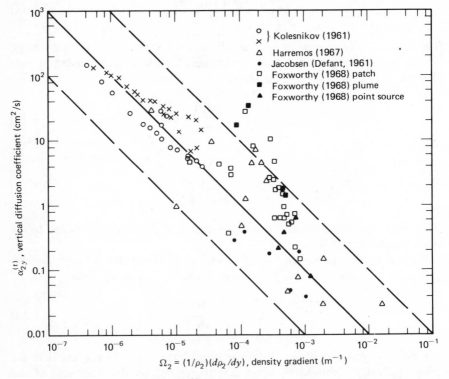

Figure 7.1-8. Correlation of $\alpha_{2y}^{(t)}$ with density gradient. (**Source.** Reference 7.)

Valisada stability frequency ν

$$\nu^2 \equiv \left| g \frac{1}{\rho_2} \frac{d\rho_2}{dy} \right| \qquad (7.1\text{-}16)$$

where g is the acceleration due to gravity and the dimension of ν is t^{-1}. It is claimed that Eq. 7.1-16 is valid for a water column up to several hundred meters deep, where the effect of adiabatic compressibility on the density gradient within the water column can be ignored. Eckart[8] has discussed the derivation and physical significance of the concept of stability frequency.

Example 7.1-1. Estimating Vertical Turbulent Diffusivities

Equation 7.1-15 and Fig. 7.1-8 provide a means of estimating $\alpha_{2y}^{(t)}$ values when no better information is available for a particular body of water. Using these means, estimate $\alpha_{2y}^{(t)}$ at the thermocline for the three water bodies of Fig. 7.1-7 and compare the results to the observed $\alpha_{2y}^{(t)}$ values.

SOLUTION

It is possible to express Ω_2 as a function of the coefficient of thermal expansion of water, β_2, and the thermal gradient by use of the chain rule to yield

$$\Omega_2 \equiv \left| -\beta_2 \frac{\partial T}{\partial y} \right| \qquad (E7.1\text{-}1)$$

where $\beta_2 \equiv \{(-1)/(\rho_2)\} \, (\partial\rho_2/\partial T)$. The density of pure water, ρ_2, as a function of temperature is given in Appendix D. From this empirical relationship $\partial\rho_2/\partial T$ can be obtained as a function of temperature:

$$\frac{\partial\rho_2}{\partial T} = 0.1546919E - 5 + 2(0.2141986E - 5)T - 3(0.6508630E - 6)T^2$$

$$+ 4(0.1975524E - 7)T^3 - 5(0.1894802E - 9)T^4 \qquad (E7.1\text{-}2)$$

Thermal gradients at the thermocline may be estimated graphically from Fig. 7.1-7. Thermocline temperature and thermal gradients appear in columns 1 and 2 of Table E7.1-1. Other pertinent calculated values appear in succeeding columns. Column 7 contains $\alpha_{2y}^{(t)}$ values calculated from Eq. 7.1-15, and column 8 contains the range of values obtained from Fig. 7.1-8. Column 9 contains the $\alpha_{2y}^{(t)}$ values observed in the field. In all cases the observed values of $\alpha_{2y}^{(t)}$ are within the lower end of the range of values

Table E7.1-1

Lake	1 T (°C)	2 $-\Delta T/\Delta Y$ (°C/m)	3 ρ_2 (g/cm³)	4 $-\partial\rho_2/\partial T$ (g/cm³·°C)	5 $-\beta_2$ (°C⁻¹)
Tahoe	12.5	0.75	0.9995	1.1879E -4	1.1885E -4
Castle	14.5	3.125	0.9992	1.14784E -4	1.1488E -4
Hungry	14.5	1.136	0.9992	1.14784E -4	1.1488E -4

Lake	6 Ω_2 (m⁻¹)	7 $\alpha_{2y}^{(t)}$ (cm²/s)	8 $\alpha_{2y}^{(t)}$ (cm²/s)	9 $\alpha_{2y}^{(t)}$ (cm²/s)	10 Estimate ÷ actual
Tahoe	8.9137E -5	1.12	0.1 -10.0	0.2	5.6
Castle	3.5899E -4	0.216	0.03 -3.0	0.05	4.3
Hungry	1.3053E -4	0.766	0.06 -6.0	0.1	7.7

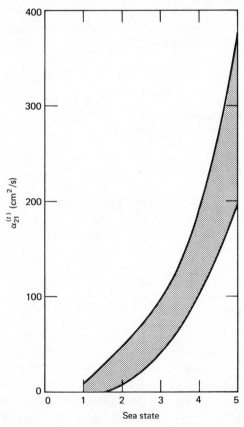

Figure 7.1-9. Dependence of $\alpha_{2y}^{(t)}$ on sea state. (**Source**. Reference 7.)

suggested by Fig. 7.1-8. Column 10 indicates that Eq. 7.1-15 overestimates the actual value of $\alpha_{2y}^{(t)}$ by a factor of approximately 6 for these lakes.

It is unlikely that the preceding empirical information concerning $\alpha_{2y}^{(t)}$ applies throughout the surface mixed layer. It has been observed that in the surface mixed layer of the ocean the density gradient is often zero. The preceding empirical relation is certainly invalid since it implies an infinite $\alpha_{2y}^{(t)}$. In this case the vertical transport is governed primarily by the vertical turbulence created by waves and wind. Koh and Fan[7] summarize equations proposed by Golubeva and Isayeva for relating the vertical diffusion coefficient in the mixed layer and the surface wave characteristics. The proposed equation is

$$\alpha_{21}^{(t)} = \frac{0.02h^2}{\theta} \tag{7.1-17}$$

where $\alpha_{21}^{(t)}$ is the vertical diffusivity at the surface, h is wave height, and θ is wave period. Thus, given the sea state, $\alpha_{21}^{(t)}$ can be estimated. Figure 7.1-9 shows the relation between $\alpha_{21}^{(t)}$ and sea state.

Problems

7.1A. THERMAL STRUCTURE OF KEOWEE RESERVOIR

Keowee Reservoir is an artificial, multipurpose lake located in the southeastern United States. Table 7.1A contains thermal profile data for the reservoir in 1972.

1. Using a single sheet of graph paper, display the annual development of the temperature profile. Plot 12 months of data on a single sheet and label each month. Place depth on the ordinate (y axis) with zero depth at the top and place temperature on the abscissa (x axis).
2. Study the graph of the thermal history for 1972 very carefully and choose a date for model time $t = 0$. Also choose a value for $T°$.
3. Based on the model time zero date and $T°$ chosen in part (2), compute values of $\alpha_2^{(t)}$ by use of Eq. 7.1-7. Compute a $\alpha_2^{(t)}$ value for each depth and time except for the months of November, December, January, and February, which are nearly isothermal. Obtain a monthly average. Obtain a yearly average. Report all values in m^2/day.
4. Review the $\alpha_2^{(t)}$ values in the table and explain the likely causes of the variations in $\alpha_2^{(t)}$ with time and with depth. Do the numerical values of $\alpha_2^{(t)}$ fall within the range for diffusion coefficients reported in Tables 3.1-4 and 3.3-1?

Table 7.1A. Keowee Reservoir Temperatures 1972

Depth (m)	Jan 27[a] 27[b]	Feb 22 53	Mar 30 89	Apr 18 108	May 16 136	Jun 22 173	Jul 19 200	Aug 22 234	Sep 17 262	Oct 27 300	Nov 22 326	Dec 20 354
						Temperature (°C) on						
0	10.4	8.0	12.4	19.1	22.2	24.4	28.4	28.3	26.8	17.8	13.9	10.8
1	10.3	8.0	12.1	18.7	21.8	24.4	27.9	28.3	26.8	17.9	13.9	10.7
2	10.1	8.0	11.8	18.0	21.4	24.2	28.0	28.0	26.6	17.9	13.9	10.5
3	10.0	7.9	11.6	17.4	21.2	24.2	27.9	28.0	26.6	17.7	13.8	10.5
4	9.9	7.9	11.4	17.0	21.1	23.7	27.4	28.0	26.5	17.8	13.8	10.5
5	9.9	7.8	11.2	16.4	20.7	23.1	26.2	27.5	26.4	17.8	13.8	10.5
6	9.8	7.8	11.0	16.0	19.3	21.8	23.6	25.6	25.0	17.8	13.8	10.5
7	9.8	7.8	10.8	15.2	17.1	20.6	21.4	22.2	22.2	17.8	13.8	10.5
8	9.8	7.8	10.6	14.7	15.6	19.2	19.3	20.0	20.1	17.8	13.8	10.5
9	9.7	7.8	10.4	13.7	14.6	17.6	17.9	18.5	18.8	17.6	13.8	10.5
10	9.7	7.8	10.2	12.9	14.1	15.8	16.8	17.6	17.8	17.4	13.7	10.5
11	9.7	7.7	10.0	12.1	13.2	14.8	15.9	16.8	17.3	16.9	13.7	10.5
12	9.7	7.7	9.7	11.6	12.7	14.1	15.0	16.1	16.6	16.5	13.7	10.5
13	9.7	7.7	9.5	11.1	12.1	13.2	14.1	15.0	15.8	16.3	13.7	10.5
14	9.6	7.7	9.2	10.7	11.6	12.6	13.2	14.0	14.8	15.6	13.7	10.4
15	9.6	7.7	8.9	10.4	11.3	12.0	12.4	13.2	14.0	14.9	13.7	10.4
20	9.5	7.7	8.3	9.3	9.9	10.6	10.6	10.9	11.4	11.8	12.5	10.3
25	9.5	7.6	8.2	8.9	9.4	9.7	9.8	9.9	10.7	10.4	10.6	10.3
30	9.5	7.6	8.1	8.7	9.1	9.5	9.3	9.6	9.8	10.0	10.0	10.1
35	9.6	7.6	7.8	8.7	8.8	–	9.1	9.3	9.5	10.0	10.0	10.3

Source. J. H. Elrod, SERT, U.S. Government, Clemson, S.C. March 1975, private communication.

[a]Date.
[b]Julian day.

7.1B. THE THERMOCLINE

A dictionary of scientific and technical terms gives the following definitions of *thermocline*: "(1) A temperature gradient as in a layer of seawater, in which the temperature decrease with depth is greater than that of the overlying and underlying water. Also known as a metalimnion; (2) A layer in a thermally stratified body of water in which such a gradient occurs."

It is not uncommon to refer to the zone between the hypolimnion and the epilimnion as the thermocline or metalimnion. A more precise definition is needed from a mathematical point of view. A common mathematical definition of the thermocline is the point where the thermal gradient $\partial T/\partial y$ is a maximum. The maximum in the gradient occurs at those points where its derivative is zero; therefore $\partial^2 T/\partial y^2 = 0$. Apply this criterion to Eq. 7.1-5 to obtain Eq. 7.1-6.

7.1C. LAKE WATER COLUMN TEMPERATURE PROFILES VIA THE SEMI-INFINITE SOLID MODEL

Using the result of the semi-infinite solid model, compute the temperature profiles for the following reservoirs and compare them with the actual profiles. Prepare a graph of each for comparison purposes. Use the following data.

Table 7.1C

Lake or Reservoir	Temperatures (°C)		Turbulent Diffusivity (cm^2/s)
	Surface Water	Bottom Water	
Tahoe	14.8	5.5	1.14
Castle	19.0	4.5	0.53
Hungry Horse	18.0	4.0	0.99

7.1D. WATER DENSITY AND STRATIFICATION

1. Make a graphical plot of the density of water (g/cm^3) as a function of temperature for the range of 0 to 30°C.

2. Study the Shagawa Lake thermal profile data for the months of December, January, February, and March of 1973 given in Problem 5.2A. Based on the effect of temperature on water density, explain how a lake can "reverse stratify" in the winter.

3. Study the Hungry Horse reservoir thermal profile data shown in Fig. 7.1-2. Based on the effect of temperature on water density, explain how a lake can stratify in the summer.

7.1E. HEAT TRANSFER RATE BETWEEN THE EPILIMNION AND THE HYPOLIMNION

Equation 7.1-5 is a first-order approximation of the temperature profile in any deep lake. The heat transfer rate through any plane parallel to the surface is

$$q_y = -\alpha_2^{(t)} \rho_2 c_{\rho_2} \frac{\partial T}{\partial y} \tag{7.1E-1}$$

1. Employing this equation and the equation for the thermocline depth, Eq. 7.1-6, show that the flux rate through the thermocline can be estimated by

$$q\Big|_{y_{tc}} = \rho_2 c_{\rho_2} \sqrt{\frac{\alpha_2^{(t)}}{2et}} \left[T(0,t) - T^\circ \right] \tag{7.1E-2}$$

2. Compute the heat flux rate through the thermocline for a day of each month from March through September in Keowee Reservoir. Use $\alpha_2^{(t)} = 1.31$ m^2/d, $t = 0$ at March 21, and $T^\circ = 7.6°C$. Give the answer in J/m^2·d.

3. Compute the heat flux rate through the air–water interface for a day of each month from March through September in Keowee Reservoir. Use Table 7.1A for the water surface temperature and the data in Table 7.1E. Give the answer in J/m^2·d.

Table 7.1E. Approximate Lake Surface Data for Keowee Reservoir

Month	Wind Speed (mi/hr)	Relative Humidity (%)	Air Temperature (°F)	Insolation (cal/cm^2·d)
March	7.5	62	58.2	303.3
April	7.3	65	60.6	356.0
May	6.5	75	66.2	373.7
June	5.1	74	71.5	375.4
July	5.4	83	77.1	342.5
August	4.9	80	77.4	351.8
September	5.1	75	75.3	320.3

4. Compare the heat flux rate through the thermocline to the heat flux rate through the air–water interface. Discuss the implications of the differing rates and how they affect lake temperature.

7.1F. AN IMPROVED MODEL FOR LAKE THERMAL HISTORY AND PROFILE

As is evident in Eq. 7.1-1, the semi-infinite slab model of a lake commenced by assuming that $\alpha_2^{(t)}$ is constant. A significantly more realistic lake thermal model results if this assumption is not applied but the equation of energy is used in the form

$$\frac{\partial T}{\partial t} = \frac{\partial}{\partial y}\left(\alpha_2^{(t)}\frac{\partial T}{\partial y}\right) \qquad (7.1F\text{-}1)$$

This model is capable of accounting for the variation of the eddy thermal diffusivity within the water column. When constructing numerical solutions of Eq. 7.1F-1 for the thermal history of particular lakes, it is convenient to relate $\alpha_2^{(t)}$ to the water temperature and/or the temperature gradient.

1. Show that by using an equation of the form of 7.1-15 along with the coefficient of thermal expansion of water that Eq. 7.1F-1 can be transformed to yield

$$\frac{\partial T}{\partial t} = a\frac{\partial}{\partial y}\left\{\frac{1}{\left|\beta_2(T)\dfrac{\partial T}{\partial y}\right|}\frac{\partial T}{\partial y}\right\} \qquad (7.1F\text{-}2)$$

where a is an empirical constant.

2. Lerman[9] suggests that $\alpha_2^{(t)}$ can be related to ν by

$$\alpha_2^{(t)} = a\exp(-b\nu) \qquad (7.1F\text{-}3)$$

where a and b are empirical constants. Show that if this equation is used, Eq. 7.1F-1 can be transformed to yield

$$\frac{\partial T}{\partial t} = a\frac{\partial}{\partial y}\left\{\exp\left(-b\left[g\beta_2(T)\frac{\partial T}{\partial y}\right]^{1/2}\right)\frac{\partial T}{\partial y}\right\} \qquad (7.1F\text{-}4)$$

Now with the appropriate empirical constants and a relationship for $\beta_2(T)$, either equation can be used to obtain more realistic predictions of thermal history and profiles in water bodies.

7.2   INTRAPHASE CHEMICAL TRANSPORT PROCESSES IN THE PRESENCE OF STRATIFICATION

In the previous section the thermal structure of lakes and the ocean was examined in detail. Based in large part on the thermal structure, these water bodies stratify vertically into a series of at least three layers. Ignoring the fine details of the thermal and density gradients, a gross description of the stratified system consists of an upper well-mixed layer, a layer with a more or less pronounced density gradient (*pycnocline*) below it, and a well-mixed layer below the pycnocline. This particular structure has important consequences for the chemodynamic behavior of lakes and the ocean. This section is devoted to the study of the rates of chemical movement, chemical lifetimes, and chemical profiles in stratified water bodies by the use of simple models.

Simple Intraphase Transport Models

MOLECULAR DIFFUSION WITHIN A HOMOGENEOUS MEDIUM

The movement of a dilute chemical species between two points within a homogeneous medium by molecular diffusion processes is well established and easily quantified. Figure 7.2-1*a* shows a section of any homogeneous medium in which a concentration gradient exists and component A is being transported by molecular processes. Species A moves from the lower plane, ①, of high concentration denoted by c_{A21}, to the upper plane, ②, of

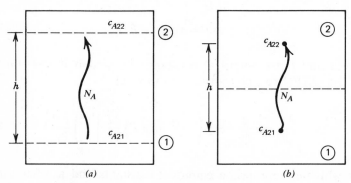

Figure 7.2-1. Basis for turbulent intraphase diffusion rate expression. (*a*) Molecular diffusion in a homogeneous medium. (*b*) Turbulent diffusion between well-mixed compartments.

lower concentration denoted by c_{A22}. The rate of movement is quantified by Fick's first law, Eq. 3.1-13:

$$N_A = \frac{\mathscr{D}_{A2}}{h}(c_{A21} - c_{A22}) \qquad (7.2\text{-}1)$$

where h is the length of the diffusion path or the distance between the planes ① and ②.

TURBULENT DIFFUSION BETWEEN WELL-MIXED CHAMBERS

Figure 7.2-1b illustrates the case for turbulent diffusion of chemical species A between two well-mixed chambers separated by a permeable plane. The lower chamber contains A at a high concentration $\bar{c}_{A21}$, and the upper chamber contains A at a lower concentration $\bar{c}_{A22}$. Since each chamber is well mixed, the average path length for movement is from the midpoint of the lower chamber to the midpoint of the upper chamber. It is desirable in many instances to have a rate expression for the movement of chemical A, and a logical expression to use is an extension of the molecular diffusion equation:

$$N_A = \frac{\mathscr{D}_{A2}^{(t)}}{h}(\bar{c}_{A21} - \bar{c}_{A22}) \qquad (7.2\text{-}2)$$

where $\mathscr{D}_{A2}^{(t)}$ is a well-chosen, average turbulent diffusion coefficient that is a characteristic of each chamber and h is the distance between the midpoints of the chambers. The movement of oxygen from the epilimnion to the hypolimnion of a lake is an example of the use of this turbulent diffusion rate equation.

The epilimnion is often referred to as a "mixed layer." This is not an altogether incorrect description. It is known that the vertical eddy thermal diffusivity in the epilimnion is large compared to the eddy thermal diffusivity in the thermocline region. It is also known that surface winds exert shear stresses upon the air–water interface that causes bulk water flow and turbulence. The constant change in direction and magnitude of the wind speed also aids in mixing the surface layer. The thermocline is a thin zone of relative calm water compared to the surface mixed layer. Below the thermocline, the turbulent diffusivity again increases dramatically, as illustrated in Fig. 7.1-7. In the light of this physical evidence, Eq. 7.2-2 is reasonable and will be used to study chemical movement rates between mixed chambers.

TURBULENT DIFFUSION FROM PHASE BOUNDARIES TO WELL-MIXED CHAMBERS

It is desirable to express the movement of a chemical from the region of a phase interface into the bulk fluid within a chamber. Examples are the movement of oxygen from the air–water interface into the epilimnion and the movement of phytoplankton-producing nutrient chemicals from the mud–water interface into the overlying hypolimnic waters. Figure 7.2-2 illustrates two such transport processes.

For the specific case of oxygen transfer shown in Fig. 7.2-2 the rate expression takes the form of

$$N_A = \frac{\mathcal{D}_{A22}^{(t)}}{h_2/2} \left(c_{A22}^* - \bar{c}_{A22} \right) \tag{7.2-3}$$

where c_{A22}^* is the oxygen concentration in water at the air–water interface and $\bar{c}_{A22}$ is the average concentration in the upper chamber. The group $\mathcal{D}_{A22}^{(t)}/(h_2/2)$ consisting of the effective turbulent diffusivity of the chamber and the diffusion path length is referred to as the *lake surface reaeration coefficient* $^1k'_{A2}$. A commonly reported range of $^1k'_{A2}$ values is 1.0 to 5.0 ft/d.

For the case of ammonium ion release from the lake mud–water interface shown in Fig. 7.2-2, the rate expression takes the form

$$N_A = \frac{\mathcal{D}_{A21}^{(t)}}{h_1/2} \left(c_{A21}^* - \bar{c}_{A21} \right) \tag{7.2-4}$$

where c_{A21}^* is the ammonia concentration at the mud interface and $\bar{c}_{A21}$ is

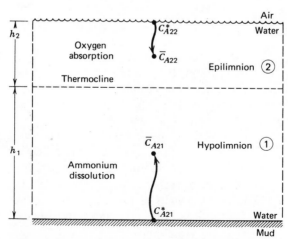

Figure 7.2-2. Turbulent diffusion from phase boundary.

the average concentration of ammonia in the overlying well-mixed chamber. The group $\mathcal{D}_{A21}^{(i)}/(h_1/2)$ is referred to as the *dissolution coefficient* $^3k'_{A2}$. (See Section 5.2 for further information concerning this coefficient.)

Time to Chemical Steady States in Lakes and Ocean

In natural systems of large dimensions, such as bodies of water, sediments, the atmosphere, many chemical processes are controlled by the transport of reacting species through the system. The distribution of chemical species in natural systems is only too often not homogeneous; concentration gradients and more or less abrupt changes in abundance from one part of an environment to another are commonplace. In general, the nonhomogeneous distributions of chemical species are a combination of (1) the geometry of the environment: its shape and location of the "sources" and "sinks" of the chemical species; (2) physics: mechanisms of transport of matter through the system; and (3) chemistry: the nature and rates of the chemical reactions in which the species enter.

Knowledge of these three facets of a natural system is indispensable when we need to understand its present chemical state and also to predict quantitatively the changes in the chemical state and their duration, as would occur when the present characteristics of the system change.

The large variation in the values of the turbulent diffusion coefficients reported in the literature for different chemical species in different environments and the laboriousness of their determination in natural environments makes it difficult in many cases to obtain accurate estimates of the time required for a certain chemical process to go to completion. However, when the turbulent diffusivities are not well known, it is still possible in some systems to choose "reasonable" lower and upper limits of the diffusion coefficients and thereby to bracket the model in short and long time estimates.

In view of the primary significance of turbulent diffusional processes in the transport of dissolved matter in a water column, these mechanisms and their bearing on a number of chemical processes are presented in this section. The effects of the magnitude of the eddy diffusivity and the presence of a relatively calm metalimnion on the transport of dissolved species in a stratified body of water are discussed in some simplified lake models.

MODEL I—TWO ADJOINING, WELL-MIXED LAYERS

The first model is highly idealized and illustrates the slowness of the intraphase diffusion process by considering chemical movement between two adjoining, well-mixed chambers separated by a plane that has no resistance. Adjoining, well-mixed layers in the atmosphere or the hydrosphere may be approximated by this crude model to obtain a first-order approximation of the real environmental exchange processes.

A general two-box model is developed so that the final result will have several applications. Figure 7.2-3 illustrates the model status at the start. A certain chemical A has been placed into the lower chamber denoted by ① and the rapid mixing processes have resulted in a uniform concentration, $\bar{c}_{A21}^0$. The chemical is also present in the upper chamber, denoted by ②, but at a low background concentration, $\bar{c}_{A22}^0$. The chambers have depths of h_1 and h_2, respectively. The quantity of chemical A placed in chamber ① is

$$m_A = h_1 A \left(\bar{c}_{A21}^0 - \bar{c}_{A22}^0 \right)$$

where A is the cross-sectional area of the layers.

The movement of chemical A between chambers is a transient process. Initially the rate is rapid because of the large concentration difference, but as time progresses the rate decreases, and c_{A21} decreases while c_{A22} increases. Eventually the concentrations are equal, and net chemical movement ceases. A simultaneous solution of component A material balances on each chamber can be manipulated to yield the concentration-time history of chamber ①;

$$t = \left[\frac{\mathcal{D}_{A22}^{(t)} h_1 + \mathcal{D}_{A21}^{(t)} h_2}{h_1 + h_2} \right] \frac{h_1 h_2}{2 \mathcal{D}_{A21}^{(t)} \mathcal{D}_{A22}^{(t)}}$$

$$\times \ln \left\{ \frac{1 - \left(c_{A22}^0 / c_{A21}^0 \right)}{\left[(h_1/h_2) + 1 \right] \left(c_{A21}/c_{A21}^0 \right) - \left[(h_1/h_2) + \left(c_{A22}^0/c_{A21}^0 \right) \right]} \right\}$$

$$(7.2\text{-}5a)$$

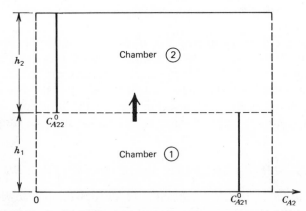

Figure 7.2-3. Chemical movement between two adjoining, well-mixed layers.

where $\mathcal{D}_{A21}^{(t)}$ and $\mathcal{D}_{A22}^{(t)}$ are turbulent diffusivities, characteristic of chambers ① and ②. The bars above the concentrations denoting average are omitted for clarity. Equation 7.2-5a gives the time required for achieving a concentration of c_{A21} in chamber ①. The only restriction on the equation is $c_{A22}^0 < c_{A21}^0$. The equation can be used for the subcases $h_1 = h_2$, $\mathcal{D}_{A21}^{(t)} = \mathcal{D}_{A22}^{(t)}$, $c_{A22}^0 = 0$, or any combination of these subcases.

The usual form of a concentration-time history equation is to have c_{A21} as the dependent variable and t as the independent variable. This equation can be obtained by solving Eq. 7.2-5a for c_{A21}. (See Example 7.2-1.) The concentration in chamber ② can be obtained from the concentration in chamber ① and the relationship

$$c_{A22} = c_{A22}^0 + \frac{h_1}{h_2}\left(c_{A21}^0 - c_{A21}\right) \tag{7.2-5b}$$

Sample calculation results using the preceding model appear in Fig. 7.2-4. The model conditions and parameters are shown in the figure. Using a broad range of turbulent diffusivities, characteristic of water environments, it appears that approximately 4 to 36 years are required before the equalization of the concentration of A occurs in well-mixed adjoining layers.

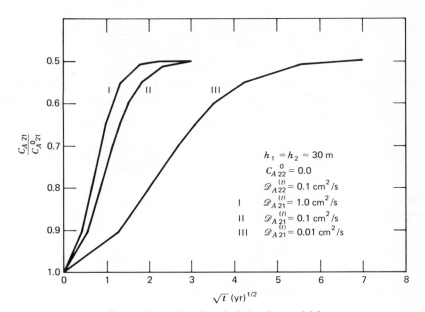

Figure 7.2-4. Sample calculation for model I.

Example 7.2-1.   Movement of a Pesticide between
Two Layers of a Reservoir

Assume that the epilimnion of Beaver Reservoir becomes uniformly contaminated at 5 ppb with a pesticide in a very short period of time because of agricultural runoff following an intense week of rain during the first week in May. It is expected that the pesticide will biodegrade to harmless products within 50 days. The regional water supply intake is below the thermocline, and there is concern about the movement of the pesticide into the hypolimnion. Estimate the maximum pesticide concentration likely to be observed in the hypolimnion. Use Fig. 7.1-4 as a source of data for Beaver Reservoir.

SOLUTION

A study of Fig. 7.1-4 indicates that during May and for a period of about 125 days the reservoir is stratified, and two layers are present. Assuming that the resistance to chemical movement through the thermocline is negligible, Eq. 7.2-5a can be used. Solving the equation for c_{A21} yields

$$c_{A21} = \frac{c_{A21}^0}{[(h_1/h_2)+1]} \left\{ \frac{h_1}{h_2} + \frac{c_{A22}^0}{c_{A21}^0} + \left(1 - \frac{c_{A22}^0}{c_{A21}^0}\right) \right.$$

$$\left. \times \exp\left(\frac{-2\mathcal{D}_{A21}^{(t)}\mathcal{D}_{A22}^{(t)}t}{h_1 h_2}\left[\frac{h_1+h_2}{\mathcal{D}_{A21}^{(t)}h_2 + \mathcal{D}_{A22}^{(t)}h_1}\right]\right)\right\} \quad \text{(E7.2-1)}$$

Let $\textcircled{1} \equiv$ epilimnion and $\textcircled{2} \equiv$ hypolimnion. From Fig. 7.1-4, $h_1 = 10$ m, $h_2 = 30$ m, $\mathcal{D}_{A22}^{(t)} = \mathcal{D}_{A21}^{(t)} = \alpha_2^{(t)} = 0.73$ m^2/d, and $t = 50$ d. Substituting into Eq. E7.2-1 with $c_{A22}^0 = 0$

$$c_{A21} = \frac{5}{(10/30)+1}\left\{\frac{10}{30} + 0 + (1-0)\exp\left(\frac{-2(0.73)50}{(10)(30)}\right)\right\} = 4.19 \text{ ppb}$$

This is the concentration of the pesticide in the upper mixed layer. The concentration in the lower mixed layer is obtained from Eq. 7.2-5b:

$$c_{A22} = 0 + \frac{10}{30}(5.00 - 4.19) = 0.270 \text{ ppb}$$

The maximum concentration likely in the hypolimnion after 50 d is 0.27 ppb.

The preceding model and further models presented in this section contain significant assumptions that allow the development of closed-form equations such as 7.2-5a and 7.2-5b. The use of these models to estimate concentrations and lifetimes of chemicals in real world situations should be tempered with a knowledge of the limitations and assumptions. The limitations and assumptions are:

1. Layers are closed; that is, there is no inflow or outflow.
2. Surface area of the air–water interface is the same as the mud–water interface.
3. Layers have constant depths.
4. There is no chemical movement through or to either interface.
5. Introduced chemicals become uniformly distributed in a short period of time.
6. No chemical or biochemical reaction consumes or produces the chemical species.

MODEL II—TWO WELL-MIXED LAYERS SEPARATED BY AN UNMIXED MIDDLE LAYER

An idealized picture of a stratified body of water is a well-mixed layer at the surface, a layer with a more or less pronounced density gradient (pycnocline) below it, and a well-mixed layer below the pycnocline. Lerman[9] developed a simple model of this situation by assuming that the

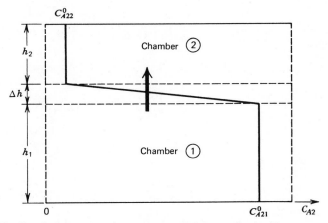

Figure 7.2-5. Chemical movement between two adjoining, well-mixed layers separated by an unmixed layer.

pycnocline, of depth Δh and eddy diffusivity $\mathcal{D}_{A2}^{(t)}$, provided all the resistance to the movement of a dissolved chemical species between two well-mixed chambers. Figure 7.2-5 illustrates the model concentration status at the start of the process.

The simplest way to estimate how long it takes for the concentrations in the lower and upper layer to become equal is to assume that the flux across the pycnocline is at all times proportional to the concentration difference between the upper and lower layer:

$$N_A = \frac{\mathcal{D}_{A2}^{(t)}}{\Delta h}\left(\bar{c}_{A21} - \bar{c}_{A22}\right) \tag{7.2-6}$$

where $\mathcal{D}_{A2}^{(t)}$ is the eddy diffusion coefficient characteristic of the middle layer and Δh is the thickness of the layer. A component balance of species A in the lower chamber yields:

$$\frac{dc_{A21}}{dt} = -\frac{\mathcal{D}_{A2}^{(t)}}{\Delta h h_1}\left(c_{A21} - c_{A22}\right) \tag{7.2-7}$$

The total amount of the chemical species in the lake is

$$m_A = c_{A21}h_1 + c_{A22}h_2 + \left(c_{A21} + c_{A22}\right)\frac{\Delta h}{2} \tag{7.2-8}$$

for a surface area of unity. The total amount may be expressed in terms of the initial concentrations in the two mixed layers and the mean concentration in the pycnocline:

$$m_A = c_{A21}^0 h_1 + c_{A22}^0 h_2 + \left(c_{A21}^0 + c_{A22}^0\right)\frac{\Delta h}{2} \tag{7.2-9}$$

Combining Eqs. 7.2-7, 7.2-8, and 7.2-9 and integrating gives

$$c_{A21} = \frac{m_A}{h_1 + h_2 + \Delta h}\left[1 - \exp\left(-\frac{\mathcal{D}_{A2}^{(t)}t}{\Delta h}(1 + \alpha)\right)\right]$$
$$+ c_{A21}^0 \exp\left(-\frac{\mathcal{D}_{A2}^{(t)}t}{\Delta h}(1 + \alpha)\right) \tag{7.2-10}$$

and

$$c_{A22} = \frac{m_A - c_{A21}(h_1 + \Delta h/2)}{h_2 + \Delta h/2} \tag{7.2-11}$$

where

$$\alpha \equiv \frac{h_1 + \Delta h/2}{h_1(h_2 + \Delta h/2)} \qquad (7.2\text{-}12)$$

The model equations are written so that they can be used for any two well-mixed chambers separated by any unmixed chamber. When the initial concentration in the upper chamber is zero, Eq. 7.2-11 may be put in the form

$$c_{A22} = c_{A21}^0 \frac{h_1 + \Delta h/2}{h_1 + h_2 + \Delta h} \left[1 - \exp\left(-\frac{\mathcal{D}_{A2}^{(t)} t}{\Delta h}(1 - \alpha) \right) \right] \qquad (7.2\text{-}13)$$

It is instructive to perform a sample calculation employing Eq. 7.2-13. Assume a dissolved species has been introduced into the lower mixed layers, and the initial concentration in the upper layer is nil. When the three-layer system remains closed and the dimensions of the water layers do not change, a conservative chemical species in one of the mixed layers will redistribute itself between the two layers because of the diffusional flux down the concentration gradient from one mixed layer into the other. For a case of transport from the lower into the upper, change in the concentration in the upper as a function of time is shown in Fig. 7.2-6. The

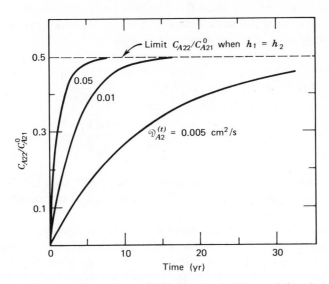

Figure 7.2-6. Sample calculation for model II. (Reprinted with permission. American Chemical Society publication: "Now Equilibrium Systems in Natural Water Chemistry" **Source**. Reference 9.)

curves have been calculated for a 60 m deep water column (lower layer $h_1 = 25$ m, pycnocline $\Delta h = 10$ m, and upper layer $h_2 = 25$ m) for three different eddy diffusion coefficients in the pycnocline (0.005, 0.01, and 0.05 cm^2/s). These values of the eddy diffusion coefficient are in the range reported for pycnoclines in stratified lakes (Table 3.1-4). The curves show that the concentrations in the two layers would equalize in 10 to 40 years. The time required to attain certain concentration levels in such a model lake depends on the eddy diffusivity in the pycnocline and on the vertical dimensions of the individual layers.

MODEL III—TWO ADJOINING, UNMIXED LAYERS[9]

In this model the two fluid layers are not well mixed; each can be characterized by a different value of the turbulent diffusivity. The initial status of this model is as shown in Fig. 7.2-3 except that the fluid in the layers is not well mixed, and concentration gradients develop within each layer. Initially, a conservative species is homogeneously distributed within the lower layer, its concentration being c_{A21}^0. Migration across the boundary between the two layers and subsequent dispersal within the upper layer would eventually equalize the concentrations. At the limit when all gradients disappear, the concentration of the species would be homogeneous throughout the two layers and equal to $c_{A21}^0 h_1/(h_1 + h_2)$.

In the two-layer system, when the diffusion coefficients in the two layers are equal, the concentration of a dissolved substance originally confined to one layer is given by the following relationship[10]:

$$c_{A2} = c_{A22}^0 + \frac{1}{2}c_{A21}^0 \sum_{n=-\infty}^{\infty} \left\{ \operatorname{erf}\frac{h_1 + 2n(h_1 + h_2) - y}{2\sqrt{\mathcal{D}_{A2}^{(i)}t}} \right.$$
$$\left. + \operatorname{erf}\frac{h_1 - 2n(h_1 + h_2) + y}{2\sqrt{\mathcal{D}_{A2}^{(i)}t}} \right\} \qquad (7.2\text{-}14)$$

where c_{A21}^0 is the initial concentration in one layer ($0 < y < h_1$), h_1 and h_2 are the boundaries of the two layers, and y is the vertical dimension ($0 \leqslant y \leqslant h_2$).

For the case when the two diffusion coefficients in the two layers are not equal, derivation of a closed-form relationship for c_{A2} is difficult. As an alternative, Lerman presents a simpler method that gives the mean concentration as a function of time in the upper layer into which the substance diffuses from the lower layer. The method for computing the mean concentration is based on the following: First, it is assumed that the upper layer is semi-infinite, extending from $y = h_1$ upward; second, the amount of dissolved matter transported from the lower layer across the plane $y = h_1$ is evaluated as a function of time; third, the amount of the matter transported from the lower into the upper (semi-infinite) layer up to some time t is divided by the height of the upper layer, $h_2 - h_1$, to obtain the mean concentration in the upper layer $\bar{c}_{A22}$. Lerman reports that the mean concentrations computed by this method are within a few percent of the values obtainable by use of a complete expression for c_{A2}, such as Eq. 7.2-14.

The concentration within a semi-infinite medium c_{A22}, into which the substance diffuses out of the lower layer of initial concentration c_{A21}^0, may be calculated by the following relationship:

$$c_{A22} = c_{A22}^0 + \tfrac{1}{2} c_{A21}^0 \left\{ -(1-p)\operatorname{erf} \frac{y - h_1}{2\sqrt{\mathcal{D}_{A22}^{(t)} t}} + (1-p^2) \sum_{n=1}^{\infty} (-p)^{n-1} \right.$$

$$\left. \cdot \operatorname{erf}\left[\frac{y - h_1}{2\sqrt{\mathcal{D}_{A22}^{(t)} t}} + \frac{n h_1}{\sqrt{\mathcal{D}_{A21}^{(t)} t}} \right] \right\} \qquad (7.2\text{-}15)$$

where $\mathcal{D}_{A21}^{(t)}$ is the diffusion coefficient in the lower layer, $0 \leqslant y \leqslant h_1$, $\mathcal{D}_{A22}^{(t)}$, is the diffusion coefficient in the upper layer, $y > h_1$, and p is the parameter defined as

$$p \equiv \frac{\left[1 - \sqrt{\mathcal{D}_{A21}^{(t)}/\mathcal{D}_{A22}^{(t)}} \right]}{\left[1 + \sqrt{\mathcal{D}_{A21}^{(t)}/\mathcal{D}_{A22}^{(t)}} \right]} \qquad (7.2\text{-}16)$$

The flux of dissolved species A across the plane $y = h_1$ may be obtained from Eq. 7.2-15 by

$$N_A = -\mathcal{D}_{A22}^{(t)} \left[\frac{\partial c_{A2}}{\partial y} \right]_{y = h_1} \qquad (7.2\text{-}17)$$

Performing the operation, we obtain

$$N_A = \tfrac{1}{2} c_{A21}^0 \sqrt{\frac{\mathcal{D}_{A22}^{(t)}}{\pi t}} \left\{ 1 - p - (1-p^2) \sum_{n=1}^{\infty} (-p)^{n-1} \exp\left[-\frac{n^2 h_1^2}{\mathcal{D}_{A21}^{(t)} t} \right] \right\} \qquad (7.2\text{-}18)$$

The amount of matter that crosses the plane $y = h_1$, up to time t, is obtained by integration of Eq. 7.2-18 to give

$$m_A = \int_0^t N_A \, dt = c_{A21}^0 \sqrt{\frac{\mathcal{D}_{A22}^{(t)} t}{\pi}} \left\{ 1 - p - (1 - p^2) \right.$$

$$\cdot \sum_{n=1}^{\infty} (-p)^{n-1} \left[\exp\left[-\frac{n^2 h_1^2}{\mathcal{D}_{A21}^{(t)} t} \right] - \frac{n h_1 \sqrt{\pi}}{\sqrt{\mathcal{D}_{A21}^{(t)} t}} \right.$$

$$\left. \left. \cdot \operatorname{erfc} \frac{n h_1}{\sqrt{\mathcal{D}_{A21}^{(t)} t}} \right] \right\} \tag{7.2-19}$$

To obtain the mean concentration within the upper layer of thickness $h_2 - h_1$, the value of m_A (in mol/L^2) in Eq. 7.2-19 is divided by $h_2 - h_1$, to give

$$\bar{c}_{A22} = c_{A22}^0 + \frac{c_{A21}^0}{h_2 - h_1} \sqrt{\frac{\mathcal{D}_{A22}^{(t)} t}{\pi}} \left\{ 1 - p - (1 - p^2) \sum_{n=1}^{\infty} (-p)^{n-1} \right.$$

$$\left. \cdot \left[\exp\left[-\frac{n^2 h_1^2}{\mathcal{D}_{A21}^{(t)} t} \right] - \frac{n h_1 \sqrt{\pi}}{\sqrt{\mathcal{D}_{A21}^{(t)} t}} \operatorname{erfc} \frac{n h_1}{\sqrt{\mathcal{D}_{A21}^{(t)} t}} \right] \right\} \tag{7.2-20}$$

Equations 7.2-18 to 7.2-20 are physically meaningful for all values of t up to the value at which the mean concentrations in the lower and upper layer become equal and the process of transport across the interface ends. At the end of the process the concentration of the diffusing substance is the same in the two layers.

The concentration-time curves for the upper layer in a two-layer model were calculated using Eq. 7.2-20 with the values of $h_1 = 30$ m, $h_2 = 30$ m, $\mathcal{D}_{A21}^{(t)} = 0.1$ cm^2/s, and $\mathcal{D}_{A22}^{(t)} = 0.01$, 0.1 and 1.0 cm^2/s as shown. As in the previous example calculations, the values of the turbulent diffusivities are taken to represent the range reported for stratified lakes. The calculated curves appear in Fig. 7.2-7. The conclusion that may be drawn from these curves is essentially the same as for the two models presented earlier: The time required to attain equal concentrations is relatively short, 1 to 20 years.

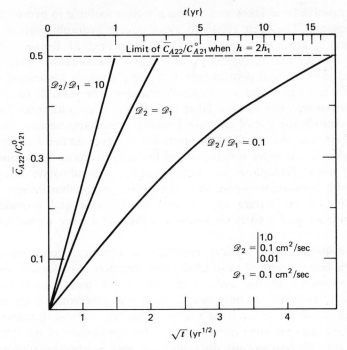

Figure 7.2-7. Sample calculation for model III. (Reprinted with permission. American Chemical Society publication: "Non Equilibrium Systems in Natural Water Chemistry" Source. Reference 9.)

CLOSURE

The preceding discussion of the three simple models suggests that in closed water bodies with stationary stratification of the water column, a change in the chemical composition of one of the layers is a transient phenomenon. The transient phenomenon is relatively short when considering the permanent thermocline in the ocean, but it is long when considering the annual thermocline in lakes. To maintain a steady concentration gradient in either water column, the water bodies must be open, such that the input of solute is balanced by its removal. An example of such a case is presented in the next section.

Steady-State Vertical Profiles of Chemicals in Lakes[11]

Because of certain similarities between lakes and oceans, the models described in this section are appropriate for studying chemical distributions in the coastal and off-shore regions of oceans. In general, chemical distribution problems in lakes and oceans are time variable in two or three space dimensions. The deep, relatively slow-moving hydraulic regime over broad horizontal areas is an important factor in this regard. Because of the greater depths, thermal differences and density structures are seasonally encountered. The vertical distribution of many chemicals associated with these regimes is a significant water quality problem. The variable nature of the wind-driven currents in both lakes and oceans is an additional factor that may preclude the use of simplified mathematical approaches. This is not to suggest that the steady-state approach is never useful, but simply that its application is more restrictive and the analysis more approximate. In spite of these limitations the one-dimensional steady-state analysis applied under appropriate conditions does offer reasonable insight and understanding of the nature of chemical distributions and provides a limiting condition and a basis on which to construct a more sophisticated analysis.

The most pronounced vertical variation in chemical concentrations occurs in the central zone of most lakes. The variation of concentration of various constituents in the vertical is one of the most characteristic chemodynamic features of the water in lakes and reservoirs. Both time-variable and steady-state analyses are valid for the vertical distribution. Although there may be some question about the application of the steady-state analysis to the temperature distribution, it may be appropriate for the dissolved oxygen analysis. During the period of maximum temperature or shortly thereafter, observations indicate in many cases a reasonable constancy of the vertical temperature gradient. Although this condition may not truly reflect a thermal steady state, it may persist for sufficiently long periods to justify the use of the steady-state analysis for other constituents.

CONSERVATIVE SUBSTANCES

A logical starting place for considering steady-state vertical profiles of chemicals in lakes is Eq. 3-1 in mass concentration form:

$$\frac{\partial \rho_{A2}}{\partial t} + \overline{\nabla} \cdot \rho_{A2}\overline{V} = \overline{\nabla} \cdot \mathcal{D}_{A2}^{(t)}\overline{\nabla}\rho_{A2} + r_{A2} \qquad (7.2\text{-}21)$$

If steady state and no advection is assumed in the lake and only the vertical profile of a conservative substance is considered, Eq. 7.2-21 simplifies to

$$0 = \frac{d}{dy}\left(\mathcal{D}_{A2}^{(t)}\frac{d\rho_{A2}}{dy}\right) \qquad (7.2\text{-}22)$$

$\mathcal{D}_{A2}^{(t)}$ is assumed to be a function of y. Two integrations yield

$$\rho_{A2} = c_1 \int_0^y \frac{dy}{\mathcal{D}_{A2}^{(t)}} + c_2$$

where c_1 and c_2 are constants of integration. To evaluate the constant c_2, the following concentration boundary condition is applied at the surface:

$$\rho_{A2} = \rho_{A20} \qquad \text{at } y = 0 \qquad\qquad (7.2\text{-}23)$$

This gives $c_2 = \rho_{A20}$.

The constant c_1 is a flux rate of emission of chemical A from either the air–water interface or the sediment–water interface. Since there are no sources or sinks for the constituent within the water column and steady state prevails, it follows that the water is of a uniform concentration ρ_{A20} and $c_1 = 0$. This simple model of the steady-state distribution of a conservative substance does not yield equations of a practical value. There are three oversimplifications that usually account for this lack of practical utility, any one of which may control.

The first factor is related to the time-variable nature of the phenomena. The time required to achieve a steady-state condition for a conservative ion may extend over a long period in many lakes. (See the preceding section.) The spatial distributions observed in this case are various stages of a slow-moving transient, which may never reach the steady state. This is identical to the question relating to the temperature distribution and the upsurge of nutrients from the mud–water interface. (See Sections 7.1 and 5.2, respectively, for a detailed analysis.)

The second factor that affects the distribution of conservative substances in lakes emanating from the bed is related to the boundary conditions at this location. The problem is slightly more complex, involving the interaction between two separate systems—the vertical water column and an associated segment of the bed—each with its own differential equation. With respect to the water column, the boundary condition at the mud–water interface is transient.

The third is the fact that a small amount of advection, either horizontal or vertical, can produce observable concentration gradients; consequently, the simplistic dispersion model may be inadequate to describe the distribution of a conservative substance, and the more encompassing advective-dispersion analysis is required. Although it may be desirable to employ the more general model for a conservative substance, it does not necessarily follow that it is required for a nonconservative substance, for which the dispersive equation subsequently described in this section may be quite adequate.

NONCONSERVATIVE SUBSTANCES

The vertical distribution of a nonconservative substance in lakes and reservoirs under steady-state conditions is similar in origin to Eq. 7.2-22:

$$0 = \frac{d}{dy}\left(\mathcal{D}_{A2}^{(t)} \frac{d\rho_{A2}}{dy} \right) + r_{A2} \qquad (7.2\text{-}24)$$

The reaction term r_{A2} accounts for sources or sinks within the vertical column.

The following analysis applies specifically to the vertical distribution of dissolved oxygen in a lake water column. Although the analysis is specific, the general technique is applicable to all nonconservative chemical species such as radon, methane, ammonia, orthophosphate, mercury, and so on. (See problems 7.2C and 7.2D.)

When applying Eq. 7.2-24 to the vertical distribution of dissolved oxygen in a lake, the internal sources and sinks must be defined and the boundary conditions specified. If one considers the more general case both a source of oxygen, caused by either a chemical or biochemical oxidation, and a sink must be assumed, both of which are functions of depth:

$$r_{A2} = r_{A2}|_c - r_{A2}|_p \qquad (7.2\text{-}25)$$

where the p and c subscripts denote oxygen production and consumption.

A HOMOGENEOUS OR UNSTRATIFIED LAKE

The boundary conditions may be specified at the surface ($y = 0$) and the bed ($y = h$). The air–water interface, which allows for the transfer of oxygen through the surface layer, is the first boundary condition:

$$\left[-\mathcal{D}_{A2}^{(t)} \frac{d\rho_{A2}}{dy} \right]_{y=0} = {}^1k'_{A2}(\rho_{A2}^* - \rho_{A2})\bigg|_{y=0} \qquad (7.2\text{-}26)$$

where ρ_{A2}^* is dissolved oxygen saturation value at a given temperature. The benthal oxygen demand at the sediment–water interface provides the second boundary condition:

$$\left[-\mathcal{D}_{A2}^{(t)} \frac{d\rho_{A2}}{dy} \right]_{y=h} = n_{Ah} \qquad (7.2\text{-}27)$$

Rearranging Eq. 7.2-24 and integrating twice yields the following:

$$\mathcal{D}_{A2}^{(t)} \frac{d\rho_{A2}}{dy} = -\int_0^y r_{A2}\,dy + C_1 \qquad (7.2\text{-}28)$$

and

$$\rho_{A2} = -\int_{y=0}^{y} \frac{dy}{\mathcal{D}_{A2}^{(t)}} \int_{y=0}^{y} r_{A2} dy + C_1 \int_{y=0}^{y} \frac{dy}{\mathcal{D}_{A2}^{(t)}} + C_2 \qquad (7.2\text{-}29)$$

The two unknown constants, C_1 and C_2, may be evaluated by applying the two boundary conditions. At the surface $\rho_{A2} = \rho_{A20}$ and Eq. 7.2-26 applies to yield

$$C_2 = \rho_{A20} \quad \text{and} \quad C_1 = -{}^1k'_{A2}(\rho^*_{A2} - \rho_{A20}). \qquad (7.2\text{-}30)$$

Substituting these into Eq. 7.2-29 yields

$$\rho_{A2} = \rho_{A20} - \int_{y=0}^{y} \frac{dy}{\mathcal{D}_{A2}^{(t)}} \int_{y=0}^{y} r_{A2} dy - {}^1k'_{A2}(\rho^*_{A2} - \rho_{A20}) \int_{0}^{y} \frac{dy}{\mathcal{D}_{A2}^{(t)}} \quad (7.2\text{-}31)$$

The remaining unknown, ρ_{A20}, may be evaluated by application of the second boundary condition, Eq. 7.2-27, at the bed, $y = h$, to Eq. 7.2-31, which yields

$$\rho_{A20} = \rho^*_{A2} - \frac{1}{{}^1k'_{A2}} \left(n_{Ah} - \int_0^h r_{A2} dy \right) \qquad (7.2\text{-}32)$$

The final equation is obtained by substituting Eq. 7.2-32 into Eq. 7.2-31 and simplifying to give

$$\rho_{A2} = \rho^*_{A2} - \left[n_{Ah} - \int_0^h r_{A2} dy \right] \left[\frac{1}{{}^1k'_{A2}} + \int_0^y \frac{dy}{\mathcal{D}_{A2}^{(t)}} \right]$$

$$- \int_0^y \frac{dy}{\mathcal{D}_{A2}^{(t)}} \int_0^y r_{A2} dy \qquad (7.2\text{-}33)$$

This equation gives the vertical distribution of dissolved oxygen or any chemical affected by comparable surface and bed conditions, such as carbon dioxide, in a lake with variable eddy diffusivity and reaction rate.

A simple application of the nonconservative, homogeneous lake model is considered. The actual conditions reflected in this subsection are admittedly not widely applicable since variable diffusivity and reaction are usually encountered in most lake environments. However, these simplified conditions provide some insight into the nature of the exchange and distribution of such constituents as dissolved oxygen. Furthermore, they may be used as an approximate solution to the more realistic situations, particularly in the central zone of broad lakes. Given the conditions that the source and sink are the surface and bed transfers and there is no reaction within the water column, Eq. 7.2-33 reduces to

$$\rho_{A2} = \rho_{A2}^* - n_{Ah}\left[\frac{1}{{}^1k_{A2}'} + \frac{y}{\mathcal{D}_{A2}^{(t)}}\right] \tag{7.2-34}$$

for constant $\mathcal{D}_{A2}^{(t)}$. A close study of this final equation yields two significant terms: one, $\rho_{A20} = \rho_{A2}^* - n_{Ah}/{}^1k_{A2}'$, which is the dissolved oxygen at the surface ($y=0$), and the other, the dimensionless number ${}^1k_{A2}'h/\mathcal{D}_{A2}^{(t)}$ in which $y=h$. Equation 7.2-34 can be transformed to

$$\rho_{A2} = \rho_{A2}^* - (\rho_{A2}^* - \rho_{A20})\left[1 + \frac{{}^1k_{A2}'y}{\mathcal{D}_{A2}^{(t)}}\right] \tag{7.2-35}$$

It can be seen from this result that if ${}^1k_{A2}'h/\mathcal{D}_{A2}^{(t)}$ is small and much less than unity, as is the case for large eddy diffusivity, the concentration is uniform and equal to ρ_{A20}. A common range for ${}^1k_{A2}'$ is 1.0 to 5.0 ft/d and for $\mathcal{D}_{A2}^{(t)}$ is 10 to 50 ft^2/d in the eqilimnion. For depths of 10 to 50 ft, the dimensionless parameter ${}^1k_{A2}'h/\mathcal{D}_{A2}^{(t)}$ has the numerical values shown in Table 7.2-1. It is apparent that in most practical cases for the conditions considered, significant gradients of dissolved oxygen develop except in the extreme case of ${}^1k_{A2}'=1$, $\mathcal{D}_{A2}^{(t)}=50$, and $h=10$. It should be emphasized that no sources such as photosynthesis were considered in this analysis.

Table 7.2-1.   Oxygen Absorption–Dispersion Parameter for a Homogeneous Lake, ${}^1k_{A2}'h/\mathcal{D}_{A2}^{(t)}$

${}^1k_{A2}'$(ft/d)	1.0		5.0	
$\mathcal{D}_{A2}^{(t)}$ (ft^2/d)	10	50	10	50
h (ft)				
10	1.0	0.2	5.0	1.0
50	5.0	1.0	25	5.0

Source. Reference 11.

Example 7.2-2.   Vertical Oxygen Profiles in Unstratified Lakes

This problem illustrates, among other things, the importance of the oxygen consumption rate at the mud–water interface on the oxygen concentration profile. Consider two unstratified lakes containing little or no phytoplank-ton that produce oxygen and little or no dissolved organic and/or in-organic material to consume dissolved oxygen within the water column (i.e., $r_{A2}=0$). One lake is a large, artificial reservoir with bottom muds consisting of a quantity of organic matter (i.e., decaying vegetation) left over from the prefilling days. The other lake is a smaller, artificial "run-of-the-river" reservoir in which the detention time is low and the flow is relatively rapid. For a water temperature of 30°C, compute and graph the vertical oxygen concentration profile as a function of depth for each lake type. Table E7.2-2 contains pertinent chemical and physical data for both lakes.

Table E7.2-2

Description	n_{Ah}^{a}	$^1k'_{A2}$	$\mathcal{D}_{A2}^{(t)}$	h
	(g O_2/m²·d)	(m/d)	(m²/d)	(m)
Large reservoir, mud bottom	0.2	0.3	0.7	20
Run-of-the-river reservoir, sandy bottom	0.05	1.5	4.0	10

^{a}Source. Reference 12.

SOLUTION

Since $r_{A2}=0$, Eq. 7.2-34 applies. At 30°C $\rho_{A2}^{*}=7.63$ mg O_2/L.
For the large reservoir:

$$\rho_{A2}=7.63-\frac{0.2\text{ g }O_2}{\text{m}^2\cdot\text{d}}\left[\frac{\text{d}}{0.3\text{ m}}+\frac{y(m)\text{ d}}{0.7\text{ m}^2}\right]$$

$$\rho_{A2}=6.96-0.286y,\ y\leqslant 20\text{ m} \qquad (E7.2-2a)$$

For the run-of-the-river reservoir:

$$\rho_{A2}=7.63-0.05\left(\frac{1}{1.5}+\frac{y}{4}\right)$$

$$\rho_{A2}=7.60-0.0125y,\ y\leqslant 10\text{ m} \qquad (E7.2-2b)$$

Figure E7.2-2 is a graph of each concentration profile. The surface oxygen concentration in both cases is less than ρ_{A2}^{*} and decreases with depth. A large reaeration coefficient and large turbulent diffusivity coupled with a low oxygen consumption rate at the mud–water interface and a shallow

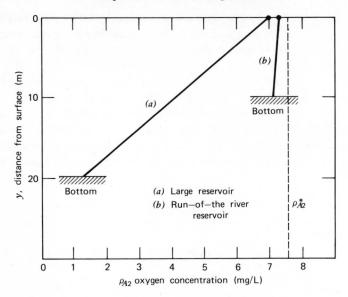

Figure E7.2-2. Oxygen concentration profiles in two reservoirs.

depth results in a near vertical profile displaying uniform oxygen concentration. The profile displayed for the large reservoir reflects a significant oxygen concentration gradient.

A TWO-LAYER OR STRATIFIED LAKE

A more realistic condition is that of a lake segmented vertically into two zones, representing the epilimnion and hypolimnion, each with its characteristic eddy diffusivity. This case is a realistic representation of a stratified lake normally encountered during the summer months. The thermocline is located at $y = h_1$ and the sediment bed at $y = h_2$. The depth of the thermocline is thus h_1 and that of the hypolimnion $h_2 - h_1$.

The solution proceeds along similar lines as that for the unstratified lake. The solution for the upper layer is identical to that determined for the previous case. The concentration profile in layer 1 can be determined by replacing the upper limit of integration h in Eq. 7.2-33 by $h_1 + h_2$. The equation can then be used for values of $0 \leqslant y \leqslant h_1$. Specifically for layer 1,

$$\rho_{A21} = \rho_{A2}^* - \left[n_{Ah_2} - \int_0^{h_1 + h_2} r_{A2}\, dy \right]\left[\frac{1}{{}^1 k'_{A2}} + \int_0^y \frac{dy}{\mathcal{D}_{A21}^{(t)}} \right]$$

$$- \int_0^y \frac{dy}{\mathcal{D}_{A21}^{(t)}} \int_0^y r_{A21}\, dy \qquad\qquad (7.2\text{-}36)$$

Developing the concentration profile for the lower layer commences with

$$0 = \frac{d}{dy}\left[\mathcal{D}_{A22}^{(t)}\frac{d\rho_{A22}}{dy}\right] + r_{A22} \tag{7.2-37}$$

where $\mathcal{D}_{A22}^{(t)}$ and r_{A22} are the eddy diffusivity and reaction rate terms characteristic of the lower layer. By specifying the flux and the concentration at the thermocline $y = h_1$, the two needed boundary conditions are obtained:

$$\mathcal{D}_{A22}^{(t)}\frac{d\rho_{A22}}{dy} = \mathcal{D}_{A21}^{(t)}\frac{d\rho_{A21}}{dy} @ y = h_1 \tag{7.2-38}$$

and

$$\rho_{A22} = \rho_{A2h_1} @ y = h_1 \tag{7.2-39}$$

Equation 7.2-37 is now integrated twice to yield

$$\mathcal{D}_{A22}^{(t)}\frac{d\rho_{A22}}{dy} = -\int_{h_1}^{y} r_{A22}\,dy + C_3 \tag{7.2-40}$$

and

$$\rho_{A22} = -\int_{h_1}^{y}\frac{dy}{\mathcal{D}_{A22}^{(t)}}\int_{h_1}^{y} r_{A22} + C_3\int_{h_1}^{y}\frac{dy}{\mathcal{D}_{A22}^{(t)}} + C_4 \tag{7.2-41}$$

Applying the boundary condition represented by Eq. 7.2-38 to Eq. 7.2-40 yields

$$\mathcal{D}_{A21}^{(t)}\frac{d\rho_{A21}}{dy}\bigg|_{h_1} = C_3 \tag{7.2-42}$$

Now the left-hand side can be obtained from Eq. 7.2-28 knowing $C_1 = -\,^1k'_{A2}(\rho_{A2}^* - \rho_{A20})$ to yield

$$-\int_0^{h_1} r_{A21}\,dy - \,^1k'_{A2}(\rho_{A2}^* - \rho_{A20}) = C_3 \tag{7.2-43}$$

Applying the boundary condition represented by Eq. 7.2-39 yields $\rho_{A2h_1} = C_4$. Substituting the appropriate expressions for C_3 and C_4 along with Eq.

7.2-32 for ρ_{A20} yields

$$\rho_{A22} = \rho_{A2h_1} - \int_{h_1}^{y} \frac{dy}{\mathscr{D}_{A22}^{(t)}} \int_{h_1}^{y} r_{A22}\,dy - n_{Ah_2} \int_{h_1}^{y} \frac{dy}{\mathscr{D}_{A22}^{(t)}} \qquad (7.2\text{-}44)$$

where $h_2 \geqslant y \geqslant h_1$.

This final relationship can be used to calculate the oxygen concentration in the hypolimnion of a lake. It should be noted that ρ_{A2h_1} is computed from Eq. 7.2-36 at $y = h_1$. Proper utilization of Eqs. 7.2-36 and 7.2-44 requires that $\mathscr{D}_{A21}^{(t)}$, $\mathscr{D}_{A22}^{(t)}$, r_{A21}, and r_{A22} be known functions of y. Problem 7.2D is realistic concentration profile calculation that includes oxygen production and utilization rates.

CLOSURE

This section introduces basic concepts one should employ when estimating concentration profiles of nonconservative chemicals in water. The specific illustration involved oxygen; however, an analogous procedure can be used to formulate steady-state models for any chemical influenced by sources or sinks within the water column or at the interfaces. Descriptions of situations involving other chemicals may be found in the problems at the end of Section 7.2.

Problems

7.2A. RATE-LIMITING DIFFUSION PROCESSES AT DEPTHS IN WATER BODIES

When the topic of evaporation of chemicals from a lake ecosystem was considered in Chapter 4, it was mentioned that other rate-limiting diffusion processes at depths in water bodies, for example, the thermocline, may be as significant as or more significant than those near the air–water interface.

1. Show for a stratified lake or ocean that

$$\frac{1}{^1K'_{A2}} = \frac{1}{^1k'_{A2}} + \frac{c_2}{c_1 H_{Ax}\,^2k'_{A1}} + \frac{h_e}{\mathscr{D}_{A2e}^{(t)}} + \frac{h_m}{\mathscr{D}_{A2m}^{(t)}} + \frac{h_h}{\mathscr{D}_{A2h}^{(t)}} \qquad (7.2\text{A-1})$$

is applicable for the transport of a chemical from the air through the water column to the water–mud interface or for transport of a chemical in the opposite direction.

2. Table 4.2-1 contains evaporation parameters for various compounds from lakes. The last column, containing the overall mass transfer coefficient for the air–water interface region, accounts for the first two terms on the right-hand side of Eq. 7.2A. Use Eq. 7.2A-1 to evaluate the significance of stratification on the total mass transfer coefficient. Do this by first computing the epilimnion resistance, the metalimnion resistance, the hypolimnion resistance (i.e., $h_e/\mathcal{D}_{A2e}^{(t)}$, $h_m/\mathcal{D}_{A2m}^{(t)}$, and $h_n/\mathcal{D}_{A2h}^{(t)}$ for n-octane, benzene, DDT, and Aroclor 1242. Use data for the three reservoirs shown in Figure 7.1-7. Report all resistances in hr/m. Are the other rate-limiting diffusion processes at depths in water bodies significant? Discuss your answer.

7.2B. MOVEMENT OF A PESTICIDE THROUGH THE THERMOCLINE OF A RESERVOIR

Employing the problem statement of Example 7.2-1 and the results of model II, reestimate the maximum concentration likely to be observed in the hypolimnion of Beaver Reservoir.

7.2C. STEADY-STATE PHOSPHORUS PROFILE IN A LAKE

Phosphorus as orthophosphate is leached out of the mud on a lake bottom, migrates through the hypolimnion, and is consumed by algal particles in the epilimnion.

1. Develop the model equations for a steady-state phosphorus profile.
2. Calculate concentration (mg/L) as orthophosphate versus distance (m) from the air–water interface for the conditions shown in Table 7.2C.

Table 7.2C

Layer	h (m)	$\mathcal{D}_{A2}^{(t)}$ (m²/d)	r_{A2} (g/m³·d)	ρ_{A2} @ $h_2=0.5$ mg/L
Epilimnion	$h_1=5$	1	0.001	
Hypolimnion	$h_2=20$	0.2	0.0	n_{Ay} @ $h_2=0.0043$ g/m²·d

7.2D. VERTICAL OXYGEN PROFILE IN A STRATIFIED LAKE

Consider a stratified lake that contains sufficient phytoplankton to produce oxygen via respiration at a rate

$$r_{A2}=0.25\exp\left(-y\right) \qquad (7.2D\text{-}1)$$

with r_{A2} in g $O_2/m^3 \cdot d$ and y in meters from the air–water interface. For a water surface temperature of 30°C compute and graph the steady-state, daylight, vertical oxygen concentration profile as a function of depth. Table 7.2D contains pertinent chemical and physical data for both layers of the lake.

Table 7.2D

Layer	$\mathcal{D}_{A2}^{(\ell)}$ (m²/d)	h (m)	Other Data
Epilimnion	1.0	5	$^1k'_{A2} = 0.2$ m/d
Hypolimnion	0.2	20	$n_{Ah_2} = 0.05$ g $O_2/m^2 \cdot d$

7.2E. VERTICAL OXYGEN PROFILE IN A STRATIFIED LAKE— A SIMPLE SYSTEM

Consider a stratified lake containing little or no phytoplankton that produce oxygen and little or no organic and/or inorganic matter that demands dissolved oxygen within the water column. For a water surface temperature of 30°C, compute and graph the vertical oxygen concentration profile as a function of depth. Table 7.2E contains pertinent chemical and physical data for both layers of the lake.

Table 7.2E

Layer	$\mathcal{D}_{A2}^{(\ell)}$ (m²/d)	h (m)	$n_{Ah_1} = 0.2$ g $O_2/m^2 \cdot d$
Epilimnion	1.0	10	
Hypolimnion	0.4	10	$^1k'_{A2} = 0.3$ m/d

7.3. INTRAPHASE CHEMICAL TRANSPORT WITHIN A HOMOGENEOUS MEDIUM

The waste products of our civilization must be disposed of. Receptacles for this debris are the earth's land masses, water bodies, and atmosphere. Waste that is released to the homogeneous fluid medium water and air is the topic of this section.

Wastes that are released to the atmosphere consist of particles and gases, and wastes released to water bodies consist of particles and substances in solution. Residence times for some of these materials may be very short— hours or even minutes. For other materials, residence times may be measured in terms of years. Regardless of the residence time, the movement of molecules and particles in the respective media is, in large measure, governed by the motions of the medium. Some motions dictate the paths to be followed by the contaminants, and other motions determine the extent to which the contaminants are diluted.

It is to be expected that in the future the designs of disposal systems for gaseous and liquid wastes will meet increasingly stricter pollution control regulations aimed at limiting specific chemical contaminants. The decisions involved will have to be based on reasonably accurate quantitative predictions of the effects of chemicals introduced into the atmosphere, ocean, lakes, and rivers.

A major practical result of this section will be methods for estimating concentrations of contaminants from continuous point sources. Calculation procedures are presented for obtaining estimates of ground level pollutant concentrations due to a pollutant plume centered at a constant height h above the surface of the earth. Oceanic and lake diffusion applications involve the use of theoretical models to predict concentrations of pollutants in sewage plumes in water. Both of these important applications involve variations of the classical Gaussian plume diffusion model.

The spread of any pollutant in the atmosphere or in various bodies of water takes place in a flow field that is almost invariably turbulent. A brief introduction to basic turbulent flow concepts was presented in Section 3.2. Without attempting here a detailed discussion, we give a brief review of the fundamental physical concepts of the statistical theory of turbulent diffusion. For a more detailed introduction to the subject, the reader is referred to Csanady[13] and Slade.[14]

In a general way, one may say that matter in the gaseous or liquid state is often subject to two kinds of random movements: one on the molecular scale, the thermal agitation of molecules, and one on a macroscopic scale, turbulence. In everyday life one may observe turbulence on a number of occasions—for example, in a billowing smoke plume.

The main differences between molecular agitation and turbulent movements are (1) a difference of scale, the typical "sweep" of a single turbulent movement being ordinarily very large compared to one molecular free path, and (2) the constraint imposed by continuity. The fairly large parcels of fluid partaking of turbulent movements can only move by displacing other fluid, which eventually has to fill in the space vacated by the moving parcel. Consequently, one may picture turbulent flow as consisting of a number of closed-flow structures of diverse shapes and sizes, called *eddies*, although it would be a mistake to think of these as resembling regular vortices or indeed regular flow structures of any kind.

Mainly because of these differences, two different theoretical models have evolved. The molecular diffusion and Brownian diffusion processes can be adequately quantified by gradient transport models such as Fick's law, whereas the turbulent diffusion processes are best quantified by statistical models. Both the molecular and the Brownian processes are Gaussian, and the experimental evidence indicates that the turbulent processes of diffusion are also Gaussian; however, each is characterized by widely different mean square displacements σ^2. Recall that the Gaussian distribution, or normal distribution, is the most commonly occurring probability distribution and has the form

$$\frac{1}{\sqrt{2\pi\sigma}} \int_{-\infty}^{u} \exp\left(\frac{-u^2}{2}\right) du \qquad (7.3\text{-}1)$$

where $u \equiv (x - e)/\sigma$, e is the mean, and σ is the variance.

Air Pollutant Diffusion Theories

The problem of turbulent diffusion in the environment has not been uniquely formulated in the sense that a single basic physical model capable of explaining all the significant aspects of the problem has not yet been proposed. Instead there are available two alternative approaches, neither of which can be categorically eliminated from consideration since each has areas of utility that do not overlap the other's. The two approaches to diffusion are the gradient transport theory and the statistical theory. Diffusion at a fixed point in the atmosphere, according to the gradient transport theory, is proportional to the local concentration gradient. Consequently, it could be said that this theory is Eulerian in that it considers properties of the fluid motion relative to the spatial fixed coordinate system. On the other hand, statistical diffusion theories consider motion following fluid particles and thus can be described as Lagrangian. Diffusion theories may be described as either continuous-motion or discontinuous-motion theories, depending on whether the particle motion is postulated to occur continuously or as discrete events. There must necessarily be a close connection among all these approaches to the diffusion problem, since obviously there is only one atmosphere.

GRADIENT TRANSPORT MODELS

This topic needs little introduction since it has been used extensively through the book in various applications. A prior introduction appears in Section 3.1. For our purposes it is sufficient to regard air and water as incompressible fluids. The mathematical statement of Fick's law has (in the three-dimensional case) the form of the classical equation of conduction:

$$\frac{\partial \rho_{A1}}{\partial t} + \bar{v} \cdot \bar{\nabla} \rho_{A1} = \mathcal{D}_{A1} \nabla^2 \rho_{A1} \qquad (7.3\text{-}2)$$

where $\mathcal{D}_{A1}$ (in the atmosphere) is the eddy diffusivity coefficient or the molecular diffusivity and ρ_{A1} refers to the value of some conservative air property per unit volume of air. For illustration we discuss here some solutions of particular relevance to atmospheric and oceanic diffusion problems.

Consider first the idealized one-dimensional case of an instantaneous plane source in a uniform medium at rest ($\bar{v}=0$). The one-dimensional equation (Eq. 3.1-3) describes a phenomenon not dependent on y or z—say, diffusion along the axis of a duct or pipe with conditions uniform across the section, or in the atmosphere along the vertical, assuming homogeneous conditions in the horizontal. The solution to the instantaneous plane source problem is

$$\rho_{A1} = \frac{m'_A}{2\sqrt{\pi \mathcal{D}_{A1} t}} \exp\left(\frac{-x^2}{4\mathcal{D}_{A2} t}\right) \qquad (7.3\text{-}3a)$$

where m'_A is an amount of material per unit area. The asymptotic behavior of $\bar{\rho}_{A1}$ as $t \to 0$ is

$$\rho_{A1} = 0, \qquad |x| > 0$$

$$\rho_{A1} = \infty, \qquad x = 0 \qquad (7.3\text{-}3b)$$

These properties show that as $t \to 0$, $\bar{\rho}_{A1}$ becomes proportional to the "delta function"

$$\rho_{A1} = m'_A \delta(x) \qquad (t=0) \qquad (7.3\text{-}3c)$$

The delta function is the mathematical description of such idealizations as the "concentrated source" such that Eq. 7.3-3a describes the diffusion of an amount of material m'_A that was initially in a thin sheet at $x=0$, in an instantaneous plane source. As time proceeds, the material spreads out as illustrated in Fig. 7.3-1.

Equation 7.3-3a is otherwise known as a *Gaussian* or *normal* curve, with the maximum concentration at the center of gravity, which remains at the origin. The movement of inertia or second moment of this distribution may

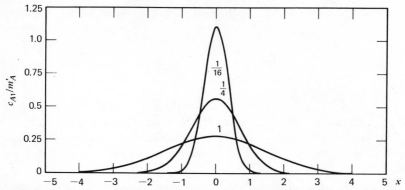

Figure 7.3-1. Concentration-distance curves for an instantaneous plane source. Numbers are values of Dt. (Reprinted by permission. **Source**. Reference 10.)

be used to characterize its spread

$$\int_{-\infty}^{\infty} \rho_{A1} x^2 \, dx = 2m'_A \mathcal{D}_{A1} t \tag{7.3-4}$$

If we divide this by the total diffusion material, we obtain a kind of mean square distance to which the particles have diffused.

$$\sigma^2 = 2\mathcal{D}_{A1} t \tag{7.3-5}$$

The length σ serves as a convenient scale of the width of the distribution, and is known as the *standard deviation*. Expressed in terms of this scale the distribution becomes

$$\rho_{A1} = \frac{m'_A}{\sqrt{2\pi}\sigma} \exp\left(\frac{-x^2}{2\sigma^2}\right) \tag{7.3-6}$$

This is perhaps the usual form in which one encounters a Gaussian distribution. An advantage of this form is that it also applies in turbulent diffusion, where σ is a different function of time since the process is not Fickian.

Similarly, a product of three one-dimensional solutions provides a description of diffusion from an "instantaneous point source":

$$\rho_{A1} = \frac{m_A}{(\sqrt{2\pi}\,\sigma)^3} \exp\left(-\frac{x^2}{2\sigma^2} - \frac{y^2}{2\sigma^2} - \frac{z^2}{2\sigma^2}\right) \tag{7.3-7}$$

where still $\sigma = \sqrt{2\mathcal{D}_{A1}t}$ for Fickian diffusion and m_A is the amount of material released at the origin at $t = 0$ in grams.

All the previous results pertain to diffusion of a chemical or particle in a uniform medium at rest. From the standpoint of environmental diffusion, a somewhat more realistic model is one in which the medium moves at a constant and uniform velocity v_x in the x-direction. Mathematically, this case differs little from the at-rest medium case in that the coordinate system moving at the wind velocity v_x is still at rest. Applying the coordinate transformation $x' = x - v_x t$, we have at once an equation of the instantaneous point source in a wind:

$$\rho_{A1} = \frac{m_A}{(\sqrt{2\pi}\,\sigma)^3} \exp\left\{ -\frac{(x - v_x t)^2 + y^2 + z^2}{2\sigma^2} \right\} \qquad (7.3\text{-}8)$$

This equation is a model of a single "puff" (point source cloud) of material moving downwind and growing in size.

A number of important problems may be modeled by the assumption that a source emits material continuously. The case of a continuous point source in a wind provides a particularly important model (a very crude model of a chimney plume). The concentration field now consists of a series of "puffs" with their centers stretched out along the x axis as the growing puffs are convected downwind. For a point source maintained indefinitely, let the rate of emission be w_A (g/s), such that in the short interval t to $t + dt$ an amount $w_A\,dt$(g) is emitted. Each "puff" generates its own cloud, and the total concentration is obtained by a summation of contributions from the individual puffs. We obtain the combined concentration field of the many puffs by integration (see Csanady[13]) to obtain

$$\rho_{A1} = \frac{w_A}{4\pi\mathcal{D}_{A1}r} \exp\left[-\frac{v_x}{2\mathcal{D}_{A1}}(r - x) \right] \qquad (7.3\text{-}9)$$

where $r^2 = x^2 + y^2 + z^2$. The concentration distribution is independent of time. This final result is not generally applicable for modeling all plumes; however, it is applicable to diffusion in uniform laminar flow and some isotropic turbulent flow cases such as the center regions of pipes. It is applicable to all cases in which $\mathcal{D}_{A1}^{(t)}$ is constant.

STATISTICAL THEORY OF TURBULENT DIFFUSION

It is apparent that the loops in the plume shown in Fig. 7.3-2 are not the result of concentration gradients but are due to eddies imbedded within

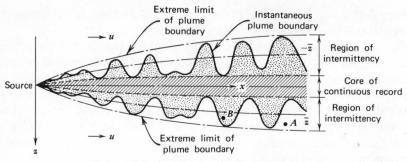

Figure 7.3-2. Continuous point source plume.

the flow structure. An instantaneous sampler located at point A would not detect the diffusing pollutant species, whereas one located at point B would. We recognize at the outset therefore that practically observable concentrations of a pollutant at some location in the environment are, in general, random variables, about which we are only able to make probabilistic predictions. Effectively the only quantity about which we have adequate evidence, both theoretical and experimental, is the first moment of the concentration probability distribution, that is, its "expectation." This expectation can be shown to be also equal to the mean concentration of an ensemble of independently diffusing particles. The statistical model results relate to the ensemble–mean concentration field.

At a fixed point in the wake of a continuous release of a tracer at a constant rate into a statistically steady and homogeneous field of turbulence of constant mean velocity, the observable ("instantaneous") concentration varies in a random manner. By the ergodic property of such processes, the ensemble–mean concentration at a given time t is equal to the time-averaged concentration $\bar{\rho}_{A1}$, obtained by averaging the concentration observed in a single experiment over a sufficiently long period of time compared to the lifetime of a typical turbulent eddy.

The time-averaged concentration field at a fixed point may be related to particle displacement probabilities. This leads us to what might be called the elementary statistical theory of turbulent diffusion. The importance of this theory rests partly on the fact that a continuous point source in a uniform wind models fairly satisfactorily such prime pollution sources as factory chimneys, although in applying the theory and obtaining the necessary parameters, many approximations have to be made.

The random movements of a diffusing particle in a field of homogeneous and stationary turbulence consist of the movements within eddies and a superimposed Brownian or molecular motion. Both components contribute to dispersion independently, the mean square displacements being simply the sum of squares arising from either source. For the mean square displacements along the x axis:

$$\sigma_x^2 = \sigma_{xt}^2 + 2\mathcal{D}_{A1}^{(l)} t \qquad (7.3\text{-}10)$$

where σ_{xt}^2 is the mean square displacement due to bulk motions alone. In most environmental applications the molecular contribution $2\mathcal{D}_{A1}^{(l)} t$ is negligible, and $\sigma_x^2 \simeq \sigma_{xt}^2$.

Experimental evidence on diffusion in a homogeneous field shows that the particle-displacement probability distribution, of which the mean square dispersion σ_x^2 is the second spatial moment, is to a high degree Gaussian. Because its second moments completely specify a Gaussian distribution, the problem of describing the mean concentration field is thus for practical purposes solved. However, the question as to why a Gaussian distribution is observed in experiments is not satisfactorily answered in theory.

On accepting that the probability distribution function of the independently diffusion particles is Gaussian, we may now write down the ensemble–average concentration field of an "instantaneous point source." Since diffusion by turbulent movement is a linear phenomenon, "continuous sources" may be built up by adding the fields of simple sources. In this instance superposition yields

$$\bar{\rho}_{A1} = \frac{w_A}{(2\pi)^{2/3}} \int_0^\infty \exp\left\{ -\frac{(x - v_x t)^2}{2\sigma_x^2} - \frac{y^2}{2\sigma_y^2} - \frac{z^2}{2\sigma_z^2} \right\} \frac{dt}{\sigma_x \sigma_y \sigma_z} \qquad (7.3\text{-}11)$$

where σ_x, σ_y, and σ_z are the mean square dispersions on the principal axes. This equation cannot be integrated until the functional dependence of the standard deviations σ_x, σ_y, σ_z on diffusion time t is specified.

At short distances from the source the instantaneous-source ensemble average cloud grows according to

$$\sigma_x = \frac{\overline{(v_x'^2)}^{1/2} x}{v_x} \qquad (7.3\text{-}12)$$

where $\overline{(v_x'^2)}^{1/2}/v_x$ are the root-mean-square velocities. (See Section 3.3.) This equation and similar relations for σ_y and σ_z are a consequence of *Taylor's theorem*. Taylor's theorem is a most important basic result in the theory of diffusion by random movements. The interested reader should consult Csanady[13] for details.

Equation 7.3-11 can be integrated once the form of the σ's are specified. One finds after a few simple approximations:

$$\bar{\rho}_{A1} = \frac{w_A}{2\pi\sigma_y\sigma_z v_x}\exp\left\{-\frac{y^2}{2\sigma_y^2}-\frac{z^2}{2\sigma_z^2}\right\} \qquad (7.3\text{-}13)$$

where we have substituted the time-dependent standard deviations back into the final result. The diffusion takes place along y and z only, and not in the x direction, the mean concentration gradients along x being negligible compared to those along y and z. This equation is exactly as found in molecular diffusion, except that the standard deviations σ_y and σ_z are now more complex functions of diffusion time $t = x/v_x$.

Beginning with the work of Sutton,[15] the elementary statistical theory of turbulent diffusion was widely utilized as a theoretical framework for the representation of data on atmospheric diffusion. The application of the theory is not straightforward and much empiricism is included. Here we use the elementary statistical theory with whatever empirical inputs are necessary to represent atmospheric observations. Practical predictions of atmospheric diffusion may be made on this basis, mainly because a good deal of empirical information may be absorbed in the functional form of the standard deviations σ_y and σ_z.

Following Sutton and others, we may write for the mean concentration field due to a continuous elevated point source of strength w_A:

$$\bar{\rho}_{A1} = \frac{w_A}{2\pi v_x\sigma_y\sigma_z}\left[\exp\left\{-\frac{y^2}{2\sigma_y^2}-\frac{(z-h)^2}{2\sigma_z^2}\right\}\right.$$
$$\left.+\exp\left\{-\frac{y^2}{2\sigma_y^2}-\frac{(z+h)^2}{2\sigma_z^2}\right\}\right] \qquad (7.3\text{-}14)$$

where the coordinate origin is vertically below the source at ground level and the source is located at height h as shown in Fig. 7.3-3. Of particular interest is the prediction of concentration at ground level, $z = 0$:

$$\bar{\rho}_{A1} = \frac{w_A}{\pi v_x\sigma_y\sigma_z}\exp\left(\frac{-y^2}{2\sigma_y^2}-\frac{h^2}{2\sigma_z^2}\right) \qquad (7.3\text{-}15)$$

The worse condition occurs along the axis of the plume, $y = 0$:

$$\bar{\rho}_{A1} = \frac{w_A}{\pi v_x\sigma_y\sigma_z}\exp\left(\frac{-h^2}{2\sigma_z^2}\right) \qquad (7.3\text{-}16)$$

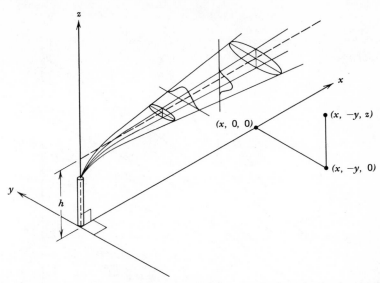

Figure 7.3-3. Coordinate system showing Gaussian distributions in the horizontal and vertical.

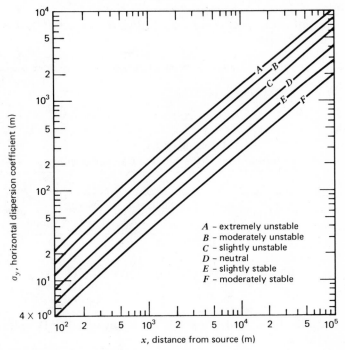

Figure 7.3-4. Horizontal dispersion coefficient as a function of distance from source. (**Source.** Reference 13.)

413

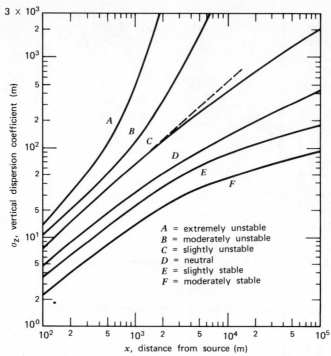

Figure 7.3-5. Vertical dispersion coefficient as a function of distance from source. (**Source.** Reference 13.)

Table 7.3-1.

Pasquill Stability Categories (They Correspond to Categories in Figures 7.3- 4 and 7.3-5)[a]

Surface Wind Speed (m/s)	Insolation			Night, Mainly Overcast or $\geqslant 4/8$ Low Cloud	$\leqslant 3/8$ Low Cloud
	Strong	Moderate	Slight		
2	A	A–B	B	–	–
2–3	A–B	B	C	E	F
3–5	B	B–C	C	D	E
5–6	C	C–D	D	D	D
6	C	D	D	D	D

[a]A, extremely unstable; B, moderately unstable; C, slightly unstable; D, neutral; E, slightly stable; F, moderately stable.

For practical use to be made of the diffusion formulas, numerical values of the diffusion coefficients σ_y and σ_z must be determined. To deal with the resulting wide variations in turbulent properties, meteorologists have introduced *stability categories* into which atmospheric conditions may be classified. Figures 7.3-4 and 7.3-5 exhibit families of curves for σ_y and σ_z for various stability categories. The manner of relating these curves to prevailing conditions of average wind speed is set out in Table 7.3-1. For concentration estimates at short distances from the source, consult Csanady.

CLOSURE

The preceding material on homogeneous atmospheric diffusion has been abstracted in large part from Csanady and Slade and has been presented as a brief introduction and for completeness. The topic of atmospheric diffusion has enjoyed a disproportionate amount of study compared to other transport topics related to chemodynamics. The serious student should consult the summary works of these authors and/or several other excellent works on this important topic.

Horizontal Diffusion in the Ocean and Large Lakes

Wastewater containing particulates and/or substances in solution must be disposed of in water bodies in a fashion similar to the disposal of contaminated air to the atmosphere. The ideas and methods developed to predict concentration of selected species dispersed into air are extended and applied to water bodies in this subsection.

The problem of dispersion arises in the context of oceanic disposal of sewage or sludge of domestic or industrial waste liquids. A usual method of disposal is to convey the liquid waste through a submarine pipe to a point some distance offshore in a large lake, sea, or ocean and release it there through a system of *diffuser ports*. The ports are designed to provide considerable mixing with sea water or lake water. (See Fig. 7.3-6a.) The net result is that the waste is distributed in some initial concentration $\bar{\rho}_{A2}^0$ over a considerable volume, which then drifts along with the lake or ocean current and mixes with its surroundings under the influence of the turbulence naturally present. A theory and mathematical models are desired to predict the dispersion or natural dilution processes of the waste plume.

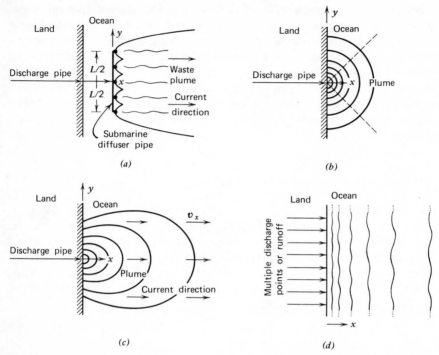

Figure 7.3-6. Pollutant dispersion models for ocean or lake outfalls. (*a*) Offshore submerged diffuser. (*b*) Point source without advection. (*c*) Point source with advection. (*d*) Distributed source.

DIFFUSION OF SEWAGE PLUMES

The diffusion of sewage plumes occurs mainly in the horizontal direction. In such applications vertical diffusion makes little or no immediate appreciable contribution to the dilution of pollutants. Such is the case, for example, when the initial cloud extends from the sea or lake bottom to the free surface. Even if the depth of water increases in the direction in which the pollutant cloud drifts, so that its vertical size increases, this does not imply vertical mixing. Often the waste liquid has some buoyancy and "boils" to the surface, effectively distributing itself over the available vertical depth during the initial mixing phase. In other situations a stable density interface, such as the thermocline, stops downward or upward vertical mixing beyond a short initial phase, much as an inversion lid does in the atmosphere. The problems to be addressed and the models presented are therefore concerned with the horizontal spread of pollutants through oceanic or lake turbulence.

Brooks[16] developed a useful theoretical model of oceanic diffusion arising out of work on sewage disposal at sea. The model assumes that the sewage "boils" to the surface, after having been mixed with seawater by a diffuser several hundred meters in length lying on the sea bottom. The effluent is forced through many ports of the diffuser, forming a number of jets that mix with the ambient fluid until the momentum is dissipated. The initial sewage field is thus fairly large horizontally and is well mixed from surface to bottom.

Since currents near the shore usually follow the depth contours, an adequate model assumes that the depth of the sewage plume remains constant and that its further dilution is caused by lateral diffusion alone. The relevant form of the diffusion equation is

$$v_x \frac{\partial \rho_{A2}}{\partial x} + k_A''' \rho_{A2} = \frac{\partial}{\partial y}\left(\mathcal{D}_{A2y}^{(t)} \frac{\partial \rho_{A2}}{\partial y} \right) \tag{7.3-17}$$

where v_x is the current velocity in the x direction and $\mathcal{D}_{A2y}^{(t)}$ is the eddy diffusivity in the y direction. An advantage of using the equation of continuity is that one may model nonconservative effects, such as die-off of bacteria (in a field of diffusing sewage) or flocculation, radioactive disintegration, and so on. In many such cases the rate of loss of the diffusing substance (bacteria, for instance) is proportional to their concentration (Eq. 1.2-2), the factor of proportionality being a *decay constant* of dimensions s^{-1} and denoted by k_A'''.

A useful equation capable of mapping the concentration field of chemical A has been presented by Csanady:

$$\rho_{A2} = \tfrac{1}{2}\rho_{A2}^0 \exp\left(\frac{-k_A''' x}{v_x} \right)\left[\mathrm{erf}\left(\frac{L/2+y}{\sqrt{2}\,\sigma_y} \right) + \mathrm{erf}\left(\frac{L/2-y}{\sqrt{2}\,\sigma_y} \right) \right] \tag{7.3-18}$$

where ρ_{A2}^0 is the constant initial concentration of chemical A after mixing by the diffuser and L is the diffuser length shown in Fig. 7.3-6a. The eddy diffusivity was converted to a variance by the relation

$$\mathcal{D}_{A2y}^{(t)} = \frac{v_x}{2}\frac{d\sigma_y^2}{dx} \tag{7.3-19}$$

$\mathcal{D}_{A2y}^{(t)}$ and hence σ_y is a function of x and not of y.

Practical estimates of the concentration field may therefore be arrived at if we have data on the decay constant k_A''' and on the point source standard deviation σ_y. In large-scale oceanic diffusion problems we may

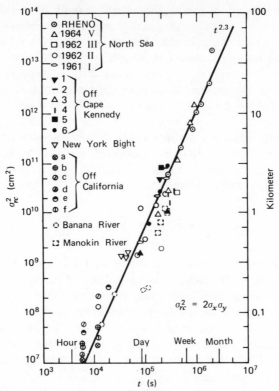

Figure 7.3-7. A diffusion diagram for variance versus diffusion time $\sigma_{rc}^2 \equiv 2\sigma_x\sigma_y$. (Reprinted by permission. **Source.** Reference 13.)

use Okubo diagrams (Fig. 7.3-7) to estimate the behavior of σ_y. The time dependence of σ_y is obtained in a continuous plume from the simple relation $x = v_x t$.

POINT SOURCES OF POLLUTANTS ENTERING LAKES OR THE OCEAN

A significant factor of water quality in lakes and similar bodies of water is the dispersion of chemicals and particulates in the vicinity of a river discharge or a wastewater discharge as shown in Fig. 7.3-6b and c. The constituents in the incoming wastewater are initially diluted by the turbulence of the discharging stream into a portion of the lake water. After this initial phase, further dilution occurs due to the natural mixing processes in the lake, as the incoming mass spreads laterally or longitudinally in the horizontal plane. As pointed out in the previous discussion, the analysis is reduced to a two-dimensional problem because of rapid but limited dispersion in the vertical direction.

POINT SOURCES WITH DIFFUSION AND ADVECTION

When a wind-driven or an offshore current is significant, an advective term must also be included for point sources. In this case the concentration pattern is distorted by a projection along the axis of the wind velocity. Thus elements of fluid containing higher concentration of pollutants are moved to larger distances from the shore. To simplify the basic differential equation, the assumption is made that dispersion is negligible along the x axis because of the advection in that direction.

Numerous experiments have been performed to observe the horizontal dispersion in large bodies of water. The usual technique is to release a marked fluid such as rhodamine B dye and follow the spreading process by crisscrossing the plume in a boat and obtaining samples that are subsequently analyzed for the dye content. In a large body of water the horizontal diffusion coefficients are generally governed by the "$\frac{4}{3}$ power law," that is, the horizontal diffusion coefficient is proportional to the $\frac{4}{3}$ power of the length scale of the diffusing patch or plume:

$$\mathcal{D}_{A2y}^{(t)} = A_L L^{4/3} \tag{7.3-20}$$

where A_L is a dissipation parameter of dimensions $cm^{2/3}/s$ or $ft^{2/3}/s$ and L is the width of the plume, usually taken to be $4\,\sigma_y$.

In the ocean numerous field experiments have been performed to estimate $\mathcal{D}_{A2y}^{(t)}$. These data are summarized in Fig. 7.3-8. It can be seen that the values of A_L are in the neighborhood of 10^{-2} to $10^{-4}\,ft^{2/3}/s$. It should be pointed out that because the effects of shear currents were not removed in the field experiments, direct use of the data requires some caution. When the effects of shear are taken into account, the lower value of $A_L = 10^{-4}$ may be more appropriate.

The effect observed that the coefficient of lateral diffusion does not remain constant with distance but varies with the scale of the diffusion phenomena was first reported by Richardson. Under these conditions, Eq. 7.3-17 is still applicable. By relating the variation of the lateral diffusion coefficient with the longitudinal dimension in accordance with the $\frac{4}{3}$ power relationships, Brooks[11] integrated the equation for $y = 0$ or the centerline of the plume:

$$\rho_{A2} = \rho_{A2}^0 \exp\left(\frac{-k_A''' x}{v_x}\right) \text{erf} \sqrt{\frac{3/2}{[1 + (8\mathcal{D}_{A20}^{(t)} x)/v_x L^2]^3 - 1}} \tag{7.3-21}$$

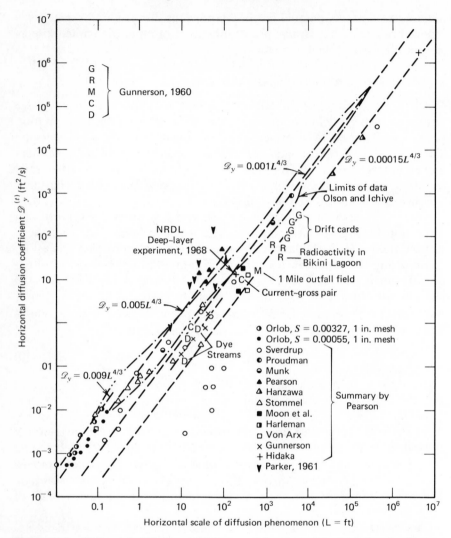

Figure 7.3-8. Horizontal diffusion coefficient as a function of horizontal scale. **(Source.** Reference 7.)

where $\mathscr{D}_{A20}^{(t)}$ is the initial diffusivity of the near shore well-mixed cloud of outlet width L.

Csanady suggests that in a nearly homogeneous field, plume growth is rather slower than suggested by Fig. 7.3-8 and may be calculated to a satisfactory approximation using a constant eddy diffusivity. He states further that experimental evidence on the behavior of sewage plumes in their early dispersal phase (which is practically the most important part) suggests an eddy diffusivity of the order 10^3 cm^2/s. It would seem to be safer to make such an assumption on the basis of engineering estimates of oceanic dispersal, rather than using the more optimistic projections that may be taken from Fig. 7.3-8.

If the diffusion coefficient is assumed constant, then Eq. 7.3-17 can be integrated and for $y = 0$ becomes

$$\rho_{A2} = \rho_{A2}^0 \exp\left(\frac{-k_A''' x}{v_x} \right) \text{erf} \sqrt{v_x L^2 / 16 \mathcal{D}_{A2y}^{(t)} x} \qquad (7.3\text{-}22)$$

where $\mathcal{D}_{A2}^{(t)}$ is some well-chosen value of the horizontal diffusivity.

POINT SOURCES WITH DIFFUSION AND NO ADVECTION[11]

During some critical periods when there are no surface winds to create a surface current and the net water flow is practically absent pollutant dispersion is solely by water turbulence. This situation is illustrated in Fig. 7.3-6b. Considering the surface plane, the steady-state distribution of concentration of a nonconservative substance may be described by the following equation:

$$0 = \mathcal{D}_{A2x}^{(t)} \frac{\partial^2 \rho_{A2}}{\partial x^2} + \mathcal{D}_{A2y}^{(t)} \frac{\partial^2 \rho_{A2}}{\partial y^2} - k_A''' \rho_{A2} \qquad (7.3\text{-}23)$$

The x and y axes are, respectively, perpendicular and parallel to the shore. Dispersion coefficients in both the longitudinal and lateral directions are assumed constant. If it is further assumed that the dispersion coefficients are equal in each direction, then Eq. 7.3-23 becomes

$$0 = \mathcal{D}_{A2}^{(t)} \left(\frac{\partial^2 \rho_{A2}}{\partial x^2} + \frac{\partial^2 \rho_{A2}}{\partial y^2} \right) - k_A''' \rho_{A2} \qquad (7.3\text{-}24)$$

In addition to being appropriate for conditions in which there is no wind, the preceding equation is appropriate for conditions in which upwelling and density difference are negligible.

Since the dispersion process is uniform in the $x - y$ plane, Eq. 7.3-24 may be transformed to polar coordinates to yield

$$\frac{\partial^2 \rho_{A2}}{\partial r^2} + \frac{1}{r} \frac{\partial \rho_{A2}}{\partial r} + \frac{1}{r^2} \frac{\partial^2 \rho_{A2}}{\partial \theta^2} - \frac{k_A''' \rho_{A2}}{\mathcal{D}_{A2}^{(t)}} = 0 \qquad (7.3\text{-}25)$$

where r is the radial distance from the source and θ is the angle on either side of the x axis. If it is assumed that ρ_{A2} is constant for a given r, then $\partial \rho_{A2}/\partial\theta$ and $\partial^2\rho_{A2}/\partial\theta^2$ are zero. This assumption does not strictly apply in the vicinity of the shore where reflection occurs and concentration patterns overlap but applies to a better degree in a zone radiating from the discharge point and 45° on either side of the x axis. For this assumption, the preceding partial differential equation reduces to the following ordinary differential equation:

$$\frac{d^2\rho_{A2}}{dr^2} + \frac{1}{r}\frac{d\rho_{A2}}{dr} - \frac{k_A''' \rho_{A2}}{\mathcal{D}_{A2}^{(t)}} = 0 \tag{7.3-26}$$

which is a Bessel equation of zero order. The solution is

$$\rho_{A2} = C_1 I_0\left[\sqrt{\frac{k_A''' r^2}{\mathcal{D}_{A2}^{(t)}}}\right] + C_2 K_0\left[\sqrt{\frac{k_A''' r^2}{\mathcal{D}_{A2}^{(t)}}}\right] \tag{7.3-27a}$$

where $I_0(x)$ and $K_0(x)$ are modified Bessel functions of the first and second kinds and C_1 and C_2 are constants.

In order to develop boundary conditions, it is assumed that the point source is replaced with an initial cloud of pollutant of concentration ρ_{A2}^0 and radius r_0. In the case of a river discharging into a lake, a reasonable estimate of r_0 is half the effective river width and ρ_{A2}^0 is the concentration of chemical A in the entering river water. The secondary boundary condition is that far away from the source the concentration of chemical A is zero. In summary, the two boundary conditions are

$$\rho_{A2} = \rho_{A2}^0 \qquad \text{at } r = r_0 \tag{7.3-27b}$$

$$\rho_{A2} = 0 \qquad \text{at } r = \infty \tag{7.3-27c}$$

The solution of Eq. 7.3-27 is

$$\rho_{A2} = \rho_{A2}^0 \frac{K_0\left(\sqrt{k_A''' r^2 / \mathcal{D}_{A2}^{(t)}}\right)}{K_0\left(\sqrt{k_A''' r_0^2 / \mathcal{D}_{A2}^{(t)}}\right)} \tag{7.3-28}$$

Values of the Bessel function $K_0(x)$ may be obtained from most mathematical handbooks.

The equations presented in the preceding development are neither necessarily definitive nor final descriptions of the various phenomena. These simple models do provide a first-order approximation of the dispersion process and provide a backdrop upon which to correlate field data in that they do indicate the significance of the various factors of diffusion, advection, and reaction.

DISTRIBUTED SOURCE WITH NO ADVECTION[11]

A distributed point source may be approached by placing multiple point sources positioned side by side as shown in Fig. 7.3-6d. The ideal distributed source consists of an infinite number of point sources. Such sources arise when chemicals such as pesticides or fertilizers are flushed into a body of water by runoff from the land surface and when chemicals enter water bodies by subterranean discharges such as pollutants in groundwater. If the extent of the source is wide enough to minimize lateral gradients, the working differential equation for a nonconservative chemical or substance is

$$0 = \mathscr{D}_{A2}^{(t)}\frac{d^2\rho_{A2}}{dx^2} - k_A'''\rho_{A2} \qquad (7.3\text{-}29a)$$

At the shoreline the concentration of chemical A is the maximum, and well away from the shoreline the concentration is zero. This yields the boundary conditions

$$\rho_{A2} = \rho_{A2}^0 \qquad \text{at } x = 0 \qquad (7.3\text{-}29b)$$

$$\rho_{A2} = 0 \qquad \text{at } x = \infty \qquad (7.3\text{-}29c)$$

The solution of Eq. 7.3-29 is

$$\rho_{A2} = \rho_{A2}^0 \exp\left[-\sqrt{\frac{k_A'''x^2}{\mathscr{D}_{A2}^{(t)}}}\right] \qquad (7.3\text{-}30)$$

This model may be used to describe the steady-state surface concentration due to a distributed input along a shore. It should be noted that ρ_{A2}^0 is the near shore concentration of A after the source has been mixed with a quantity of lake water and not the concentration of the source.

Example 7.3-1. Bacterial Distribution in Lake Michigan in the Vicinity of the Indiana Harbor

Brooks et al.[11] obtained field observations of bacterial concentrations in Lake Michigan in the vicinity of the point source of the Indiana Harbor.

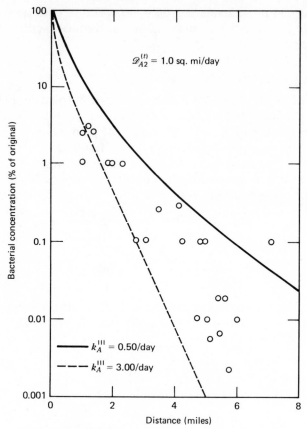

Figure E7.3-1. (Source. Reference 11.)

Three sets of data taken along radii located on the x axis and 45° on either side of this axis for the months of June, July, and September appear in Fig. E7.3-1.

The diameter of the initial field is taken at 300 ft, which is the order of the outlet width. The dispersion coefficient has a value of 1 mi^2/d. Available data on the value of the bacteria decay coefficient k_A''' range from 0.5 to 3.0 per day. Since the wind velocity during this period was less than 1 mi/hr, use the appropriate model and calculate relative bacteria concentrations for the range of decay constants given.

SOLUTION

This particular situation can be modeled best with the point source model without advection. The radius of the initial field is 150 ft. Equation 7.3-28 for $\rho_{A2}^0 = 100$ and $k_A''' = 0.5$ per day becomes

$$\rho_{A2} = 100 \frac{K_0\left(\sqrt{\dfrac{0.5}{d} \left|\dfrac{r^2\,mi^2}{}\right| \dfrac{d}{1.0\,mi^2}}\right)}{K_0\left(\sqrt{0.5 \left|\left(\dfrac{150}{5280}\right)^2\right| \dfrac{1}{1.0}}\right)}$$

$$\rho_{A2} = 100 \frac{K_0(\sqrt{0.5 r^2})}{K_0(0.02)}$$

From a table of Bessel functions $K_0(0.02) = 3.908$. The following calculations appear as a function of distance from the source, r miles.

r (mi)	$\sqrt{0.5 r^2}$	$K_0(\sqrt{0.5 r^2})$	ρ_{A2}
1	0.707	0.654	16.7
2	1.41	0.240	6.14
4	2.83	0.0425	1.09
6	4.24	0.00852	.22
8	5.66	0.00181	.05

For $k_A''' = 3.0$ per day we have the following calculations: $K_0(0.0492) = 3.012$

r (mi)	$\sqrt{3.0 r^2}$	$K_0(\sqrt{3 r^2})$	ρ_{A2}
1	1.73	.159	5.29
2	3.46	.0204	.68
4	6.93	.459E − 3	.15E − 1
6	10.4	.119E − 4	.40E − 3

The model results are shown plotted in Fig. E7.3-1. It is apparent that no significant error is introduced by omitting the advective component, and the range of k_A''' values is sufficient to bracket the observations.

Problems

7.3A. BACTERIAL DISTRIBUTION IN LAKE MICHIGAN[11]

Additional data were obtained on the distribution of bacteria in Lake Michigan in the vicinity of the Indiana Harbor and these data appear in Table 7.3A. During the periods of observation, the winds were reasonably

Table 7.3A.   Bacterial Concentration in Lake Michigan

x (mi)	ρ_{A2}	x (mi)	ρ_{A2}
0	100	2.2	1.2
0.4	80	4.0	1.1
0.5	90	2.8	1.0
0.6	15	6.0	0.8
1.6	15	5.5	0.2
1.8	10	7.2	0.13
3.8	8.5	5.2	0.11
4.3	2.0		

uniform and constant and the wind direction was along the x axis for the periods. The water velocity was taken as 2.5% of the wind velocity and was 6.0 mi/d. Using the range of bacterial decay constants given in Example 7.3-1, calculate the concentration profile for the following model cases:

1. Point source with diffusion and advection and variable $\mathcal{D}_{A2}^{(t)}$. Use $\mathcal{D}_{A20}^{(t)} = 0.0063$ mi^2/d.
2. Point source with diffusion and advection and constant diffusion. Use $\mathcal{D}_{A2}^{(t)} = 1.0$ mi^2/d.
3. Plot the computed profiles and the observed data in the manner of Fig. E7.3-1.

REFERENCES

1. G. T. Orlob, L. A. Roesner, and W. R. Norton, "Mathematical Models for the Prediction of Thermal Energy Changes in Impoundments," Water Pollution Control Research Series, 16130EXT12/69, Washington, D.C., 1969.

2. L. J. Thibodeaux, "Semi-Infinite Solid Model for Prediction of Temperature in Deep Reservoirs and Lakes," *Water Resour. Bull.*, **11** (3), 449–454 (1975).

3. H. S. Carslaw and J. C. Jaeger, *Conduction of Heat in Solids*, 2nd ed., Oxford University Press, London, 1958, p. 76.

4. F. L. Parker, B. A. Benedict, and C. Tsai, "Evaluation of Mathematical Models for Temperature Prediction in Deep Reservoirs," U.S. Environmental Protection Agency, EPA-660/3-75-38, Corvallis, Oreg., 1975.

5. D. N. Lapedes, *Encyclopedia of Environmental Science*, McGraw-Hill, New York, 1974, pp. 619–620.

6. G. Neumann and W. J. Pierson, Jr., *Principles of Physical Oceanography*, Prentice-Hall, Englewood Cliffs, N.J., 1966, pp. 445–450.

7. C. Y. Koh and L.-N. Fan, "Mathematical Models for the Prediction of Temperature Distributions Resulting from the Discharge of Heated Water into Large Bodies of

Water," Water Pollution Control Research Series, 16130DW010/70, Washington, D.C., 1970.

8. C. Eckart, *Hydrodynamics of Oceans and Atmosphere*, Pergamon, New York, 1960, pp. 57–71.

9. A. Lerman, "Time to Chemical Steady-States in Lakes and Oceans," in J. D. Hem, Ed., *Non Equilibrium Systems in Natural Water Chemistry*, Advances in Chemistry Series 106, American Chemical Society, Washington, D.C., 1971, pp. 30–76.

10. J. Crank, *The Mathematics of Diffusion*, Oxford University Press, Oxford, 1956, pp. 9–61, 121–46.

11. D. J. O'Connor, R. V. Thomann, D. M. Ditoro, and N. H. Brooks, "Mathematical Modeling of Natural Systems," Summer Institute in Water Pollution Control, Manhattan College, New York, 1974.

12. R. V. Thomann, *Systems Analysis and Water Quality Management*, McGraw-Hill, New York, 1972, p. 104.

13. G. T. Csanady, *Turbulent Diffusion in the Environment*, Reidel, Boston, 1973.

14. D. H. Slade, Ed., *Meteorology and Atomic Energy*, U.S. Atomic Energy Commission, Oak Ridge, Tenn., 1968.

15. O. G. Sutton, *Micrometeorology*, McGraw-Hill, New York, 1953.

16. N. H. Brooks, in Proceedings of the 1st International Conference on Waste Disposal in the Marine Environment, Pergamon, New York, p. 246.

CHAPTER 8

REVIEW OF ADDITIONAL
CHEMICAL TRANSPORT
PROCESSES IN THE
ENVIRONMENT

In the previous chapter numerous chemical transport processes were presented and analyzed. In a generic sense the term *transport* should be interpreted to include all processes that move chemicals from one place to another within the exterior environment. Major additional chemodynamic processes are introduced briefly in this chapter. Unlike previous chapters, this one is entirely qualitative and does not contain example problems or exercise problems. These additional transport processes are treated briefly to restrict the size of the book, and it therefore should not be inferred that the processes are of minor importance from a chemodynamic standpoint.

The processes are divided into two categories. One category involves intraphase processes that affect the degree of transport to a greater or lesser extent but are not transport processes per se. The second category involves those natural environmental processes that are directly responsible for moving chemicals and particles to and from interfaces.[1-3]

8.1 PROCESSES WITHIN THE GEOSPHERES

Degradation or alteration of a chemical contaminant may disallow or at least change the transport pathways. The word *degradation* is used here in a broad sense. This includes either transformation to other less complex products or complete breakdown to carbon dioxide, water, and minerals. In the environment the chemical may be altered or degraded by a number of processes that can be divided into the categories biodegradation, photochemical degradation, and chemical degradation.

The importance of the three mechanisms varies considerably depending on the particular media in which a chemical resides. For example, photochemical processes are very significant in the atmosphere but are of minimum importance in soil. In contrast, biodegradation is frequently the most important mechanism in soil and water.

Biological Transformations

Biological transformations of organic chemicals occur via relatively specific processes. Since the soil and water contain a host of microorganisms, there is sufficient latitude or nonspecificity in these organisms to allow them to perform a significant role in the degradation and/or removal of various synthetic chemicals that are introduced into the environment. Because of their high catabolic versatility, species diversity, and metabolic rate per unit weight, microorganisms play a dominant role in chemical degradation.

Microorganisms are exceedingly versatile in the metabolism of synthetic products. When confronted with a toxic substance, the microorganisms attempt to detoxify their environment and usually do; however, the product of this detoxification may be more toxic to higher organisms than the original chemical.

A consideration of the heavy metals indicates the complexity of the biological interconversions possible. For example, with mercury there is the following disproportionation:

$$Hg_2^{2+} \rightleftharpoons Hg^{2+} + Hg^0$$

Volatilization of mercury metal (Hg^0) shifts the equilibrium to the right, but return of Hg^0 from the atmosphere to the earth's crust shifts the equilibrium to the left. Some bacteria detoxify their environment of Hg^{2+} by converting it to methylmercury and dimethylmercury. Other bacteria detoxify their environment of methylmercury by reducing it to Hg^0 and methane. Dimethylmercury is volatile, but in the atmosphere photolysis yields Hg^0, methane, and formaldehyde. These interconversions in the mercury cycle lead to steady-state concentrations of methylmercury in sediments.

Chemical Degradations

Ubiquitous substances such as air, water, and soil may also catalyze environmental degradation. Many chemicals are sensitive to autooxidation with air or hydrolysis with water. The free radical character and solubility in water make oxygen a most effective environmental reactant. Water, besides being an important medium for chemical reactions, is reactive itself through its ability to form hydroxyl and hydrogen ions; thus, pH of the media can be important.

Water solubility appears to be important to breakdown processes. Water-insoluble chemicals have frequently been found to endure longer in the environment than water-soluble chemicals. It is thought that the resistance of water-insoluble compounds to degradation may be linked to failure of the chemical to penetrate to the reaction site, or inaccessibility of the chemical because of adsorption or entrapment in inert material.

Photochemical Degradation and or Transformation

Considering that the atmosphere is a highly reactive medium and that many chemicals undergo sunlight-induced photochemical reactions under laboratory conditions, photochemical reactions cannot be neglected when describing chemical transport by air.

The extensive recent literature on the photochemistry of pesticides has shown that it is possible for these chemicals to react photochemically to yield complex mixtures of products. Sunlight-induced photochemical reactions have been demonstrated in the field. Laboratory studies have shown, for example, that intramolecular eliminations and isomerizations tend to dominate when pesticides in molecular vapor form are irradiated, whereas other reactions, such as oxidation or condensation, can occur with pesticides dissolved in water.

Photochemical reactions of gaseous species, especially of the nitrogen oxides, promote oxidation of naturally occurring forest terpenes and waste industrial organic molecules, which produce the hygroscopic and polymeric species of smog. Sulfur dioxide from smelting waste and fuel combustion—again, with the assistance of nitrogen oxides—oxidizes to sulfur trioxide, which hydrolyzes to sulfuric acid mists and ammonium sulfate aerosols.

The preceding are but a few of many photochemically induced, directly or indirectly, reactions that destroy and create chemical species within the atmosphere. The concentration of the species in air is usually directly related to transport rate in either the interphase or intraphase mode.

Photochemical degradation is an important process with atmospheric contaminants and some chemicals that reside on surfaces such as pesticides on leaves and vegetation or in water. Chemicals in the environment can react photochemically either by reaching an excited state themselves or by reacting with another chemical that has reached an excited state. A chemical can reach an excited state either by direct adsorption of light (sunlight contains wavelengths greater than 290 nm) or by accepting energy from an excited donor molecule (sensitizer).

Decomposition of Chemicals—Closure

Chemicals degrade in the environment through chemical, photochemical, and microbial processes. Examples of chemical degradation include air oxidation, neutralization, and hydrolysis. Many organic compounds decompose in the presence of sunlight.

Microorganisms are known to metabolize a wide variety of chemicals. Microbial activity is extremely important in soils. Although some industrial chemicals are not readily degraded by microorganisms, sufficient information is generally available to predict, on the basis of chemical structure, which compounds are likely to biodegrade. The substitution of biodegradable compounds for persistent ones may affect microbial activity in localized areas. Increased activity could either enhance detoxification or aggravate problems akin to methylmercury formation.

On the basis of existing knowledge, it is generally possible for microbiologists, biochemists, and chemists working together to determine metabolic sequences for both natural and totally synthetic compounds and to identify those that might pose environmental problems.

An appreciation for the interdependency of reaction and transport as related to fate can be obtained from a study of such classic pollutants as DDT[4] and other synthetic chemicals.[5] These aspects of chemical degradation and transformations are presented in textbook form by Tinsley.[6] Tinsley's book provides a complementary view of chemodynamics from the viewpoint of environmental chemistry and is an excellent companion to this book.

8.2   TRANSPORT PROCESSES TO AND FROM INTERFACES

Particles in the Atmosphere

Conceptually, a particle is considered to include spray droplets, dust particles, atmospheric aerosols, wind-blown soil particles, and the like. For some of the physical processes in the atmosphere, the behavior of molecular vapor is indistinguishable from that of other small particles.

Aerosol particles interact strongly with one another, with water and ice clouds, and with dispersed chemicals. From these interactions, a complicated and time-varying distribution of aerosol sizes that extend over many decades result from ionic clusters containing as few as 10 to 10^4 molecules and having radii of 10^{-3} to 10^{-2} μm, to soil particles and droplets only briefly suspended in the air by violent winds.

Particles descend through the atmosphere under the influence of gravity by well-known physical laws that govern the resistance to motion of particles moving in viscous media. The descent is described by the Stokes equation, commonly found in physics texts. Large particles, as is evident, tend to settle to the surface rather rapidly, but as particle size decreases, the effect of gravity becomes less, owing to buoyancy, viscous forces, and the random turbulent motions of the air in which the particles are suspended. Transport potential then depends on the balance achieved between particle size, gravity, and the strength of the random turbulent motions, so that long-range aerial transport of particles is probably associated with particles finer than 10 μm in diameter. Particles larger than 100 μm in diameter fall relatively rapidly and are largely unaffected by air turbulence. Particles of 10 μm or less are dispersed primarily by turbulence under all meteorological conditions, molecular vapor is returned to the surface with difficulty, and long-range transport is indeed very probable.

Dry Deposition

The deposition of particles from the atmosphere onto exposed surfaces in the absence of rain or snowfall, so-called dry deposition, occurs as a long-term process for virtually all airborne materials. It will continue as long as the air mass containing the material remains in contact with the ground. Thus, surface deposition may be a very effective mechanism for removing airborne chemical residues even though particle fall velocity is negligible. Empirical observations allow the rate of particle removal by surface deposition to be quantified by a mass transfer coefficient type of rate expression. The proportionality factor between flux and concentration is called the *deposition velocity*. Deposition velocities depend primarily on particle transport and hence are a function basically of particle size rather than of composition of the particle. Molecular vapor is an exception in that deposition velocities, which may vary widely, depend on the reactivity of the molecule with the contact surface. Besides particle size, dry deposition is also apparently strongly dependent on the nature of the ground surface and aerodynamic factors of the atmospheric flow over the surface. Once a chemical pollutant particle is deposited on the ground, it is usually not resuspended in the atmosphere and thus should be considered as entering the soil or water transport systems. An excellent review of deposition velocity theory and dry removal processes appears in a National Academy of Sciences report.[7]

Precipitation Scavenging

Precipitation provides a major mechanism by which pollutant materials are removed from the atmosphere. Although gaseous materials may be dissolved in clouds and raindrops, aerosol particles are especially susceptible to precipitation scavenging mechanisms. Small particles can be incorporated in cloud drops and ultimately in rain by providing initial condensation or freezing nuclei. At later stages in the precipitation growth process, particles may be brought by Brownian motion into contact with cloud drops or ice crystals. If the pollutant and precipitation particles are in the proper size range, the precipitation particles may also capture the pollutants as they fall to the ground. Studies of nuclear bomb fallout have shown that precipitation scavenging processes account for 90% of the atmospheric aerosol removal in regions with moderate rainfall.

The inclusion of airborne pesticide residues in falling rain or snow is competitive with dry deposition as a removal mechanism. A spray droplet or dust particle that is in the atmosphere for an appreciable period of time as a discernible entity may serve as a nucleating particle in the growth of a raindrop and will soon fall to earth.

Rainfall may also remove molecular vapor. The solubilization of molecular vapor in a raindrop is essentially a partitioning process in which a given concentration of gaseous chemical establishes an equilibrium concentration in solution. The reasonably good agreement of the washout ratios for chemicals such as DDT, dieldrin, and lindane as measured in the laboratory and in the field has been purported to show that these chemicals exist in the atmosphere essentially as unassociated molecular vapor.

Wind Erosion

Wind erosion is the detachment, transportation, and deposition of loose topsoil or sand by the action of wind. Transport by wind erosion of contaminated surface dust to high altitudes and for long distances has been well documented. This transport mechanism is a well-known route for pesticides to enter the atmosphere, and results in a spectrum of particle sizes entering the air. Part of the eroded material returns quickly to the surface a short distance downwind; part is deposited more slowly further downwind; part becomes airborne for a much longer time, diluting rapidly in the atmosphere and returning to the surface very slowly.

Volatilization of Chemicals from Vegetation Surfaces

Air–soil volatilization is important; however, considering the fraction of the surface of the earth covered with vegetation, this aspect cannot be overlooked. The volatilization of chemicals deposited directly or indirectly on vegetation surfaces has not been studied as extensively as that from soil. Volatilization is reduced, in the case of pesticides, by surface solubilization in plant waxes and soils, penetration to internal portions of the plant, and translocation. Chemical transformation occurring on the surface caused by the plant or by sunlight may be important. The effect of plant transpiration on pesticide volatilization is unclear, although water evaporation from plant leaves is known to change the energy budget. Plants with adequate moisture are relatively cool because the incoming solar radiation is dissipated as latent heat of evaporation. Water evaporation is obviously restricted under dry conditions, and the plants become warmer through insolation. Cooler plant surfaces may restrict chemical volatilization.

Bursting Bubbles at the Sea Surface

Water, salts, organic materials, and a net electric charge are transferred to the air through the ejection of droplets by bubbles bursting at the sea surface. The exchange of chemicals by the sea and the atmosphere is of importance to chemical resuspension from water surfaces. On evaporation of the water, the droplet residues are carried great distances by winds. These particles become nuclei for cloud and raindrop formations.

The lipid-soluble, hydrophobic pesticides may preferentially dissolve in the natural organic surface films associated with all water bodies. They then may become airborne by the action of spray or bursting bubbles at the surface, at a rate exceeding that predicted by volatilization rate equations for the small amounts in true solution.

Other Processes

Other processes include suspended particulate matter sedimentation in water. Waterborne particulates consisting of living or dead organisms and organic films preferentially absorb organic chemical residues. Heavy metal ions, for example, are sorbed to suspended inorganic matter owing to ion-exchange equilibriums. Exchange of pollutants between solid phases and water depends on such factors as oxygen content, ionic strength, and composition of the water. Much of this suspended material, along with attached pollutants, is deposited in lacustrine and marine sediments or ephemerally in river sediments.

An important process that controls the transport of chemicals in soil is its movement with water. Although downward movement is most common, lateral and even upward movement are sometimes significant. The upward movement, which is a result of evaporation from the surface, is effective in removing chemicals from the topmost zone.

Where water percolation is rapid, the downward movement of the chemical in the direction of flow predominates, but as percolation becomes slower and slower, diffusion into the pores of soil particles becomes important in determining the distribution of the chemical. As a chemical is carried through the soil by water movement, a localized equilibrium exists between the dissolved phase and the adsorbed phase. As a consequence, a chemical that is tightly adsorbed will be leached slowly. Water movement and leaching are important aspects of nuclear waste disposal as related to mechanisms of migration of radionuclides in geologic formations.

8.3   CLOSURE ON CHEMODYNAMICS AND A LIST OF RELATED WORKS

The subject of chemodynamics in the exterior environment is a rapidly growing new field of specialization. The subject matter is presently in a state of transition. A few textbooks are now appearing and attempts are being made to organize the ever-expanding quantity of material on the subject into some logical format for inclusion into existing or evolving curriculums. In this transition period, much of the material, particularly at the upper graduate level, will be offered in the form of conference proceedings and published symposium papers.

Chemodynamics cuts across many disciplinary lines as is readily apparent from scanning the reference lists in each chapter. There is a large body of knowledge concerning "natural" chemicals in the exterior environment that can be drawn on to address problems concerned with the movement of synthetic chemicals. In this book an attempt was made to bring together this knowledge and organize it so that it could be easily used to address specifically the transport and related fate aspects of chemicals in the exterior environment.

The book is an introduction to the subject. Readers wishing to dig deeper into individual topics or professors searching for supplemental and/or reference material to accompany the textbook for a one-semester course are referred to the list of readings at the end of this chapter. This list, presented in alphabetical order by name of the primary author or editor, contains relevant books and significant reports directly related to the subject of chemodynamics.

REFERENCES

1. N. Norton, Chairman, *Principles for Evaluating Chemicals in the Environment*, National Academy of Sciences, Washington, D.C., 1975.
2. D. E. Glotfelty and J. H. Caro, "Introduction, Transport, and Fate of Persistent Pesticides in the Atmosphere," in V. R. Deitz, Ed., *Removal of Trace Contaminants from the Air*, Am. Chem. Soc. Symp. Ser. 17, American Chemical Society, Washington, D.C., 1975, p. 42–62.
3. P. H. Howard, J. Saxena, and H. Sikka, "Determining the Fate of Chemicals," *Environ. Sci. Technol.*, 12 (4), 398–407, (1978).
4. N. Lee Wolfe, R. G. Zepp, D. F. Paris, G. L. Baughman, and B. C. Hollis, "Methoxychlor and DDT Degradation in Water: Rates and Products," *Environ. Sci. Technol.*, 11 (12), 1077–1081 (1977).
5. R. G. Zepp, N. Lee Wolfe, J. A. Gordon, and G. L. Baughman, "Dynamics of 2,4-D Esters in Surface Waters, Hydrolysis, Photolysis and Vaporization," *Environ. Sci. Technol.*, 9 (13), 1144–1150 (1975).
6. I. J. Tinsley, *Behavior of Chemicals in the Environment*, Wiley-Interscience, New York, 1979.
7. J. M. Prospero, Chairman, *The Tropospheric Transport of Pollutants and Other Substances to the Oceans*, National Academy of Science, Washington, D.C., 1978, Chap. 4.

A LIST OF SUPPLEMENTAL READING AND REFERENCE MATERIAL

J. O'M. Bockris, Ed., *Environmental Chemistry*, Plenum, New York, 1977.

F. A. Brooks and W. O. Pruitt, "Investigations of Energy, Momentum and Mass Transfer Near the Ground," U.S. Army Electronics Command, Atmospheric Sciences Laboratory, Research Division, Fort Huachuca, Ariz., Final Report, 1965.

G. S. Campbell, *An Introduction to Environmental Biophysics*, Springer-Verlag, New York, 1977.

T. M. Church, Ed., *Marine Chemistry in the Environment*, Am. Chem. Soc. Symp. Ser. 18, American Chemical Society, Washington, D.C., 1975.

G. T. Csanady, *Turbulent Diffusion in the Environment*, Reidel, Boston, 1973.

D. A. de Vries and N. H. Afgan, Eds., *Heat and Mass Transfer in the Biosphere, Part 1—Transfer Processes in Plant Environment*, Scripta, Washington, D.C., 1975.

G. Eglinton, Ed., *Environmental Chemistry*, Vol. 1, The Chemical Society, Burlington House, London, 1975.

S. Eskinazi, *Fluid Dynamics and Thermodynamics of Our Environment*, Academic, New York, 1975.

A. M. Friedman, Ed., *Actinides in the Environment*, Am. Chem. Soc. Symp. Ser. 35, American Chemical Society, Washington, D.C., 1976.

R. Haque and V. H. Freed, Eds., *Environmental Dynamics of Pesticides*, Plenum, New York, 1975.

R. A. Horne, *The Chemistry of Our Environment*, Wiley-Interscience, New York, 1978.

I. N. McCave, Ed., *The Benthic Boundary Layer*, Plenum, New York, 1976.

N. Nelson, Chairman, *Principles for Evaluating Chemicals in the Environment*, National Academy of Sciences, Washington, D.C., 1975.

J. M. Prospero, Chairman, *The Tropospheric Transport of Pollutants and Other Substances to the Oceans*, National Academy of Sciences, Washington, D.C., 1978.

W. O. Pruitt, D. L. Morgan, F. J. Lourence, and F. V. Jones, Jr., "Energy, Momentum and Mass Transfer above Vegetative Surfaces," U.S. Army Electronics Command, Atmospheric Sciences Laboratory, Research Division, Fort Huachuca, Ariz., 1967.

N. J. Rosenberg, *Microclimate: The Biological Environment*, Wiley-Interscience, New York, 1974.

I. H. Suffet, Ed., *Fate of Pollutants in the Air and Water Environments*, Wiley-Interscience, New York, 1977, Parts 1 and 2.

O. G. Sutton, *Micrometeorology*, McGraw-Hill, New York, 1953.

I. J. Tinsley, *Behavior of Chemicals in the Environment*, Wiley-Intersience, New York, 1979.

H. L. Windom and R. A. Duce, Eds., *Marine Pollutant Transfer*, Lexington Books, Lexington, Mass., 1976.

T. F. Yen, Ed., *Chemistry of Marine Sediments*, Ann Arbor Science Publishers Inc., Ann Arbor, Mich., 1977.

APPENDIX A. **THE METRIC SYSTEM OF MEASUREMENT AND CONVERSION TABLE**

A.1. THE METRIC SYSTEM OF MEASUREMENT: Interpretation and Modification of the International System of Units for the United States*

Section 3 of Pub. L. 94–168, the Metric Conversion Act of 1975, declares that the policy of the United States shall be to coordinate and plan the increasing use of the metric system in the United States. Section 403 of Pub. L. 93–380, the Education Amendments of 1974, states the policy of the United States to encourage educational agencies and institutions to prepare students to use the metric system of measurement as part of the regular education program. Under both these acts, the "metric system of measurement" is defined as the International System of Units as established by the General Conference of Weights and Measures in 1960 and nterpreted or modified for the United States by the Secretary of Commerce (subsec. 4(4), Pub. L. 94–168; subsec. 403 (a) (3), Pub. L. 93–380). The Secretary has delegated his authority under these subsections to the Assistant Secretary for Science and Technology. Accordingly, in implementation of this authority, the following tables and associated materials set forth the interpretation and modification of the International System of Units (hereinafter "SI") for the United States.

    This notice supersedes the notice of the National Bureau of Standards published in the *Federal Register* of June 19, 1975 (40 FR 25837).

Source. Office of the Secretary for Science and Technology, Federal Register Doc. 76-36414, November 9, 1976.

Table A.1-1. *SI Base and Supplementary Units*

Quantity	Name	Symbol
SI base Units:		
Length	meter	m
Mass[a]	kilogram	kg
Time	second	s
Electric current	ampere	A
Thermodynamic tempera-ture[b]	kelvin	K
Amount of substance	mole	mol
Luminous intensity	candela	cd
SI supplementary units:		
Plane angle	radian	rad
Solid angle	steradian	sr

[a]"Weight" is the commonly used term for "mass."
[b]Wide use is made of "Celsius temperature" (symbol t) defined by

$$t = T - T_0$$

where T is the thermodynamic temperature, expressed in kelvins, and $T_0 = 273.15$ K by definition. The unit "degree Celsius" is thus equal to the unit "kelvin," but the degree Celsius (symbol °C) is a special name used instead of kelvin for expressing Celsius temperature. A temperature interval or a Celsius temperature difference may be expressed in degrees Celsius as well as in kelvins.

The SI is constructed from seven base units for independent quantities plus two supplementary units for plane angle and solid angle, listed in Table A.1.

Units for all other quantities are derived from these nine units. In Table A.2 are listed 17 SI derived units with special names which were derived from the base and supplementary units in a coherent manner, which means, in brief, that they are expressed as products and ratios of the nine base and supplementary units without numerical factors.

All other SI derived units, such as those in Tables A.3 and A.4, are similarly derived in a coherent manner from the 26 base, supplementary, and special-name SI units.

For use with the SI units there is a set of 16 prefixes (see Table A.5) to form multiples and submultiples of these units. It is important to note that the kilogram is the only SI unit with a prefix. Because double prefixes are not to be used, the prefixes of Table A.5, in the case of mass, are to be used with gram (symbol g) and not with kilogram (symbol kg).

Table A.1-2. *SI Derived Units with Special Names*

Quantity	SI unit		
	Name	Symbol	Expression in terms of other units
Frequency	hertz	Hz	s^{-1}
Force	newton	N	$kg \cdot m/s^2$
Pressure, stress	pascal	Pa	N/m^2
Energy, work, quantity of heat.	joule	J	$N \cdot m$
Power, radiant flux	watt	W	J/s
Quantity of electricity, electric charge.	coulomb	C	$A \cdot s$
Electric potential, potential difference, electromotive force.	volt	V	W/A
Capacitance	farad	F	C/V
Electric resistance	ohm	Ω	V/A
Conductance	siemens	S	A/V
Magnetic flux	weber	Wb	$V \cdot s$
Magnetic flux density	tesla	T	Wb/m^2
Inductance	henry	H	Wb/A
Luminous flux	lumen	lm	$cd \cdot sr$
Illuminance	lux	lx	lm/m^2
Activity (of ionizing radiation source).	becquerel	Bq	s^{-1}
Absorbed dose	gray	Gy	J/kg

Table A.1-3. *Examples of SI Derived Units Expressed in Terms of Base Units*

Quantity	SI unit	Unit symbol
Area	square meter	m^2
Volume	cubic meter	m^3
Speed, velocity	meter per second	m/s
Acceleration	meter per second squared.	m/s^2
Wave number	1 per meter	m^{-1}
Density, mass density.	kilogram per cubic meter.	kg/m^3
Current density	ampere per square meter.	A/m^2

Table A.1-3. (Continued)

Quantity	SI unit	Unit symbol
Magnetic field strength.	ampere per meter	A/m
Concentration (of amount of substance).	mole per cubic meter	mol/m^3
Specific volume	cubic meter per kilogram.	m^3/kg
Luminance	candela per square meter.	cd/m^2

Table A.1-4. *Examples of SI Derived Units Expressed by Means of Special Names*

Quantity	Name	Unit symbol
Dynamic viscosity	pascal second	$Pa \cdot s$
Moment of force	newton meter	$N \cdot m$
Surface tension	newton per meter	N/m
Power density, heat flux density, irradiance.	watt per square meter.	W/m^2
Heat capacity, entropy.	joule per kelvin	J/K
Specific heat capacity, specific entropy.	joule per kilogram kelvin.	$J/(kg \cdot K)$
Specific energy	joule per kilogram	J/kg
Thermal conductivity.	watt per meter kelvin.	$W/(m \cdot K)$
Energy density	joule per cubic meter.	J/m^3
Electric field strength	volt per meter	V/m
Electric charge density.	coulomb per cubic meter.	C/m^3
Electric flux density	coulomb per square meter.	C/m^2
Permittivity	farad per meter	F/m
Permeability	henry per meter	H/m
Molar energy	joule per mole	J/mol
Molar entropy, molar heat capacity	joule per mole kelvin.	$J/(mol \cdot K)$

Table A.1-5.   *SI Prefixes*

Factor	Prefix	Symbol
10^{18}	exa	E
10^{15}	peta	P
10^{12}	tera	T
10^{9}	giga	G
10^{6}	mega	M
10^{3}	kilo	k
10^{2}	hecto	h
10^{1}	deka	da
10^{-1}	deci	d
10^{-2}	centi	c
10^{-3}	milli	m
10^{-6}	micro	μ
10^{-9}	nano	n
10^{-12}	pico	p
10^{-15}	femto	f
10^{-18}	atto	a

Certain units which are not part of the SI are used so widely that it is impractical to abandon them. The units that are accepted for continued use in the United States with the International System are listed in Table A.6.

In those cases where their usage is already well established, the use, for a limited time, of the units in table 7 is accepted, subject to future review.

Table A.1-6.   *Units in Use with the International System*

Name	Symbol	Value in SI unit
Minute (time)	min	1 min = 60 s
Hour	h	1 h = 60 min = 3 600 s
Day	d	1 d = 24 h = 86 400 s
Degree (angle)	°	$1° = (\pi/180)$ rad
Minute (angle)	′	$1' = (1/60)° = (\pi/10\ 800)$ rad
Second (angle)	″	$1'' = (1/60)'$
		$= (\pi/648\ 000)$ rad
Liter	L^a	1 L = 1 dm^3 = 10^{-3} m^3
Metric ton	t	1 t = 10^3 kg
Hectare (land area)	ha	1 ha = 10^4 m^2

[a]The international symbol for liter is the lowercase "1", which can easily be confused with the numeral "1." Accordingly, the symbol "L" is recommended for United States use.

Table A.1-7. *Units to be used for a limited time*

Nautical mile	angstrom	gal[a]
Knot	barn	curie
Standard atmosphere	bar	roentgen
		rad[b]

[a]Unit of acceleration.
[b]Unit of absorbed dose.

Metric units, symbols, and terms that are not in accordance with the foregoing Interpretation and Modification are no longer accepted for continued use in the United States with the International System of Units. Accordingly, the following units and terms listed in the table of metric units in section 2 of the act of July 28, 1866, that legalized the metric system of weights and measures in the United States, are no longer accepted for use in the United States:

myriameter
stere
millier or tonneau
quintal
myriagram
kilo (for kilogram)

For more information regarding the International System of Units, contact the Office of Technical Publications, National Bureau of Standards, U.S. Department of Commerce, Washington, D.C. 20234.

BETSY ANCKER-JOHNSON, Ph.D.,
Assistant Secretary for
Science and Technology.

A.2. CONVERSION TABLE

Directions For Use

- Avoid use of prefixes in denominators (except kg).
- The use of hecto, deka, deci, and centi prefixes should be avoided except when used in areas and volumes.
- SI symbols are not capitalized unless the unit is derived from a proper name, e.g., Hz for H. R. Hertz. Unabbreviated units are not capitalized, e.g., hertz, newton, kelvin. Only T, G, M prefixes are capitalized.

- Except at the end of a sentence, SI units are not to be followed by periods.
- Four or more digits in a group should be separated in groups of three with no comma, e.g., 1 983 212.322 7, not 1,983,212.3227.
- With derived unit abbreviations, use center dot to denote multiplication and a slash for division, e.g., newton-second/meter^2 = $N \cdot s/m^2$.

Table A.2

To Convert From	To	Multiply by
angstrom	meter (m)	1.000 000*E−10
atmosphere (normal)	newton/meter ^{2}(N/m^2)	1.013 250*E +05
barrel (for petroleum, 42 gal)	meter^3 (m^3)	1.589 873E −01
British thermal unit (International Table)	joule (J)	1.055 056E +03
Btu/lbm-deg F (c, heat capacity)	joule/kilogram-kelvin (J/kg·K)	4.186 800*E +03
Btu/hour	watt (W)	2.930 711E −01
Btu/second	watt (W)	1.055 056E +03
Btu/ft^2-hr-deg F (heat transfer coefficient)	joule/meter^2-second-kelvin (J/m^2·s·K)	5.678 264E +00
Btu/ft^2-hr (heat flux)	joule/meter^2-second (J/m^2·s)	3.154 591E +00
Btu/ft-hr-deg F (thermal conductivity)	joule/meter-second-kelvin (J/m·s·K)	1.730 735E +00
calorie (International Table)	joule (J)	4.186 800*E +00
cal/g-deg C	joule/kilogram-kelvin (J/kg·K)	4.186 800*E +03
centimeter	meter (m)	1.000 000*E −02
centimeter of mercury (0 C)	pascal (Pa) (N/m^2)	1.333 22E +03
centimeter of water (4 C)	pascal (Pa) (N/m^2)	9.806 38E +01
centipoise	pascal·second (Pa·s) (N·s/m^2)	1.000 000*E −03
centistoke	meter^2/second (m^2/s)	1.000 000*E −06
degree Celsius	kelvin (K)	$t_K = t_C + 273.15$
degree Fahrenheit	kelvin (K)	$t_K = (t_F + 459.67)/1.8$
degree Rankine	kelvin (K)	$t_K = t_R/1.8$
dyne	newton (N)	1.000 000*E −05
erg	joule (J)	1.000 000*E −07

To Convert From	To	Multiply by
farad (international of 1948)	farad (F)	9.995 05E−01
fluid ounce (U.S.)	meter³ (m³)	2.957 353E−05
foot	meter (m)	3.048 000*E−01
foot (U.S. survey)	meter (m)	3.048 006E−01
foot of water (39.2 F)	pascal (Pa)	
	(N/m²)	2.988 98E+03
foot²	meter² (m²)	9.290 304*E−02
foot/second²	meter/second² (m/s²)	3.048 000*E−01
foot²/hour	meter²/second (m²/s)	2.580 640*E−05
foot-pound-force	joule (J)	1.355 818E+00
foot²/second	meter²/second (m²/s)	9.290 304*E−02
foot³	meter³ (m³)	2.831 685E−02
gallon (U.S. liquid)	meter³ (m³)	3.785 412E−03
gram	kilogram (kg)	1.000 000*E−03
horsepower (550 ft·lbf/s)	watt (W)	7.456 999E+02
hour (mean solar)	second (s)	3.600 000E+03
inch	meter (m)	2.540 000*E−02
inch of mercury (60 F)	pascal (Pa)	
	(N/m²)	3.376 85E+03
inch of water (60 F)	pascal (Pa)	
	(N/m²)	2.488 4E+02
inch²	meter² (m²)	6.451 600*E−04
inch³	meter³ (m³)	1.638 706E−05
kilocalorie	joule (J)	4.186 800*E+03
kilogram-force (kgf)	newton (N)	9.806 650*E+00
knot (international)	meter/second (m/s)	5.144 444E−01
liter	meter³ (m³)	1.000 000*E−03
micron	meter (m)	1.000 000*E−06
mil	meter (m)	2.540 000*E−05
mile (U.S. statute)	meter (m)	1.609 344*E+03
mile/hour	meter/second (m/s)	4.470 400*E−01
millimeter of mercury (0 C)	pascal (Pa)	
	(N/m²)	1.333 224E+02
minute (angle)	radian (rad)	2.908 882E−04
minute (mean solar)	second (s)	6.000 000E+01
ohm (international of 1948)	ohm (Ω)	1.000 495E+00
ounce-mass (avoirdupois)	kilogram (kg)	2.834 952E−02
ounce (U.S. fluid)	meter³ (m³)	2.957 353E−05
pint (U.S. liquid)	meter³ (m³)	4.731 765E−04
poise (absolute viscosity)	pascal·second (Pa·s)	
	(N·s/m²)	1.000 000*E−01

To Convert From	To	Multiply by
poundal	newton (N)	1.382 550E−01
pound-force (lbf avoirdupois)	newton (N)	4.448 222E+00
pound-force-second/foot^2	pascal·second (Pa·s)	
	(N·s/m^2)	4.788 026E+01
pound-mass (lbm avoirdupois)	kilogram (kg)	4.535 924E−01
Pound-mass/foot^3	kilogram meter^3	
	(kg/m^3)	1.601 846E+01
pound-mass/foot-second	pascal·second (Pa·s)	
	(N·s/m^2)	1.488 164E+00
psi	pascal (Pa)	
	(N/m^2)	6.894 757E+03
quart (U.S. liquid)	meter^3 (m^3)	9.463 529E−04
second (angle)	radian (rad)	4.848 137E−06
slug	kilogram (kg)	1.459 390E+01
stoke (kinematic viscosity)	meter^2/second (m^2/s)	1.000 000*E−04
ton (long, 2240 lbm)	kilogram (kg)	1.016 047E+03
ton (short, 2000 lbm)	kilogram (kg)	9.071 847E+02
torr (mm Hg, 0 C)	pascal (Pa)	
	(N/m^2)	1.333 22E+02
volt (international of 1948)	volt (absolute) (V)	1.000 330E+00
watt (international of 1948)	watt (W)	1.000 165E+00
watt-hour	joule (J)	3.600 000*E+03
yard	meter (m)	9.144 000*E−01

Source. *Am. Inst. Chem. Eng. J.*, **17** (2), 511, 512 (March 1971).

*An asterisk after the sixth decimal place indicates the conversion factor is exact and all subsequent digits are zero.

PHYSICAL CONSTANTS, MATHEMATICAL CONSTANTS, AND MATHEMATICAL TABLE

PHYSICAL CONSTANTS

Universal gas constants:
   0.0821 atm L/mol K
   8.319 J/mol K
Acceleration of gravity:
   980.7 cm/s^2
Avogadro's number:
   6.023E 23 molecules/mol
Planck's constant:
   6.624E−34 J/s
Stefan-Boltzmann constant:
   5.673E−12 J/s·cm^2 K^4
Boltzmann's constant:
   1.3805E−23 J/molecule K
Heat of fusion of water at 1 atm, 0°C
   334 J/g
Heat of vaporization of water at 1 atm, 100°C
   2260 J/g

Molecular weight of dry air
    28.97 g/mol
One mole of an ideal gas at 0°C and 1 atm occupies
    22.4 L.
One atmosphere of pressure is
    760 mm Hg = 1.013E 5 N/M^2

MATHEMATICAL CONSTANTS

$\pi = 3.1416$, $e = 2.71828$

MATHEMATICAL TABLE

Table B.1. The Error Function

ϕ	erf ϕ	ϕ	erf ϕ
0	0.0	0.85	0.7707
0.025	0.0282	0.90	0.7970
0.05	0.0564	0.95	0.8209
0.10	0.1125	1.0	0.8427
0.15	0.1680	1.1	0.8802
0.20	0.2227	1.2	0.9103
0.25	0.2763	1.3	0.9340
0.30	0.3286	1.4	0.9523
0.35	0.3794	1.5	0.9661
0.40	0.4284	1.6	0.9763
0.45	0.4755	1.7	0.9838
0.50	0.5205	1.8	0.9891
0.55	0.5633	1.9	0.9928
0.60	0.6039	2.0	0.9953
0.65	0.6420	2.2	0.9981
0.70	0.6778	2.4	0.9993
0.75	0.7112	2.6	0.9998
0.80	0.7421	2.8	0.9999

CHEMICAL DATA

Table C.1. Relative Atomic Weights for the Chemical Elements (1971) (based on an assigned relative atomic weight for $^{12}C = 12$)[a]

Element	Symbol	Atomic Number	Atomic Weight	Element	Symbol	Atomic Number	Atomic Weight
Actinium	Ac	89		Mercury	Hg	80	200.59
Aluminum	Al	13	26.98154	Molybdenum	Mo	42	95.94
Americium	Am	95		Neodymium	Nd	60	144.24
Antimony	Sb	51	121.75	Neon	Ne	10	20.179
Argon	Ar	18	39.948	Neptunium	Np	93	237.0482
Arsenic	As	33	74.9216	Nickel	Ni	28	58.71
Astatine	At	85		Niobium	Nb	41	92.9064
Barium	Ba	56	137.34	Nitrogen	N	7	14.0067
Berkelium	Bk	97		Nobelium	No	102	
Beryllium	Be	4	9.01218	Osmium	Os	76	190.2
Bismuth	Bi	83	208.9804	Oxygen	O	8	15.9994
Boron	B	5	10.81	Palladium	Pd	46	106.4
Bromine	Br	35	79.904	Phosphorus	P	15	30.97376
Cadmium	Cd	48	112.40	Platinum	Pt	78	195.09
Calcium	Ca	20	40.08	Plutonium	Pu	94	
Californium	Cf	98		Polonium	Po	84	
Carbon	C	6	12.011	Potassium	K	19	39.098
Cerium	Ce	58	140.12	Praseodymium	Pr	59	140.9077
Cesium	Cs	55	132.9054	Promethium	Pm	61	
Chlorine	Cl	17	35.453	Protactinium	Pa	91	231.0359
Chromium	Cr	24	51.996	Radium	Ra	88	226.0254
Cobalt	Co	27	58.9332	Radon	Rn	86	
Copper	Cu	29	63.546	Rhenium	Re	75	186.2

Table C.1. (Continued)

Element	Symbol	Atomic Number	Atomic Weight	Element	Symbol	Atomic Number	Atomic Weight
Curium	Cm	96		Rhodium	Rh	45	102.9055
Dysprosium	Dy	66	162.50	Rubidium	Rb	37	85.4678
Einsteinium	Es	99		Ruthenium	Ru	44	101.07
Erbium	Er	68	167.26	Samarium	Sm	62	150.4
Europium	Eu	63	151.96	Scandium	Sc	21	44.9559
Fermium	Fm	100		Selenium	Se	34	78.96
Fluorine	F	9	18.99840	Silicon	Si	14	28.086
Francium	Fr	87		Silver	Ag	47	107.868
Gadolinium	Gd	64	157.25	Sodium	Na	11	22.98977
Gallium	Ga	31	69.72	Strontium	Sr	38	87.62
Germanium	Ge	32	72.59	Sulfur	S	16	32.06
Gold	Au	79	196.9665	Tantalum	Ta	73	180.9479
Hafnium	Hf	72	178.49	Technetium	Tc	43	98.9062
Helium	He	2	4.00260	Tellurium	Te	52	127.60
Holmium	Ho	67	164.9304	Terbium	Tb	65	158.9254
Hydrogen	H	1	1.0079	Thallium	Tl	81	204.37
Indium	In	49	114.82	Thorium	Th	90	232.0381
Iodine	I	53	126.9045	Thulium	Tm	69	168.9342
Iridium	Ir	77	192.22	Tin	Sn	50	118.69
Iron	Fe	26	55.847	Titanium	Ti	22	47.90
Krypton	Kr	36	83.80	Tungsten	W	74	183.85
Lanthanum	La	57	138.9055	Uranium	U	92	238.029
Lawrencium	Lr	103		Vanadium	V	23	50.9414
Lead	Pb	82	207.2	Xenon	Xe	54	131.30
Lithium	Li	3	6.941	Ytterbium	Yb	70	173.04
Lutetium	Lu	71	174.97	Yttrium	Y	39	88.9059
Magnesium	Mg	12	24.305	Zinc	Zn	30	65.38
Manganese	Mn	25	54.9380	Zirconium	Zr	40	91.22
Mendelevium	Md	101					

[a]The values apply to the elements as they exist in materials of terrestrial origin and to certain aritificial elements. Weights are reliable to ±3 if the last digit is in small type. See *Pure Appl. Chem.*, **21** (1), 105 (1970). (Reprinted by permission of Pergamon Press Ltd.)

Table C.2. Dissolved-Oxygen Solubility Data
(dissolved oxygen mg/L)

Tempera-ture (°C)	Chloride Concentration (mg/L)				
	0[a]	5,000[b]	10,000[b]	15,000[b]	20,000[b]
0	14.16	13.79	12.97	12.14	11.32
1	13.77	13.41	12.61	11.82	11.03
2	13.40	13.05	12.28	11.52	10.76
3	13.05	12.72	11.98	11.24	10.50
4	12.70	12.41	11.69	10.97	10.25
5	12.37	12.09	11.39	10.70	10.01
6	12.06	11.79	11.12	10.45	9.78
7	11.76	11.51	10.85	10.21	9.57
8	11.47	11.24	10.61	9.98	9.36
9	11.19	10.97	10.36	9.76	9.17
10	10.92	10.73	10.13	9.55	8.98
11	10.67	10.49	9.92	9.35	8.80
12	10.43	10.28	9.72	9.17	8.62
13	10.20	10.05	9.52	8.98	8.46
14	9.98	9.85	9.32	8.80	8.30
15	9.76	9.65	9.14	8.63	8.14
16	9.56	9.46	8.96	8.47	7.99
17	9.37	9.26	8.78	8.30	7.84
18	9.18	9.07	8.62	8.15	7.70
19	9.01	8.89	8.45	8.00	7.56
20	8.84	8.73	8.30	7.86	7.42
21	8.68	8.57	8.14	7.71	7.28
22	8.53	8.42	7.99	7.57	7.14
23	8.38	8.27	7.85	7.43	7.00
24	8.25	8.12	7.71	7.30	6.87
25	8.11	7.96	7.56	7.15	6.74
26	7.99	7.81	7.42	7.02	6.61
27	7.86	7.67	7.28	6.88	6.49
28	7.75	7.53	7.14	6.75	6.37
29	7.64	7.39	7.00	6.62	6.25
30	7.53	7.25	6.86	6.49	6.13

[a]Solubility of oxygen from a wet atmosphere at a pressure of 760 mm Hg.
Source. G. E. Hutchinson, *A Treatise on Limnology*, Wiley, New York, 1957.
[b]Saturation values of dissolved oxygen in fresh and seawater exposed to dry air containing 20.90% oxygen under a total pressure of 760 mm of mercury.

Source: G. C. Whipple and M. C. Whipple, "Solubility of Oxygen in Sea Water," *J. Am. Chem. Soc.*, **33**, (1911), 362.

Table C.3. The Solubility of Carbon Dioxide in Seawater (in $mol/L \times 10^4$, STP)

Chlorinity (Cl ‰)	0°C	4°C	10°C	16°C	20°C	26°C
0	770	662	536	442	394	332
15	674	578	472	393	351	299
16	667	573	468	390	348	297
17	660	567	464	387	346	294
18	653	562	460	384	343	292
19	646	557	456	381	340	289
20	640	551	452	377	337	287
21	633	546	448	374	335	285

Source. R. A. Horne, *The Chemistry of Our Environment*, Wiley-Interscience, New York, (1978), p. 734. Copyright © (John Wiley and Sons, Inc. 1978). Reprinted by permission of John Wiley and Sons, Inc.

Table C.4. The Solubility of Nitrogen in Seawater (mL N_2/mL H_2O, STP)

Chlorinity (Cl ‰)	0°C	4°C	10°C	16°C	20°C	26°C
15	0.0193	0.0177	0.0155	0.0140	0.0131	0.0120
16	0.0190	0.0174	0.0154	0.0138	0.0129	0.0119
17	0.0188	0.0172	0.0152	0.0136	0.0128	0.0117
18	0.0185	0.0170	0.0150	0.0135	0.0126	0.0116
19	0.0182	0.0167	0.0148	0.0133	0.0125	0.0115
20	0.0180	0.0165	0.0146	0.0131	0.0123	0.0114
21	0.0177	0.0163	0.0145	0.0130	0.0122	0.0113

Source. R. A. Horne, "The Chemistry of Our Environment," Wiley-Interscience, (1978), p. 734. Copyright © (John Wiley and Sons, Inc. 1978). Reprinted by permission of John Wiley and Sons, Inc.

Table C.5. Henry's Law Constants for Gases in Water ($H \times 10^{-4}$)

T, °C	CO_2	CO	C_2H_6	C_2H_4	He	H_2	H_2S	CH_4	N_2	O_2
0	0.0728	3.52	1.26	0.552	12.9	5.79	2.68	2.24	5.29	2.55
10	0.104	4.42	1.89	0.768	12.6	6.36	3.67	2.97	6.68	3.27
20	0.142	5.36	2.63	1.02	12.5	6.83	4.83	3.76	8.04	4.01
30	0.186	6.20	3.42	1.27	12.4	7.29	6.09	4.49	9.24	4.75
40	0.233	6.96	4.23		12.1	7.51	7.45	5.20	10.4	5.35

Source. National Research Council, *International Critical Tables, Vol. III*, McGraw-Hill, New York, 1929.

$p_A = Hx_A$
p_A = partial pressure of A in the gas, atm
x_A = mole fraction of A in the liquid
H = Henry's law constant, atm/mol fraction

Table C.6. Apparent Henry's Constants[a] of Pesticides and Related Materials

Compound	P_A Vapor Pressure at 20–25°C (mm Hg)	ρ_{A2} Water Solubility at 20–25°C (μg/L)	$-\text{Log } H_p$
Toxaphene	3×10^{-1}	10^2–10^3	2.5–3.5
Aroclor 1242-1260	4×10^{-4} to 4×10^{-5}	2.7–240	4.8–5.8
Heptachlor	3×10^{-4}	56	5.3
2,4-D esters	10^{-3}–10^{-4}	$(10^3)^b$	6.0–7.0
γ-chlordane	10^{-4}–10^{-5}	(10^2)	6.0–7.0
Aldrin	2.3×10^{-5}	27	6.1
Trifluralin	10^{-4}	580	6.8
p,p'-DDT	1.5×10^{-7}	1.2	6.9
o,p'-DDT[c]	5.5×10^{-6}	85	7.2
p,p'-DDE[c]	6.5×10^{-6}	120	7.3
EPTC	1.6×10^{-2}	3.8×10^5	7.4
Dieldrin	2.8×10^{-6}	140	7.7
Heptachlor epoxide	$(10^{-4}$–$10^{-6})$	350	6.5–8.5
Diazinon	2.8×10^{-4}	4×10^4	8.2
Lindane	3.3×10^{-5}	7×10^3	8.3
Endrin	2×10^{-7}	2×10^2	9.0
Parathion	2.3×10^{-5}	2.4×10^4	9.0
Carbofuran	(10^{-5})	2.5×10^5	10.4
2,4-D salts	$\sim 10^{-10}$	Soluble	–

Source. D. E. Glotfelty and J. H. Caro, "Introduction, Transport, and Fate of Persistent Pesticides in the Atmosphere," in Victor R. Deitz, Ed., ACS Symposium Series, 17, *Removal of Trace Contaminants from the Air*, Washington, D.C., (1975). Reprinted with permission. Copyright by the American Chemical Society.

[a]Henry's law: $p_A = H_p \rho_{A2}$
[b]Values in brackets are estimated.
[c]Data at 30°C.

Table C.7a. Properties of Selected Hydrocarbons

Hydrocarbon n-Alkane	ρ_{A2} Solubility in Water at 25°C and 1 atm (g/m³)	p_A Vapor Pressure at 25°C (atm)	γ_{A2} Activity Coefficient in Water	$\gamma_{A2}f_A^0$ Coefficient Fugacity Product (atm)	H_A Henry's Law Constant (atm·mol/m³)
Methane (g)[a]	24.1	269	$1.37\,5 \times 10^2$	3.70×10^4	0.665
Ethane (g)	60.4	39.4	7.02×10^2	2.77×10^4	0.499
Propane (g)	62.4	9.29	4.23×10^3	3.93×10^4	0.707
n-Butane (g)	61.4	2.40	2.19×10^4	5.26×10^4	0.947
n-Pentane (l)	38.5	0.675	1.04×10^5	7.02×10^4	1.26
n-Hexane (l)	9.5	0.205	5.04×10^5	9.59×10^4	1.85
n-Heptane (l)	2.93	0.0603	1.90×10^6	1.15×10^5	2.07
n-Octane (l)	0.66	0.0186	9.62×10^6	1.79×10^5	3.22
n-Nonane (l)	0.22	5.64×10^{-3}	3.24×10^7	1.83×10^5	3.29
Decane (l)	0.052	1.73×10^{-3}	1.58×10^8	2.74×10^5	4.93
Dodecane (l)	0.0037	1.55×10^{-4}	2.56×10^9	3.96×10^5	7.12
Tetradecane (l)	0.0022	1.26×10^{-4}	5.01×10^9	6.31×10^4	1.14

Source. D. Mackay and Wan-Ying Shin, "The Aqueous Solubility and Air–Water Exchange Characteristics of Hydrocarbons Under Environmental Conditions," in *Chemistry and Physics of Aqueous Gas Solutions*, American Society Testing Materials, Philadelphia, Pa., (1974), pp. 104–108. Reprinted by permission.

[a] Here g denotes gas, l denotes liquid, and s denotes solid.

Table C.7b. Properties of Selected Hydrocarbons

Cycloalkanes Branched-Chain Alkanes	ρ_{A2} Solubility in Water at 25°C and 1 atm (g/m³)	p_A Vapor Pressure at 25°C (atm)	γ_{A2} Activity Coefficient in Water	$\gamma_{A2}f_A^0$ Coefficient Fugacity Product (atm)	H_A Henry's Law Constant (atm·mol/m³)
Cyclopentane (l)	156	0.418	2.50×10^4	1.04×10^4	0.187
Cyclohexane (l)	55	0.128	8.50×10^4	1.09×10^4	0.196
Methyl-cyclopentane (l)	42	0.181	1.11×10^5	2.01×10^4	0.362
Methyl-cyclohexane (l)	14	0.0610	3.90×10^5	2.38×10^4	0.428
Propyl-cyclopentane (l)	2.04	0.0162	3.06×10^6	4.96×10^4	0.893
Isobutane (g)	48.9	3.52	1.96×10^4	6.88×10^4	1.24
Isopentane (l)	47.8	0.904	8.39×10^4	7.58×10^4	1.364
2-Methyl-pentane (l)	13.8	0.276	3.47×10^5	9.59×10^4	1.73
2-Methyl-hexane (l)	4.40	0.0867	2.19×10^6	1.90×10^5	3.42
2,2-Dimethyl-pentane (l)	4.40	0.138	1.27×10^6	1.75×10^5	3.15
3-Methyl-heptane (l)	0.792	0.0257	8.01×10^6	2.06×10^5	3.71
2,2,4-Trimethyl-pentane (l)	2.44	0.0649	2.60×10^6	1.69×10^5	3.04
4-Methyl-octane (l)	0.115	8.91×10^{-3}	6.20×10^7	5.52×10^7	9.936

Table C.7c. Properties of Selected Hydrocarbons

Aromatics	ρ_{A2} Solubility in Water at 25°C and 1 atm (g/m³)	p_A Vapor Pressure at 25°C (atm)	γ_{A2} Activity Coefficient in Water	$\gamma_{A2}f_A^0$ Coefficient Fugacity Product (atm)	H_A Henry's Law Constant (atm·mol/m³)
Benzene (l)	1780	0.125	2.4×10^3	3.05×10^2	5.49×10^{-3}
Toluene (l)	515	0.0374	9.9×10^3	3.71×10^2	6.66×10^{-3}
Ethyl benzene (l)	152	0.0125	3.88×10^4	4.85×10^4	8.73×10^{-3}
O-Xylene (l)	175	8.71×10^{-3}	3.37×10^4	2.93×10^2	5.27×10^{-3}
Isopropylbenzene (l)	50	6.03×10^{-3}	1.34×10^5	8.05×10^5	1.45×10^{-2}
Naphthalene (s)	34.4	1.14×10^{-4}	7.69×10^4	23.60	4.25×10^{-4}
Biphenyl (s)	7.48	7.45×10^{-5}	4.75×10^5	35.34	6.36×10^{-4}
Acenaphthene (s)	3.88	3.97×10^{-5}	3.19×10^5	12.67	2.28×10^{-4}
Fluorene (s)	1.90	1.64×10^{-5}	7.95×10^5	13.07	2.35×10^{-4}
Anthracene (s)	0.075	5.04×10^{-5}	1.814×10^6	91.40	1.65×10^{-3}
Phenanthrene (s)	1.18	4.53×10^{-6}	1.82×10^6	8.242	1.48×10^{-4}

Table C.7d. Properties of Selected Hydrocarbons

Olefins and Acetylenes	ρ_{A2} Solubility in Water at 25°C and 1 atm (g/m³)	p_A Vapor Pressure at 25°C (atm)	γ_{A2} Activity Coefficient in Water	$\gamma_{A2} f_A^0$ Coefficient Fugacity Product (atm)	H_A Henry's Law Constant (atm·mol/m³)
Ethene (g)	131	59.91	1.99×10^2	1.19×10^4	0.214
Propene (g)	200	11.29	1.15×10^3	1.29×10^4	0.232
1-Butene (g)	222	2.933	5.07×10^{13}	1.49×10^4	0.268
1-Pentene (l)	148	0.839	2.6×10^4	2.21×10^4	0.398
1-Hexene (l)	50	0.245	9.35×10^4	2.29×10^4	0.412
2-Heptene (l)	15	0.0637	3.64×10^5	2.32×10^4	0.418
1-Octene (l)	2.7	0.0229	2.31×10^6	5.28×10^4	0.905
Propyne (g)	3640	5.505	1.01×10^2	6.12×10^2	0.0110
1-Butyne (g)	2870	1.858	5.8×10^2	1.08×10^3	0.0194

Table C.7e. Solubility of Selected Hydrocarbons in Water and Seawater (g/m^3)

Hydrocarbons	Water	Seawater	Ratio (%)
n-Pentene	38.5	27.6	71.5
Dodecane	0.0037	0.0029	78
Tetradecane	0.0022	0.0017	77.5
Benzene	1780	1391	78
Toluene	515	402	78

Table C.8. Diffusivities and Schmidt Numbers of Gases in Air

Substance	Temperature (°C)	$\mathcal{D}_{A1}$ (cm^2/s)	Sc $= \mu_1/\rho_1\mathcal{D}_{A1}$
Acetic acid	0	0.1064	
	25	0.133	1.16
Acetone	0	0.109	
Ammonia	0	0.216	0.61
	25	0.28	0.78
n-Amyl alcohol	0	0.0589	
	25	0.07	2.21
sec-Amyl alcohol	30	0.072	
Amyl butyrate	0	0.040	
Amyl formate	0	0.0543	
i-Amyl formate	0	0.058	
Amyl isobutyrate	0	0.0419	
Amyl propinate	0	0.046	
Aniline	0	0.0610	
	25	0.072	2.14
	30	0.075	
Anthracene	0	0.0421	
Benzene	0	0.077	
	25	0.088	1.76
Benzidine	0	0.0298	
Benzyl chloride	0	0.066	
n-Butyl acetate	0	0.058	
i-Butyl acetate	0	0.0612	
n-Butyl alcohol	0	0.0703	
	25.9	0.097	1.72
	30	0.088	
	59	0.104	
i-Butyl alcohol	0	0.0727	
Butyl amine	0	0.0821	
	25	0.101	1.53
i-Butyl amine	0	0.0853	

Table C.8. (Continued)

Substance	Temperature (°C)	$\mathscr{D}_{A1}$ (cm^2/s)	Sc $= \mu_1/\rho_1\mathscr{D}_{A1}$
i-Butyl butyrate	0	0.0468	
i-Butyl formate	0	0.0705	
i-Butyl isobutyrate	0	0.0457	
i-Butyl proprionate	0	0.0529	
i-Butly valerate	0	0.0424	
Butyric acid	0	0.067	
i-Butyric acid	0	0.0679	
	25	0.081	1.91
Caproic acid	0	0.050	
i-Caproic acid	0	0.0513	
	25	0.060	2.58
Carbon dioxide	0	0.138	
Carbon dioxide	25	0.164	0.94
	44	0.177	
Carbon disulfide	0	0.0892	
	25	0.107	1.45
Chlorine	0	0.093	1.42
Chlorobenzene	0	0.062	2.13
	30	0.075	
Chloroform	0	0.091	
Chloropicrin	25	0.088	
m-Chlorotoluene	0	0.054	
o-Chlorotoluene	0	0.059	
p-Chlorotoluene	0	0.051	
Chlorotoluene	25	0.065	2.38
Cyanogen chloride	0	0.111	
Cyclohexane	45	0.086	
Diethylamine	0	0.0884	
	25	0.105	1.47
Diphenyl	0	0.0610	
	25	0.068	2.28
Ethane	0	0.108	1.22
Ether (diethyl)	0	0.0778	
Ethyl acetate	0	0.0715	
	30	0.089	
Ethyl alcohol	0	0.102	
	25	0.119	1.30
	42	0.145	
Ethyl benzene	0	0.0658	
	25	0.077	2.01
Ethyl *n*-butyrate	0	0.0579	

Table C.8 (Continued)

Substance	Temperature (°C)	$\mathcal{D}_{A1}$ (cm^2/s)	$Sc = \mu_1/\rho_1\mathcal{D}_{A1}$
Ethyl *i*-butyrate	0	0.0591	
Ethyl ether	25	0.093	1.66
	0	0.0779	1.70
Ethyl formate	0	0.0840	
Ethyl propionate	0	0.068	
Ethyl valerate	0	0.0512	
Eugenol	0	0.0377	
Formic acid	0	0.1308	
	25	0.159	0.97
Hexane	21	0.080	
Hexyl alcohol	0	0.0499	
	25	0.059	2.60
Hydrogen	0	0.611	
	25	0.410	0.22
Hydrogen cyanide	0	0.173	
Hydrogen peroxide	60	0.188	
Iodine	0	0.07	
Mercury	0	0.112	
Mesitylene	0	0.056	
	25	0.067	2.31
Methyl acetate	0	0.084	
Methyl alcohol	0	0.132	
	25	0.159	0.97
Methyl butyrate	0	0.0633	
Methyl *i*-butyrate	0	0.0639	
Methyl formate	0	0.0872	
Methyl propionate	0	0.0735	
Methane	0	0.16	0.84
Methyl valcrate	0	0.0569	
Naphthalene	0	0.0513	
Nitrogen	0	0.13	0.98
n-Octane	0	0.0505	
	25	0.060	2.58
Oxygen	0	0.178	
	25	0.206	0.75
n-Pentane	21	0.071	
Phosgene	0	0.095	
Propane	0	0.088	1.51
Propionic acid	0	0.0829	
	25	0.099	1.56

Table C.8 (Continued)

Substance	Temperature (°C)	$\mathcal{D}_{A1}$ (cm^2/s)	$Sc = \mu_1 / \rho_1 \mathcal{D}_{A1}$
Propyl acetate	0	0.067	
n-Propyl alcohol	0	0.085	
	25	0.100	1.55
i-Propyl alcohol	0	0.0818	
	30	0.101	
n-Propyl benzene	0	0.0481	
	25	0.059	2.62
i-Propyl benzene	0	0.0489	
n-Propyl bromide	0	0.085	
	25	0.105	1.47
i-Propyl bromide	0	0.0902	
Propyl butyrate	0	0.0530	
Propyl formate	0	0.0712	
n-Propyl iodide	0	0.079	
	25	0.096	1.61
i-Propyl iodide	0	0.0802	
n-Propyl isobutyrate	0	0.0549	
i-Propyl isobutyrate	0	0.059	
Propyl propionate	0	0.057	
Propyl valerate	0	0.0466	
Safrol	0	0.0434	
i-Safrol	0	0.0455	
Sulfur dioxide	0	0.103	1.28
Toluene	0	0.076	
	30	0.088	
	59.0	0.104	
Trimethyl carbinol	0	0.087	
n-Valeric acid	0	0.050	
	25	0.067	2.31
i-Valeric acid	0	0.0544	
Water	0	0.220	
	25	0.256	0.60
Xylene	25	0.071	2.18

Compiled from the following sources. C. O. Bennett and J. E. Myers, *Momentum, Heat and Mass Transfer*, 2nd ed., McGraw-Hill, New York, 1974, pp. 787–788. T. K. Sherwood and R. L. Pigford, *Absorption and Extraction*, 2nd ed., McGraw-Hill, New York, 1952, p. 20. C. J. Geankoplis, *Mass Transport Phenomena*, Holt, Rinehart & Winston, New York, 1972, pp. 22, 70, 478. J. H. Perry, Ed., *Chemical Engineer's Handbook*, 4th Ed., McGraw-Hill, New York, 1964, pp. 14–22, 14–23.

Table C.9. Diffusivities and Schmidt Numbers of Chemicals in Water

Substance	Temperature (°C)	$\mathcal{D}_{A2} \times 10^5$ (cm^2/s)	Sc $= \mu_2/\rho_A \mathcal{D}_{A2}$
Acetamide	25	1.19	
Acetic acid	20	0.88	1140
Acetonitrile	25	1.66	
Acetylene	20	1.56	645
Allyl alcohol	20	0.93	1080
Ammonia	20	1.76	570
i-amyl alcohol	25	1.0	
Bromine	20	1.2	840
n-butanol	25	0.96	
Butanol	20	0.77	1310
Caffeine	25	0.63	
Carbon dioxide	20	1.77	559
Chloral hydrate	25	0.77	
Chlorine	20	1.22	824
Ethanol	20	1.00	1005
Formic acid	25	1.37	
Glucose	20	0.6	
	25	0.69	
Gylcerol	20	0.72	1400
Hydrogen	25	5.85	
	30	5.42	
Hydrogen sulfide	20	1.41	712
Hydroquinone	20	0.77	1300
Lactose	20	0.43	2340
Maltose	20	0.43	2340
Mannitol	20	0.58	1730
Methanol	20	1.28	785
Nitric acid	20	2.6	390
Nitrogen	10	1.29	
	20	1.64	613
	25	1.9, 2.01	
	40	2.83	
	55	3.80	
Oxygen	10	1.54	
	20	1.80	558
	25	2.5, 2.2	
	30	3.49	
	40	3.33	
	55	4.50	
Phenol	20	0.84	1200
n-Propanol	25	1.1	
	20	0.87	1150

Table C.9 (continued)

Substance	Temp. (°C)	$\mathcal{D}_{A2} \times 10^5$ (cm^2/s)	$Sc = \mu_2/\rho_2 \mathcal{D}_{A2}$
Pyridine	25	0.76	
Pyrogallol	20	0.70	1440
Ratlinose	20	0.37	2720
Resorcinol	20	0.80	1260
Saccharose	25	0.49	
Sodium chloride	20	1.35	745
Sodium hydroxide	20	1.51	665
Succinic acid	25	0.94	
Sucrose	20	0.45	2230
Sulfur dioxide	25	1.7	
Sulfuric acid	20	1.73	580
Tartaric acid	25	0.80	
Urea	20	1.06	946
Urethane	20	0.92	1090

Compiled from the following sources. C. O. Bennett and J. E. Myers, *Momentum, Heat and Mass Transfer*, 2nd ed., McGraw-Hill, New York, (1974), pp. 787–788. R. A. Horne, *Marine Chemistry*, Wiley-Interscience, New York, (1969). J. H. Perry, Ed., *Chemical Engineers' Handbook*, 4th ed., McGraw-Hill, New York, (1964), pp. 14-25, 14-26.

Table C.10. Diffusivities in the Solid State

System	T, °C	$\mathcal{D}_{A3}$(cm^2/s)
He in SiO_2[a]	20	2.4–5.5E−10
He in pyrex[a]	20	4.5E−11
He in pyrex[a]	500	2E−8
H_2 in SiO_2[a]	500	0.6–2.1E−8
elements in glass (borosilicate)[b]	100	1.0–100.0E−14

[a]Values taken from R. M. Barrer, *Diffusion in and through Solids*, Macmillan, New York, 1941, pp. 141, 222, 275.
[b]Range of values reported by G. deMarsily, E. Ledoux, and J. Margat, "Nuclear Waste Disposal: Can Geologists Guarantee Isolation," Science, **197** (Aug. 5, 1977), 197.

PHYSICAL PROPERTY DATA

Table D.1. Physical Properties of Dry Air at Atmospheric Pressure

T (°C)	ρ (g/L)	c_p (J/kg·K)	ν (cm²/s)	k (J/m·s·K)	α (cm²/s)
−17.8	1.380	1.005E3	0.117	0.0228	0.165
− 1.1	1.297	1.005E3	0.132	0.0241	0.184
15.6	1.224	1.005E3	0.148	0.0252	0.206
26.7	1.177	1.005E3	0.157	0.0263	0.221
37.8	1.137	1.005E3	0.168	0.0270	0.237

Table D.2. Physical Properties of Water

T (°C)	ρ (g/cm²)	c_p (J/kg · K)	$\nu \times 10^{+5}$ (cm²/s)	k (J/m·s·K)	$\alpha \times 10^3$ (cm²/s)
− 1.1	1.00	4230	1790	0.552	1.31
15.6	0.999	4190	1130	0.588	1.41
26.7	0.997	4180	863	0.611	1.46
37.8	0.993	4180	684	0.630	1.51

Table D.3. Comparison of Transport Phenomena in Pure Water and in Seawater at 1 atm

Name, Symbol, Units	Pure Water		Seawater Salinity 35/mille‰	
	0°C	20°C	0°C	20°C
Dynamic viscosity, η, g/cm/ sec = poise	0.01787	0.01022	0.01877	0.01075
Thermal conductivity, k, W/ cm/°C	0.00566	0.00599	0.00563	0.00596
Kinematic viscosity, $\nu = \eta/\rho$, cm^2/sec	0.01787	0.01004	0.01826	0.01049
Thermal diffusivity, $\kappa = k/c_P\rho$ cm^2/sec	0.00134	0.00143	0.00138	0.00149
Diffusivity, D, cm^2/sec NaCl	0.0000074	0.0000141	0.0000068	0.0000129
N$_2$	0.0000106	0.0000169	–	–
O$_2$		0.000021	–	–
Prandtl number, $N_p = \nu/\kappa$	13.3	7.0	13.1	7.0

Source. R. B. Montgomery in D. E. Gray, Ed., *American Institute of Physics Handbook*, 1957. Used with permission of McGraw-Hill Book Co.

Table D.4. Specific Volume of Seawater (cm^3/gm)

		0°C	10°C	20°C
Pure water	1 bar	1.0000	1.0003	1.0017
	500 bar	0.9767	0.9781	0.9804
	1000 bar	0.9567	0.9590	0.9618
10‰ Cl	1 bar	0.9921	0.9926	0.9942
	500 bar	0.9695	0.9711	0.9735
	1000 bar	0.9501	0.9526	0.9554
20‰ Cl	1 bar	0.9842	0.9850	0.9868
	500 bar	0.9623	0.9642	0.9666
	1000 bar	0.9435	0.9461	0.9491
30‰ Cl	1 bar	0.9764	0.9775	0.9794
	500 bar	0.9552	0.9572	0.9598
	1000 bar	0.9370	0.9397	0.9428
40‰ Cl	1 bar	0.9687	0.9770	0.9720
	500 bar	0.9481	0.9504	0.9530
	1000 bar	0.9305	0.9334	0.9365

Source. R. A. Horne, *Marine Chemistry*, Wiley-Interscience, New York, 1969. Copyright © (John Wiley and Sons, Inc.). Reprinted by permission of John Wiley and Sons, Inc.

Table D.5 Physical Properties of Solids at 20°C

Material	ρ (g/cm^3)	c_P (J/kg·K)	k (J/m·s·K)	ϵ (cm^3/cm^3)
Brick, fire clay	2.31	920	1.12	
Clay, in water	1.28			
Concrete	2.31	880	1.21	
Diatomaceous, Earth powder	0.224	837	0.52	
Earth, mud, flowing	1.7			
Mud, packed	1.8			
Earth's crust	2.67		1.67	
Glass, silica	2.72	840	0.78	
Ice @ 0°C	0.90	2040	2.09	
River mud, in water	1.44			
Rock				
Granite	2.59–2.76	804	0.167–4.06	0.004–0.0384
Marble	2.64–2.87	879	0.502–2.09	0.004–0.021
Limestone	1.87–2.80	904		0.011–0.310
Sandstone	1.91–2.69	921	1.00–2.51	0.019–0.273
Slate	2.69–2.88		1.97	0.001–0.017
Sand			3.9	
Soil[a]				
Sandy loam, 4% H$_2$O	1.66			0.43
Sandy loam, 10% H$_2$O	1.94			
Clay, clay loam, silt loam	1.00–1.60		0.99	0.51
Sands and sandy loam	1.20–1.80		1.98–2.44	0.35–0.50
Compact subsoils	$\geqslant 2.0$		1.74	
Soil minerals, dry	2.60–2.75	837		
Soil, 20% H$_2$O		1380		
Soil, 30% H$_2$O		1590		

Compiled primarily from the following sources: A. R. Jumikis, *Thermal Geotechnics*, Rutgers University Press, New Brunswick, N.J., 1977. J. H. Perry Ed., "Chemical Engineers' Handbook," 3rd Ed. McGraw-Hill, New York (1950).

[a]See Table 6.4-2 for thermal properties of soils.

SPECIFIC GRAVITY OF WATER VS. TEMPERATURE

The following equation* approximating the curve shown in Fig. D.1 has been developed through a computer application of the least squares technique:

$$
\begin{aligned}
\text{Specific gravity factor} = \\
-0.2260569(10^{-5}) + 0.1546919(10^{-5})T \\
+0.2141968(10^{-5})T^2 - 0.6508630(10^{-6})T^3 \\
+0.1975524(10^{-7})T^4 - 0.1894802(10^{-9})T^5
\end{aligned}
$$

where T = temperature in °C, $0° \leqslant T \leqslant 50°$

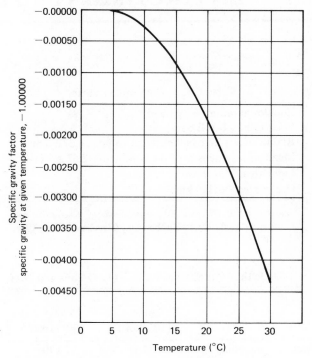

Figure D.1. Specific gravity of water versus temperature.

*From Joseph W. Dewitt, Civil Engineer, U.S. Army Corps of Engineers District Office, Savannah, Georgia.

ENVIRONMENTAL DATA

Table E.1a.   Data on the Earth

Equatorial radius, 6.38E 6 m
Polar radius, 6.36E 6 m
Radius of a sphere having same volume, 6.37E 6 m
Mean surface density of the continents, 2.67 g/cm^3
Land area, 149E 6 km^2
Ocean area, 361E 6 km^2

Source. R. C. Weast, Ed., *Handbook of Chemistry and Physics*, 49th ed., Chem-Rubber Co., Cleveland, Ohio, 1968.

Table E.1b.   The Sizes of the World's Oceans and Seas

Body of Water		Area (10^6/km^2)	Volume (10^6/km^3)	Mean Depth (m)
Pacific Ocean	including	174.68	723.70	4028
Atlantic Ocean	adjacent	106.46	354.68	3332
Indian Ocean	seas	74.92	291.95	3897
Total		361.06	1370.32	3795
Pacific Ocean	excluding	165.25	707.56	4282
Atlantic Ocean	adjacent	82.44	323.61	3926
Indian Ocean	seas	73.44	291.03	3963
Total		321.13	1322.20	4117

Body of Water	Area $(10^6/\text{km}^2)$	Volume $(10^6/\text{km}^3)$	Mean Depth (m)
Baltic Sea	0.42	0.02	55
Red Sea	0.44	0.22	491
North Sea	0.58	0.05	94
Irish Sea	0.10	0.006	60
Bering Sea	2.27	3.26	1437
Mediterranean and Black Seas	2.97	4.24	1429

Source. R. A. Horne, *The Chemistry of Our Environment*, Wiley-Interscience, New York, 1978. Copyright © (John Wiley and Sons, Inc.) Reprinted by permission of John Wiley and Sons, Inc.

Table E.2a. Manning Roughness Coefficients for Clean, Straight Channels

Type of Lining	Condition	n^a
Glazed coating or enamel	In perfect order	0.010
Timber	Planed boards, carefully laid	0.010
	Planed boards, inferior workmanship or aged	0.012
	Unplaned boards, carefully laid	0.012
	Unplaned boards, inferior workmanship or aged	0.014
Metal	Smooth	0.010
	Riveted	0.015
	Slightly tuberculated	0.020
Masonry	Neat cement plaster	0.010
	Sand and cement plaster	0.012
	Concrete, steel troweled	0.012
	Wood troweled	0.013
	Brick, in good condition	0.013
	Rough	0.015
Masonry in bad condition		0.020
Stonework	Smooth, dressed ashlar	0.013
	Rubble set in cement	0.017
	Fine, well-packed gravel	0.020

Type of Lining	Condition	n^a
Earth	Regular surface in good condition	0.020
	In ordinary condition	0.0225
	With stones and weeds	0.025
	In poor condition	0.035
	Partially obstructed with debris or weeds	0.050

Source. C. V. Davis, Ed., *Handbook of Applied Hydraulics*, McGraw-Hill, New York, 1952, p. 1204. Reprinted by permission.

[a]*Manning* formula:

$$V = \frac{1.486}{n} R^{2/3} S^{1/2} \quad \text{or} \quad V = \frac{1.486}{n} R^{1/6}\sqrt{RS_1}$$

where V = mean velocity, fps
R = hydraulic radius, ft
S = hydraulic gradient
n = coefficient of roughness in $(ft)^{1/6}$

Table E.2b. Manning Roughness Coefficients for Natural Channels

Condition	n
Rivers and earth canals in fair condition—some growth	0.025
Winding natural streams and canals in poor condition—considerable moss growth	0.035
Mountain streams with rocky beds and rivers with variable sections and some vegetation along banks	0.040–0.050
Alluvial channels, sand bed, no vegetation	
1. Tranquil flow, $Fr < 1$	
Plane bed	0.014 – 0.02
Ripples	0.018–0.028
Dunes	0.018–0.035
Washed-out dunes or transition	0.014–0.024
Plane bed	0.012–0.015
2. Rapid flow $Fr > 1$	
Standing waves	0.011–0.015
Anti-dunes	0.012–0.020

Reprinted by permission. **Source.** Richard J. Chorley, Editor, *Water, Earth and Man*, from Chapter 7. "Channel Flow" by D. B. Simmons, Methuen and Co. Ltd., London (1969), p. 309.

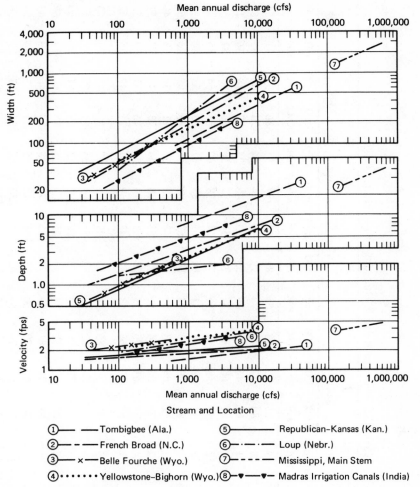

Figure E.1. River systems—width, depth, and velocity. (**Source.** L. B. Leopold, "Rivers," *Am. Sci.*, **50**, No. 4 (December 1962), 511.)

Table E.3. Average Relative Humidity (%)—Selected Cities

State	Station	Length of Record (yr.)	Jan.		Feb.		Mar.		Apr.		May		June	
			7:00 A.M.	1:00 P.M.	7:00 A.M.	1:00 P.M.	7:00 A.M.	1:00 P.M.	7:00 A.M.	1:00 P.M.	7:00 A.M.	1:00 P.M.	7:00 A.M.	1:00 P.M.
Ala.	Mobile	9	81	63	78	55	82	54	87	54	86	52	87	54
Ak.	Juneau	28	80	77	83	77	79	70	74	64	74	62	75	63
Ariz.	Phoenix	11	67	44	61	39	59	33	46	24	37	18	36	18
Ark.	Little Rock	11	81	60	80	57	77	54	81	56	86	56	87	54
Calif.	Los Angeles	12	66	53	71	56	76	58	78	60	81	64	85	70
	Sacramento	11	91	87	88	81	84	69	83	61	83	53	80	50
	San Francisco	12	86	79	83	74	80	69	82	66	85	65	86	65
Colo.	Denver	11	62	44	68	45	70	43	68	37	71	39	73	40
Conn.	Hartford	12	72	58	74	58	73	53	70	44	73	46	77	51
Del.	Wilmington	24	75	61	75	59	74	53	74	51	76	53	79	53
D.C.	Washington	11	68	53	68	53	68	48	69	47	71	49	75	52
Fla.	Jacksonville	35	87	57	85	52	85	49	85	47	83	48	86	55
	Miami	6	84	61	82	57	82	56	81	55	81	60	86	67
Ga.	Atlanta	11	79	59	76	55	78	50	80	52	83	54	86	60
Hawaii	Honolulu[a]	16	80	80	78	78	77	75	75	71	75	69	74	69
Idaho	Boise	32	81	74	81	69	74	55	71	48	70	46	69	43
Ill.	Chicago	8	70	64	70	60	73	58	72	55	71	51	74	54
	Peoria	12	77	68	79	65	82	64	80	58	81	57	82	56
Ind.	Indianapolis	12	79	68	78	65	79	62	77	55	81	57	82	57

Iowa	Des Moines	10	79	72	81	69	81	64	80	57	80	57	83	59
Kans.	Wichita	18	79	63	79	60	75	52	77	50	82	55	83	54
Ky.	Louisville	11	76	63	77	62	76	59	75	53	84	56	85	58
La.	New Orleans	23	86	67	85	63	84	60	88	60	89	60	90	62
Me.	Portland	31	78	62	75	60	76	59	74	55	75	57	78	59
Md.	Baltimore	18	72	57	72	56	71	51	72	49	76	51	79	52
Mass.	Boston	7	66	57	70	60	69	58	68	54	70	57	72	58
Mich.	Detroit	38/34	79	69	79	66	78	60	74	53	71	51	74	53
	Sault Ste. Marie	30	82	76	83	73	84	68	80	61	79	56	85	61
Minn.	Duluth	10	74	67	73	61	76	60	77	57	76	54	80	61
	Minneapolis– St. Paul	12	74	68	75	65	78	64	77	55	77	54	81	56
Miss.	Jackson	8	86	64	85	58	87	57	92	58	92	55	91	55
Mo.	Kansas City	11	75	61	73	58	73	53	72	51	77	54	82	60
	St. Louis	11	79	61	76	58	78	55	76	52	80	55	83	57
Mont.	Great Falls	10	67	63	67	59	65	54	68	49	68	46	71	46
Nebr.	Omaha	8	75	65	75	59	75	53	74	50	76	53	80	57
Nev.	Reno	8	72	65	70	56	65	45	65	38	66	35	70	37
N.H.	Concord	6	70	57	75	60	77	56	77	46	77	46	86	54
N.J.	Atlantic City	7	75	57	75	56	77	53	74	49	75	52	82	56
N. Mex.	Albuquerque	11	66	47	62	43	53	32	45	25	42	23	44	24
N.Y.	Albany	6	79	64	76	59	73	54	69	46	75	51	77	55
	Buffalo	11	77	72	79	71	81	68	76	58	75	56	77	57
	New York	52	68	60	68	58	67	55	68	52	71	53	73	55
N.C.	Charlotte	11	78	55	76	52	79	49	80	48	84	52	87	57
	Raleigh	7	76	53	72	47	78	46	80	45	85	53	87	58
N. Dak.	Bismarck	12	71	66	75	68	77	62	77	52	77	49	84	55

Table E.3. (Continued)

State	Station	Length of Record (yr.)	Jan. 7:00 A.M.	Jan. 1:00 P.M.	Feb. 7:00 A.M.	Feb. 1:00 P.M.	Mar. 7:00 A.M.	Mar. 1:00 P.M.	Apr. 7:00 A.M.	Apr. 1:00 P.M.	May 7:00 A.M.	May 1:00 P.M.	June 7:00 A.M.	June 1:00 P.M.
Ohio	Cincinnati	9	76	64	78	62	77	59	76	53	77	52	79	53
	Cleveland	11	75	69	78	69	78	65	74	57	75	56	78	57
	Columbus	12	76	67	77	65	74	59	76	53	79	55	81	53
Okla.	Oklahoma City	6	81	63	79	56	76	52	77	52	81	56	83	56
Oreg.	Portland	31	86	82	86	80	86	72	86	68	86	66	85	65
Pa.	Philadelphia	12	73	59	71	57	71	53	69	48	75	52	77	53
	Pittsburgh	12	76	66	76	64	77	60	73	51	76	51	79	52
R.I.	Providence	8	69	55	71	56	69	54	68	47	70	49	74	54
S.C.	Columbia	19	85	54	85	52	83	48	81	45	82	46	83	49
S. Dak.	Sioux Falls	8	76	67	78	65	80	61	82	55	81	52	83	58
Tenn.	Memphis	32	80	64	79	60	77	56	79	54	81	54	83	56
	Nashville	6	80	64	78	58	76	50	80	52	85	54	87	54

474

Tex.	Dallas	12	77	59	75	56	74	54	79	56	82	57	81	55
	El Paso	11	59	41	50	35	44	30	33	21	35	20	41	25
	Houston	8	84	64	84	60	86	59	89	63	90	61	91	61
Utah	Salt Lake City	12	77	69	77	64	70	50	66	44	64	37	63	33
Vt.	Burlington	7	68	61	71	61	73	57	75	52	74	52	79	56
Va.	Norfolk	23	75	60	75	57	73	53	74	51	77	56	79	57
	Richmond	37	81	57	79	52	78	48	75	45	78	49	81	53
Wash.	Seattle–Tacoma	12	82	80	80	76	82	75	85	73	84	70	83	67
	Spokane	12	83	81	83	78	78	67	75	56	74	51	73	48
W. Va.	Charleston	24	77	62	76	59	74	53	75	48	82	50	86	53
Wis.	Milwaukee	11	75	69	75	67	80	66	79	61	77	59	81	61
Wyo.	Cheyenne	12	53	43	61	46	62	45	65	40	68	40	71	42
P.R.	San Juan	16	84	65	82	63	79	61	78	64	80	67	81	68

Source. U.S. National Oceanic and Atmospheric Administration, *Local Climatological Data*. Monthly with annual summary.

Table E.4. Average Relative Humidity (%)—Selected Cities

Station	State	July 7:00 A.M.	July 1:00 P.M.	Aug. 7:00 A.M.	Aug. 1:00 P.M.	Sept. 7:00 A.M.	Sept. 1:00 P.M.	Oct. 7:00 A.M.	Oct. 1:00 P.M.	Nov. 7:00 A.M.	Nov. 1:00 P.M.	Dec. 7:00 A.M.	Dec. 1:00 P.M.	Annual 7:00 A.M.	Annual 1:00 P.M.
Mobile	Ala.	89	62	90	62	87	59	84	53	85	54	83	63	85	57
Juneau	Ak.	81	70	84	73	88	78	87	81	86	82	82	81	81	73
Phoenix	Ariz.	47	29	57	36	55	33	53	28	61	38	70	49	54	32
Little Rock	Ark.	88	58	89	57	90	58	86	50	83	57	80	61	84	57
Los Angeles	Calif.	86	68	85	68	82	64	78	58	75	59	70	55	78	61
Sacramento		76	48	76	50	75	50	78	57	87	77	91	87	83	64
San Francisco		88	65	88	67	83	65	81	67	84	74	86	79	84	70
Denver	Colo.	72	36	71	37	72	39	65	36	70	44	67	45	69	40
Hartford	Conn.	79	50	83	52	86	55	85	51	80	58	79	62	78	53
Wilmington	Del.	80	54	84	56	85	55	85	54	81	56	77	60	79	55
Washington	D.C.	75	52	78	53	79	54	79	49	73	52	71	57	73	51
Jacksonville	Fla.	87	58	90	59	91	62	90	58	88	55	87	57	87	55
Miami		85	64	85	65	89	67	88	66	85	59	84	58	84	61
Atlanta	Ga.	90	64	91	62	89	59	84	53	82	54	80	60	83	57
Honolulu	Hawaii	74	69	75	70	74	70	76	72	77	75	77	76	76	73
Boise	Idaho	53	33	51	34	59	40	68	49	79	66	83	75	70	53
Chicago	Ill.	77	55	79	55	80	55	76	52	77	63	78	70	75	58
Peoria		86	59	87	58	88	59	84	57	82	65	83	71	83	62
Indianapolis	Ind.	87	60	89	59	89	58	86	56	85	66	82	71	83	61
Des Moines	Iowa	83	58	85	57	86	61	78	54	79	63	81	71	81	62
Wichita	Kans.	79	50	79	49	81	55	80	53	78	55	79	62	79	55
Louisville	Ky.	86	58	88	56	89	59	86	54	79	61	77	64	82	59
New Orleans	La.	91	66	91	66	89	65	87	59	86	59	86	67	88	63

476

City	State														
Portland	Me.	60	80	62	80	64	85	60	86	60	86	59	84	59	80
Baltimore	Md.	53	77	58	74	53	77	52	82	54	85	55	83	52	80
Boston	Mass.	58	72	61	71	64	76	58	77	61	79	56	74	55	73
Detroit	Mich.	58	78	69	80	64	79	55	81	54	83	53	80	51	75
Duluth	Minn.	62	79	73	79	69	80	62	81	63	86	61	86	57	82
Minneapolis–St. Paul		61	80	72	79	67	82	59	84	61	88	55	85	56	83
Jackson	Miss.	59	90	67	89	56	91	53	93	59	94	61	94	60	93
Kansas City	Mo.	57	77	64	77	57	76	51	74	57	82	55	80	57	79
St. Louis		57	82	66	82	59	82	53	82	58	89	55	87	58	85
Great Falls	Mont.	49	65	62	66	54	64	44	61	45	64	34	58	37	65
Omaha	Nebr.	58	79	66	79	60	80	54	80	60	86	58	85	58	82
Reno	Nev.	45	70	65	73	58	74	40	71	35	70	33	70	30	68
Concord	N.H.	55	82	64	79	63	84	55	88	59	93	53	89	53	87
Atlantic City	N.J.	56	81	59	79	57	83	56	87	57	87	57	87	59	86
Albuquerque	N. Mex.	37	57	52	69	43	63	36	57	41	57	40	66	36	61
Albany	N.Y.	57	78	67	81	63	82	55	84	57	86	53	81	54	77
Buffalo		63	80	73	81	71	82	61	82	60	83	59	83	56	79
New York		56	72	60	70	59	73	55	76	58	79	57	78	55	75
Charlotte	N.C.	54	84	55	79	51	84	53	88	56	90	59	89	59	89
Raleigh		53	84	52	78	49	83	55	90	58	93	61	92	61	91
Bismarck	N. Dak.	56	78	66	74	61	78	47	77	49	82	43	80	48	83
Cincinnati	Ohio	57	80	68	78	61	78	52	82	56	86	54	86	55	84
Cleveland		62	78	70	75	66	78	57	80	59	83	59	84	57	81
Columbus		59	81	69	79	65	82	55	83	58	88	57	88	56	84
Oklahoma City	Okla.	54	80	60	80	52	78	52	80	57	86	50	81	49	80
Portland	Oreg.	73	86	84	87	82	88	79	90	67	87	65	85	62	83
Philadelphia	Pa.	55	76	60	73	56	77	53	82	56	84	55	81	54	80
Pittsburgh		57	79	68	78	62	80	52	81	56	86	55	86	52	83
Providence	R.I.	54	74	60	75	59	78	53	80	55	80	53	77	56	76

Table E.4. (Continued)

	July		Aug.		Sept.		Oct.		Nov.		Dec.		Annual		State	Station
	7:00 A.M.	1:00 P.M.	7:00 A.M.	1:00 P.M.	7:00 A.M.	1:00 P.M.	7:00 A.M.	1:00 P.M.	7:00 A.M.	1:00 P.M.	7:00 A.M.	1:00 P.M.	7:00 A.M.	1:00 P.M.		
	86	52	89	53	90	55	90	49	88	47	85	52	86	50	S.C.	Columbia
	84	54	84	50	87	58	81	56	83	63	81	71	82	59	S. Dak.	Sioux Falls
	85	57	86	56	86	55	84	50	80	55	80	62	82	56	Tenn.	Memphis
	90	58	92	61	91	58	86	55	80	57	80	64	84	57		Nashville
	76	50	77	51	82	56	80	53	79	55	77	60	78	55	Tex.	Dallas
	58	39	59	41	63	45	56	34	57	38	64	45	52	34		El Paso
	91	59	91	59	88	60	86	52	85	60	84	64	87	60		Houston
	51	26	56	30	62	35	68	42	74	58	79	72	67	46	Utah	Salt Lake City
	79	54	84	58	88	63	83	63	81	70	77	70	78	60	Vt.	Burlington
	82	60	85	62	84	61	84	61	79	56	76	58	79	58	Va.	Norfolk
	85	56	88	57	89	56	89	53	84	50	81	55	82	53		Richmond
	84	66	83	69	87	76	87	80	84	81	83	82	84	75	Wash.	Seattle–Tacoma
	60	37	58	41	68	49	79	66	87	82	87	84	75	62		Spokane
	90	61	92	57	91	54	88	53	80	56	78	61	82	55	W. Va.	Charleston
	82	60	87	60	88	63	82	63	80	67	81	73	81	64	Wis.	Milwaukee
	69	34	65	33	65	37	58	36	57	40	56	43	62	40	Wyo.	Cheyenne
	81	68	81	67	81	68	82	67	83	66	83	66	81	66	P.R.	San Juan

Source. U.S. National Oceanic and Atmospheric Administration, *Local Climatological Data.* Monthly with annual summary.

478

Table E.5. Normal Monthly and Annual Precipitation—Selected Cities (in inches.)

State	Station	Jan.	Feb.	Mar.	Apr.	May	June	July	Aug.	Sept.	Oct.	Nov.	Dec.	Annual
Ala.	Mobile	4.64	4.59	7.23	6.36	4.88	6.23	9.67	6.44	6.25	3.03	3.35	5.46	68.13
Ak.	Juneau	4.00	3.06	3.27	2.87	3.24	3.39	4.49	5.02	6.67	8.33	6.06	4.22	54.62
Ariz.	Phoenix	0.73	0.85	0.66	0.32	0.13	0.09	0.77	1.12	0.73	0.46	0.49	0.85	7.20
Ark.	Little Rock	5.22	4.33	4.81	4.93	5.28	3.61	3.34	2.82.	3.23	2.88	4.12	4.09	48.66
Calif.	Los Angeles	2.66	2.88	1.79	1.05	0.13	0.05	0.01	0.02	0.17	0.39	1.09	2.39	12.63
	Sacramento	3.18	2.99	2.36	1.40	0.59	0.10	T	0.02	0.19	0.77	1.45	3.24	16.29
	San Francisco	4.01	3.48	2.69	1.30	0.48	0.11	0.01	0.02	0.19	0.74	1.57	4.09	18.69
Colo.	Denver	0.55	0.69	1.21	2.11	2.70	1.44	1.53	1.28	1.13	1.01	0.69	0.47	14.81
Conn.	Hartford	3.58	2.94	3.80	3.73	3.41	3.70	3.61	1.01	3.65	3.18	3.84	3.47	42.92
Del.	Wilmington	3.40	2.95	4.02	3.33	3.53	4.07	4.25	5.59	3.95	2.91	3.53	3.03	44.56
D.C.	Washington	3.03	2.47	3.21	3.15	4.14	3.21	4.15	4.90	3.83	3.07	2.84	2.78	40.78
Fla.	Jacksonville	2.45	2.91	3.49	3.55	3.47	6.33	7.68	6.85	7.56	5.16	1.69	2.22	53.36
	Miami	2.03	1.87	2.27	3.88	6.44	7.37	6.75	6.97	9.47	8.21	2.83	1.67	59.76
Ga.	Atlanta	4.44	4.51	5.37	4.47	3.16	3.83	4.72	3.60	3.26	2.44	2.96	4.38	47.14
Hawaii	Honolulu	3.76	3.30	2.89	1.31	0.99	0.33	0.44	0.89	0.99	1.84	2.16	2.99	21.89
Idaho	Boise	1.32	1.33	1.32	1.16	1.29	0.89	0.21	0.16	0.39	0.84	1.20	1.32	11.43
Ill.	Chicago	1.86	1.60	2.74	3.04	3.73	4.07	3.37	3.16	2.73	2.78	2.20	1.90	33.18
	Peoria	1.88	1.71	2.85	3.97	4.27	4.08	3.54	2.88	3.05	2.53	2.14	1.94	34.84
Ind.	Indianapolis	3.05	2.28	3.41	3.74	3.99	4.62	3.50	3.03	3.24	2.62	3.09	2.68	39.25
Iowa	Des Moines	1.30	1.10	2.09	2.53	4.07	4.71	3.06	3.67	2.88	2.06	1.76	1.14	30.37
Kans.	Wichita	0.81	0.92	1.64	2.30	3.97	4.21	3.64	2.87	3.22	2.40	1.49	0.94	28.41
Ky.	Louisville	4.10	3.29	4.59	3.82	3.90	3.99	3.96	2.97	2.63	2.25	3.20	3.22	41.32
La.	New Orleans	3.84	3.99	5.34	4.55	4.38	4.43	6.72	5.34	5.03	2.84	3.34	4.10	53.90
Me.	Portland	4.37	3.80	4.34	3.73	3.41	3.18	2.86	2.42	3.52	3.20	4.17	3.85	42.85

Table E.5. (Continued)

State	Station	Jan.	Feb.	Mar.	Apr.	May	June	July	Aug.	Sept.	Oct.	Nov.	Dec.	Annual
Md.	Baltimore	3.43	2.89	3.82	3.60	3.98	3.29	4.22	5.19	3.33	3.18	3.13	2.99	43.05
Mass.	Boston	3.94	3.32	4.22	3.77	3.34	3.48	2.88	3.66	3.46	3.14	3.93	3.63	42.77
Mich.	Detroit	2.05	2.08	2.42	3.00	3.53	2.83	2.82	2.86	2.44	2.63	2.21	2.08	30.95
	Sault Ste. Marie	2.07	1.50	1.81	2.16	2.77	3.30	2.48	2.89	3.81	2.82	3.33	2.28	31.22
Minn.	Duluth	1.15	0.96	1.62	2.36	3.29	4.27	3.54	3.81	2.86	2.17	1.78	1.16	28.97
	Minn.–St. Paul	0.70	0.78	1.53	1.85	3.19	4.00	3.27	3.18	2.43	1.59	1.40	0.86	24.78
Miss.	Jackson	5.18	4.96	5.74	4.91	4.38	3.79	4.76	3.33	2.53	2.04	3.90	5.30	50.82
Mo.	Kansas City	1.41	1.24	2.49	3.56	4.40	4.57	3.19	3.77	3.25	2.86	1.80	1.53	34.07
	St. Louis	1.98	2.04	3.08	3.71	3.73	4.29	3.30	3.02	2.76	2.86	2.57	1.97	35.31
Mont.	Great Falls	0.61	0.74	0.92	0.98	2.10	2.90	1.28	1.26	1.20	0.73	0.75	0.60	14.07
Nebr.	Omaha	0.82	0.95	1.45	2.56	3.48	4.53	3.37	3.98	2.63	1.73	1.26	0.80	27.56
Nev.	Reno	1.19	1.02	0.68	0.54	0.52	0.37	0.27	0.17	0.23	0.51	0.57	1.08	7.15
N.H.	Concord	3.23	2.48	3.26	3.31	3.17	3.60	3.41	2.96	3.75	2.66	3.72	3.25	38.80
N.J.	Atlantic City	3.56	3.13	3.91	3.41	3.51	2.83	3.72	4.90	3.31	3.20	3.66	3.22	42.36
N. Mex.	Albuquerque	0.41	0.38	0.48	0.47	0.75	0.57	1.20	1.33	0.95	0.75	0.38	0.46	8.13
N.Y.	Albany	2.47	2.20	2.72	2.77	3.47	3.25	3.49	3.07	3.58	2.77	2.70	2.59	35.08
	Buffalo	2.84	2.72	3.24	3.01	2.95	2.54	2.57	3.05	3.13	3.00	3.60	3.00	35.65
	New York	3.31	2.84	4.01	3.43	3.67	3.31	3.70	4.44	3.87	3.14	3.39	3.26	42.37
N.C.	Charlotte	3.53	3.55	4.39	3.49	3.11	3.61	4.88	4.22	3.49	2.96	2.53	3.62	43.38
	Raleigh	3.22	3.23	3.35	3.52	3.52	3.70	5.49	5.20	3.85	2.71	2.77	3.02	43.58
N. Dak.	Bismarck	0.44	0.43	0.78	1.22	1.97	3.40	2.19	1.73	1.19	0.85	0.59	0.36	15.15
Ohio	Cincinnati	3.67	2.80	3.89	3.63	3.80	4.18	3.59	3.28	2.71	2.24	2.95	2.77	39.51
	Cleveland	2.67	2.33	3.13	3.41	3.52	3.43	3.31	3.28	2.90	2.42	2.61	2.34	35.35
	Columbus	3.16	2.31	3.16	3.49	4.00	4.16	3.93	2.86	2.65	2.11	2.50	2.34	36.67
Okla.	Oklahoma City	1.31	1.37	1.97	3.12	5.19	4.47	2.37	2.52	3.02	2.51	1.56	1.41	30.82
Oreg.	Portland	5.37	4.22	3.83	2.09	1.99	1.67	0.41	0.65	1.63	3.61	5.33	6.38	37.18

Pa.	Philadelphia	3.32	2.80	3.80	3.40	3.74	4.05	4.16	4.63	3.46	2.78	3.40	2.94	42.48
	Pittsburgh	2.97	2.19	3.32	3.08	3.91	3.78	3.88	3.31	2.54	2.52	2.24	2.40	36.14
R.I.	Providence	3.81	3.10	4.14	3.75	3.35	2.76	2.91	3.96	3.52	3.10	4.11	3.62	42.13
S.C.	Columbia	3.02	3.74	4.26	4.01	3.54	3.85	6.09	5.74	4.31	2.38	2.36	3.52	46.82
S. Dak.	Sioux Falls	0.62	0.93	1.54	2.31	3.38	4.35	2.84	3.59	2.61	1.25	1.00	0.74	25.16
Tenn.	Memphis	6.07	4.69	5.07	4.63	4.23	3.68	3.54	2.97	2.82	2.72	4.38	4.93	49.73
	Nashville	5.49	4.51	5.19	3.74	3.72	3.25	3.72	2.86	2.87	2.32	3.28	4.19	45.15
Tex.	Dallas	2.32	2.55	2.85	4.00	4.83	3.24	1.94	1.93	2.82	2.70	2.70	2.67	34.55
	El Paso	0.46	0.41	0.35	0.29	0.40	0.69	1.29	1.19	1.14	0.85	0.33	0.49	7.89
	Houston	3.78	3.44	2.67	3.24	4.32	3.69	4.29	4.27	4.26	3.77	3.86	4.36	45.95
Utah	Salt Lake City	1.35	1.18	1.56	1.76	1.40	0.98	0.58	0.87	0.53	1.15	1.30	1.24	13.90
Vt.	Burlington	1.95	1.79	2.11	2.63	2.99	3.49	3.85	3.37	3.31	2.97	2.62	2.13	33.21
Va.	Norfolk	3.33	3.21	3.45	3.16	3.36	3.61	5.92	5.97	4.22	2.92	3.05	2.74	44.94
	Richmond	3.46	2.90	3.42	3.15	3.72	3.75	5.61	5.54	3.65	3.00	3.04	2.97	44.21
Wash.	Seattle–Tacoma	5.73	4.24	3.79	2.40	1.73	1.58	0.81	0.95	2.05	4.02	5.35	6.29	38.94
	Spokane	2.44	1.86	1.50	0.91	1.21	1.49	0.38	0.41	0.75	1.57	2.24	2.43	17.19
W. Va.	Charleston	4.32	3.53	4.34	3.68	3.71	3.69	5.67	3.95	2.92	2.58	2.79	3.25	44.43
Wis.	Milwaukee	1.83	1.40	2.31	2.53	3.16	3.64	2.95	3.06	2.72	2.10	2.18	1.63	29.51
Wyo.	Cheyenne	0.52	0.56	1.21	1.88	2.52	2.11	1.82	1.44	1.10	0.83	0.62	0.45	15.06
P.R.	San Juan	4.70	2.90	2.20	3.72	7.12	5.66	6.25	7.13	6.76	5.83	6.49	5.45	64.21

Source. U.S. National Oceanic and Atmospheric Administration, *Local Climatological Data*. Monthly with annual summary.

481

Table E.6. Sunshine: Average Percentage of Possible Sunshine—Selected Cities.

State	Station	Length of Record (yr.)	Jan.	Feb.	Mar.	Apr.	May	June	July	Aug.	Sept.	Oct.	Nov.	Dec.	Annual
Ala.	Mobile	48	49	51	57	65	69	67	61	63	64	72	62	48	61
Ak.	Juneau	26	33	31	37	39	38	35	30	30	24	19	23	20	31
Ariz.	Phoenix	76	78	80	83	88	93	94	84	85	89	88	84	77	86
Ark.	Little Rock	29	46	53	56	60	67	72	70	71	67	69	56	47	62
Calif.	Los Angeles	31	71	72	73	69	66	65	82	83	79	73	74	71	73
	Sacramento	23	45	61	71	80	86	92	98	96	94	86	65	46	79
	San Francisco	35	56	63	69	73	72	72	66	66	73	71	63	54	67
Colo.	Denver	22	72	71	71	67	64	70	71	72	75	74	66	68	70
Conn.	Hartford	17	57	56	57	56	58	59	61	64	60	58	45	50	58
Del.	Wilmington	25	50	54	57	57	59	64	63	61	60	60	54	51	58
D.C.	Washington	23	49	51	56	56	58	65	62	63	63	60	50	48	57
Fla.	Jacksonville	21	58	61	66	71	70	61	60	58	52	55	61	56	61
	Miami	22	68	74	74	72	68	62	62	63	58	59	66	65	66
Ga.	Atlanta	37	48	52	57	65	69	67	62	66	64	67	60	50	61
Hawaii	Honolulu	19	63	65	69	69	71	73	75	77	75	67	61	60	69
Idaho	Boise	32	42	52	62	68	71	75	89	86	82	68	46	40	68
Ill.	Chicago	29	44	47	51	53	61	67	69	68	64	61	41	40	57
	Peoria	28	46	50	52	56	60	66	69	69	66	64	46	41	58
Ind.	Indianapolis	29	41	51	52	56	62	68	70	72	68	65	43	40	59
Iowa	Des Moines	21	51	55	56	56	60	67	71	70	65	63	51	45	60
Kans.	Wichita	18	58	59	61	62	63	68	73	73	68	67	62	57	65

State	City														
Ky.	Louisville	24	41	46	51	56	63	67	66	69	67	63	47	41	58
La.	New Orleans	46	49	51	57	65	69	67	61	63	64	72	62	48	61
Me.	Portland	31	55	59	58	57	57	61	65	65	62	58	46	54	59
Md.	Baltimore	21	52	55	55	55	58	64	66	63	62	59	51	49	58
Mass.	Boston	36	53	57	58	56	58	63	65	66	64	61	51	54	60
Mich.	Detroit	32	32	43	49	52	59	65	70	65	61	56	35	32	54
	Sault Ste. Marie	30	34	45	54	53	55	58	63	59	45	42	23	27	47
Minn.	Duluth	21	50	56	59	55	55	59	67	63	52	49	34	40	55
	Minneapolis–St. Paul	33	50	57	54	56	58	62	70	67	61	57	39	40	58
Miss.	Jackson	7	46	56	61	60	66	70	63	61	61	64	55	43	60
Mo.	Kansas City	38	53	56	58	60	63	68	76	73	68	67	57	50	64
	St. Louis	12	52	51	54	57	63	68	70	67	66	63	50	43	59
Mont.	Great Falls	29	51	59	68	63	64	65	81	78	67	61	46	47	64
Nebr.	Omaha	36	54	55	55	59	62	67	76	71	67	67	53	48	62
Nev.	Reno	29	65	69	74	79	79	83	92	93	91	82	70	62	80
N.H.	Concord	30	51	54	52	52	54	58	62	60	55	54	41	48	54
N.J.	Atlantic City	11	53	49	54	52	55	61	58	62	60	57	51	44	55
N. Mex.	Albuquerque	32	73	74	74	77	80	83	76	76	81	80	78	71	77
N.Y.	Albany	33	46	51	53	53	55	60	64	62	57	54	37	40	54
	Buffalo	28	33	40	47	52	59	67	69	67	61	54	30	29	53
	New York	95	51	55	57	59	62	65	65	64	63	61	52	50	59
N.C.	Charlotte	21	56	59	64	69	70	71	69	71	68	69	63	60	66
	Raleigh	17	56	59	65	63	60	61	62	61	62	62	64	58	61
N. Dak.	Bismarck	32	55	56	60	59	63	63	77	74	65	60	45	48	62
Ohio	Cincinnati	56	42	45	51	56	61	68	69	67	67	60	45	39	54
	Cleveland	30	31	37	45	53	60	66	67	65	60	55	31	27	52
	Columbus	20	38	42	46	53	60	64	66	67	65	59	40	32	55

Table E.6. (Continued)

State	Station	Length of Record (yr.)	Jan.	Feb.	Mar.	Apr.	May	June	July	Aug.	Sept.	Oct.	Nov.	Dec.	Annual
Okla.	Oklahoma City	19	58	61	64	63	65	73	76	78	72	68	62	58	67
Oreg.	Portland	22	23	36	41	48	53	50	69	63	58	40	29	20	47
Pa.	Philadelphia	29	50	53	57	56	58	64	62	62	61	60	53	51	58
	Pittsburgh	19	37	39	47	50	55	61	63	63	63	58	41	32	52
R.I.	Providence	18	57	56	57	56	58	59	59	59	59	59	49	54	57
S.C.	Columbia	18	58	60	65	67	67	65	64	66	65	65	65	62	64
S. Dak.	Sioux Falls	27	54	59	55	60	63	67	77	74	67	65	52	48	63
Tenn.	Memphis	21	48	53	57	63	70	73	73	76	72	72	59	50	65
	Nashville	30	40	47	52	59	63	68	65	66	64	64	51	41	58
Tex.	Dallas	31	51	53	58	57	61	73	77	76	70	66	62	55	65
	El Paso	29	78	82	84	87	89	89	79	81	82	84	83	77	83
	Houston	7	47	56	57	54	62	72	74	70	66	73	60	50	62
Utah	Salt Lake City	34	47	54	64	67	73	78	83	82	83	74	54	44	69
Vt.	Burlington	28	42	48	54	51	57	62	66	63	56	50	31	34	53
Va.	Norfolk	18	57	58	63	66	67	68	65	65	65	60	61	58	63
	Richmond	21	51	54	59	62	65	67	65	64	64	59	55	51	60
Wash.	Seattle–Tacoma	6	19	46	49	49	56	54	66	66	57	37	27	15	48
	Spokane	24	25	41	54	60	63	66	81	78	70	51	29	20	57
W. Va.	Charleston[a]	69	31	36	42	49	56	60	62	60	60	54	37	29	48
Wis.	Milwaukee	31	44	46	51	54	59	64	71	68	61	57	41	39	56
Wyo.	Cheyenne	36	62	64	64	60	58	65	69	67	69	69	61	59	64
P.R.	San Juan	16	65	68	74	69	60	58	66	68	63	63	61	58	64

Source. U.S. National Oceanic and Atmospheric Administration, *Local Climatological Data.* Monthly with annual summary.

Table E.7. Temperature: Normal Monthly Average Temperature—Selected Cities (in F°.)

State	Station	Jan.	Feb.	Mar.	Apr.	May	June	July	Aug.	Sept.	Oct.	Nov.	Dec.	Annual
Ala.	Mobile	53.0	55.2	60.3	67.6	75.6	81.5	82.6	82.1	77.9	69.9	58.9	54.1	68.2
Ak.	Juneau	25.1	26.8	30.4	38.0	45.6	52.3	55.3	54.1	48.9	41.6	34.3	28.4	40.1
Ariz.	Phoenix	49.7	53.5	59.0	67.2	75.0	83.6	89.8	87.5	82.8	70.7	58.1	51.6	69.0
Ark.	Little Rock	40.6	44.4	51.8	62.4	70.5	78.9	81.9	81.3	74.3	63.1	49.5	41.9	61.7
Calif.	Los Angeles	54.4	55.2	57.0	59.4	62.0	64.8	69.1	69.1	68.5	64.9	61.1	56.9	61.9
	Sacramento	45.2	49.2	53.4	58.4	64.0	70.5	75.4	74.1	71.6	63.5	52.9	46.4	60.4
	San Francisco	50.7	53.0	54.7	55.7	57.4	59.1	58.8	59.4	62.0	61.4	57.4	52.5	56.8
Colo.	Denver	28.5	31.5	36.4	46.4	56.2	66.5	72.9	71.5	63.0	51.4	37.7	31.6	49.5
Conn.	Hartford	26.0	27.1	36.0	48.5	59.9	68.7	73.4	71.2	63.3	53.0	41.3	28.9	49.8
Del.	Wilmington	33.4	33.8	41.3	52.1	62.7	71.4	76.0	74.3	67.6	56.6	45.4	35.1	54.1
D.C.	Washington	36.9	37.8	44.8	55.7	65.8	74.2	78.2	76.5	69.7	59.0	47.7	38.1	57.0
Fla.	Jacksonville	55.9	57.5	62.2	68.7	75.8	80.8	82.6	82.3	79.4	71.0	61.7	56.1	69.5
	Miami	66.9	67.9	70.5	74.2	77.6	80.8	81.8	82.3	81.3	77.8	72.4	68.1	75.1
Ga.	Atlanta	44.7	46.1	51.4	60.2	69.1	76.6	78.9	78.2	73.1	62.4	51.2	44.8	61.4
Hawaii	Honolulu	72.5	72.4	72.8	74.2	75.9	77.9	78.8	79.4	79.2	78.2	75.9	73.6	75.9
Idaho	Boise	29.1	34.5	41.7	50.4	58.2	65.8	75.2	72.1	62.7	51.6	38.6	32.2	51.0
Ill.	Chicago	26.0	27.7	36.3	49.0	60.0	70.5	75.6	74.2	66.1	55.1	39.9	29.1	50.8
	Peoria	25.7	28.4	37.6	50.8	61.5	71.7	76.0	74.3	66.4	55.3	39.7	29.1	51.4
Ind.	Indianapolis	29.1	31.1	38.9	50.8	61.4	71.1	75.2	74.3	66.5	55.4	40.9	31.1	52.1
Iowa	Des Moines	19.9	23.4	33.8	48.7	60.6	71.0	76.3	74.1	65.4	54.2	37.1	25.3	49.2
Kans.	Wichita	32.0	36.3	44.5	56.7	66.0	76.5	80.9	80.8	71.3	59.9	44.4	35.8	57.1
Ky.	Louisville	35.0	35.8	43.3	54.8	64.4	73.4	77.6	76.2	69.5	57.9	44.7	36.3	55.7
La.	New Orleans	54.6	57.1	61.4	67.9	74.4	80.1	81.6	81.9	78.3	70.4	60.0	55.4	68.6

485

Table E.7. (Continued)

State	Station	Jan.	Feb.	Mar.	Apr.	May	June	July	Aug.	Sept.	Oct.	Nov.	Dec.	Annual
Me.	Portland	21.8	22.8	31.4	42.5	53.0	62.1	68.1	66.8	58.7	48.6	38.1	25.8	45.0
Md.	Baltimore	34.8	35.7	43.1	54.2	64.4	72.5	76.8	75.0	68.1	57.0	45.5	35.8	55.2
Mass.	Boston	29.9	30.3	37.7	47.9	58.8	67.8	73.7	71.7	65.3	55.0	44.9	33.3	51.4
Mich.	Detroit	26.9	27.2	34.8	47.6	59.0	69.7	74.4	72.8	65.1	53.8	40.4	29.9	50.1
	Sault Ste. Marie	15.8	15.7	23.8	38.0	49.6	59.0	64.6	64.0	55.8	46.3	33.3	20.9	40.6
Minn.	Duluth	8.7	10.8	21.3	37.0	49.2	58.8	65.5	63.8	54.2	44.6	27.3	14.0	37.9
	Minneapolis–													
	St. Paul	12.4	15.7	27.4	44.3	57.3	66.8	72.3	70.0	60.4	48.9	31.2	18.1	43.7
Miss.	Jackson	47.9	50.5	56.5	64.9	73.1	79.8	82.3	82.0	76.5	67.0	55.5	49.4	65.5
Mo.	Kansas City	31.7	35.8	43.3	55.7	65.6	75.9	81.5	79.8	71.3	60.2	44.6	35.8	56.8
	St. Louis	31.9	34.7	42.6	54.9	64.2	74.1	78.1	76.8	69.5	58.4	44.1	34.8	55.3
Mont.	Great Falls	22.1	23.8	30.7	43.6	53.0	59.9	69.4	66.8	57.4	47.5	34.3	27.3	44.7
Nebr.	Omaha	22.3	26.5	36.9	51.7	63.0	73.1	78.5	76.2	66.9	55.7	38.9	28.2	51.5
Nev.	Reno	30.4	35.6	41.5	48.0	53.9	60.1	67.7	65.5	58.8	49.2	39.3	31.9	48.4
N.H.	Concord	21.2	22.7	31.7	43.8	55.5	64.5	69.6	67.4	59.3	48.7	37.6	25.0	45.6
N.J.	Atlantic City	34.8	34.7	41.1	51.0	61.3	70.0	75.1	73.7	67.2	57.2	46.7	36.6	54.1
N. Mex.	Albuquerque	35.0	39.9	45.8	55.7	65.1	74.9	78.5	76.2	70.0	58.0	43.6	37.0	56.6
N.Y.	Albany	22.7	23.7	33.0	46.2	57.9	67.3	72.1	70.0	61.6	50.8	39.1	26.5	47.6
	Buffalo	24.5	24.1	31.5	43.5	54.8	64.8	69.8	68.4	61.4	50.8	39.1	27.7	46.7
	New York	32.2	33.4	40.5	51.4	62.4	71.4	76.8	75.1	68.5	58.3	47.0	35.9	54.5
N.C.	Charlotte	42.7	44.2	50.0	60.3	69.0	77.1	79.2	78.7	72.9	62.5	50.4	42.7	60.8
	Raleigh	41.6	43.0	49.5	59.3	67.6	75.1	77.9	76.9	71.2	60.5	50.0	41.9	59.5
N. Dak.	Bismarck	9.9	13.5	26.2	43.5	55.9	64.5	71.7	69.3	58.7	46.7	28.9	17.8	42.2

State	City													
Ohio	Cincinnati	33.7	35.1	42.7	54.2	64.2	73.4	76.9	75.7	69.0	57.9	44.6	35.3	55.2
	Cleveland	28.4	28.5	35.1	47.0	58.0	67.8	71.9	70.4	64.2	53.4	41.3	30.5	49.7
	Columbus	29.9	31.1	38.9	50.8	61.5	70.8	74.8	73.2	65.9	54.2	41.2	31.5	52.0
Okla.	Oklahoma City	37.0	41.3	48.5	59.9	68.4	78.0	82.5	82.8	73.8	62.9	48.4	40.3	60.3
Oreg.	Portland	38.4	42.0	46.1	51.8	57.4	62.0	67.2	66.6	62.2	54.2	45.1	41.3	52.9
Pa.	Philadelphia	32.3	33.2	41.0	52.0	62.6	71.0	75.6	73.6	66.7	55.7	44.3	33.9	53.5
	Pittsburgh	28.9	29.2	36.8	49.0	59.8	68.4	72.1	70.8	64.2	53.1	40.8	30.7	50.3
R.I.	Providence	29.2	29.7	37.0	47.2	57.5	66.2	72.1	70.5	63.2	53.2	43.0	32.0	50.1
S.C.	Columbia	46.9	48.4	54.4	63.6	72.2	79.7	81.6	80.5	75.3	64.7	53.7	46.4	64.0
S. Dak.	Sioux Falls	15.2	19.1	30.1	45.9	58.3	68.1	74.3	71.8	61.8	50.3	32.6	21.1	45.7
Tenn.	Memphis	41.5	44.1	51.1	61.4	70.3	78.5	81.3	80.5	73.9	63.1	50.1	42.5	61.5
	Nashville	39.9	42.0	49.1	59.6	68.6	77.4	80.2	79.2	72.8	61.5	48.5	41.4	60.0
Tex.	Dallas	45.9	49.5	56.1	65.0	72.9	81.3	84.9	85.0	77.9	67.8	54.9	48.1	65.8
	El Paso	42.9	49.1	54.9	63.4	71.9	81.0	81.9	80.4	74.5	64.4	51.2	44.1	63.3
	Houston	53.6	55.8	61.3	68.5	76.0	81.6	83.0	83.2	79.2	71.4	60.8	55.7	69.2
Utah	Salt Lake City	27.2	32.5	40.4	49.9	58.9	67.4	76.9	74.5	64.4	51.7	36.7	30.1	50.9
Vt.	Burlington	16.2	17.4	26.7	41.2	53.8	64.2	69.0	66.7	58.4	47.6	35.3	21.5	43.2
Va.	Norfolk	41.2	41.6	48.0	58.0	67.5	75.6	78.8	77.5	72.6	62.0	51.4	42.5	59.7
	Richmond	38.7	39.9	47.7	58.1	67.0	75.1	78.1	76.0	70.2	58.7	48.5	39.7	58.1
Wash.	Seattle–Tacoma	38.3	40.8	43.8	49.2	55.5	59.8	64.9	64.1	59.9	52.4	43.9	40.8	51.1
	Spokane	25.3	30.0	38.1	47.3	56.2	61.9	70.5	68.0	60.9	49.1	35.7	30.1	47.8
W. Va.	Charleston	36.6	37.5	44.4	55.3	64.8	72.0	74.9	73.8	68.2	57.3	45.3	37.1	55.6
Wis.	Milwaukee	20.6	22.4	31.0	43.6	53.4	63.3	68.7	67.8	60.3	50.0	35.8	24.6	45.1
Wyo.	Cheyenne	25.4	27.3	32.4	42.6	52.9	63.0	70.0	67.7	58.6	47.5	34.2	29.5	45.9
P.R.	San Juan	74.4	74.4	75.3	76.6	78.7	80.0	80.4	80.9	80.5	80.0	78.2	76.2	78.0

Source. U.S. National Oceanic and Atmospheric Administration, *Local Climatological Data.* Monthly with annual summary.

487

Table E.8. Average Wind Speed—Selected Cities (in mi/hr.)

State	Station	Length of Record (yr)	Jan.	Feb.	Mar.	Apr.	May	June	July	Aug.	Sept.	Oct.	Nov.	Dec.	Annual
Ala.	Mobile	23	11.0	11.4	11.5	10.8	9.2	8.0	7.2	7.1	8.3	8.6	9.7	10.5	9.4
Ak.	Juneau	28	8.5	8.6	9.0	8.9	8.4	7.9	7.7	7.7	8.2	8.8	8.9	9.5	8.6
Ariz.	Phoenix	26	4.9	5.5	6.2	6.5	6.6	6.6	6.9	6.3	6.0	5.5	5.0	4.8	5.9
Ark.	Little Rock	29	8.9	9.4	10.2	9.7	8.2	7.7	7.0	6.7	7.0	7.0	8.2	8.5	8.2
Calif.	Los Angeles	23	6.7	7.3	7.9	8.4	8.2	7.8	7.5	7.4	7.1	6.7	6.6	6.6	7.4
	Sacramento	23	8.2	8.1	9.1	9.1	9.6	10.1	9.3	8.8	8.0	7.0	6.6	7.3	8.4
	San Francisco	44	7.1	8.5	10.3	12.1	13.2	14.0	13.7	12.9	11.1	9.3	7.1	6.8	10.5
Colo.	Denver	23	9.3	9.3	10.0	10.4	9.4	9.0	8.5	8.2	8.2	8.2	8.7	9.0	9.0
Conn.	Hartford	17	9.8	10.0	10.5	10.8	9.6	8.5	7.9	7.8	7.8	8.3	8.9	9.2	9.1
Del.	Wilmington	23	9.8	10.5	11.2	10.5	9.0	8.3	7.7	7.3	7.8	8.1	9.1	9.3	9.0
D.C.	Washington	23	10.2	10.6	11.1	10.6	9.3	8.8	8.3	8.1	8.3	8.6	9.3	9.4	9.4
Fla.	Jacksonville	22	8.7	9.8	9.8	9.5	9.1	8.7	7.9	7.7	8.8	9.0	8.6	8.4	8.8
	Miami	14	9.4	10.1	10.3	10.4	9.4	8.1	7.8	7.6	8.2	9.1	9.2	8.8	9.0
Ga.	Atlanta	33	10.6	11.1	11.0	10.2	8.7	7.9	7.4	7.1	8.0	8.4	9.2	9.8	9.1
Hawaii	Honolulu	22	9.9	10.6	11.2	11.8	12.1	12.8	13.5	13.6	11.7	10.7	11.1	11.2	11.7
Idaho	Boise	32	8.7	9.4	10.4	10.2	9.6	9.2	8.5	8.3	8.4	8.7	8.7	8.6	9.1
Ill.	Chicago	29	11.5	11.7	11.9	11.8	10.5	9.3	8.3	8.1	9.0	9.8	11.4	11.2	10.4
	Peoria	28	11.3	11.6	12.3	12.4	10.6	9.2	8.1	7.9	8.9	9.6	11.4	11.0	10.4
Ind.	Indianapolis	23	11.1	11.1	11.8	11.5	9.7	8.4	7.4	7.1	8.1	8.9	10.8	10.6	9.7
Iowa	Des Moines	22	11.8	11.8	13.3	13.5	11.9	10.6	9.1	8.8	9.8	10.8	12.0	11.7	11.2
Kans.	Wichita	18	12.7	13.0	14.6	14.7	13.4	12.9	11.3	11.3	11.9	12.4	12.3	12.2	12.7

Ky.	Louisville	24	9.6	9.7	10.4	10.0	8.0	7.2	6.6	6.2	6.8	7.0	9.0	9.2	8.3
La.	New Orleans	23	9.6	10.2	10.2	9.6	8.3	6.9	6.3	6.2	7.5	7.8	8.9	9.2	8.4
Me.	Portland	31	9.2	9.5	10.0	9.9	9.2	8.2	7.7	7.5	7.8	8.5	8.8	9.0	8.8
Md.	Baltimore	21	10.1	10.8	11.4	11.2	9.8	8.8	8.3	8.5	8.6	9.2	9.7	9.6	9.7
Mass.	Boston	14	14.6	14.5	14.5	13.7	12.7	11.8	11.3	11.3	11.5	12.3	13.2	14.1	13.0
Mich.	Detroit	38	11.5	11.5	11.4	11.1	9.8	9.0	8.2	8.0	8.8	9.5	11.4	11.3	10.1
	Sault Ste. Marie	30	10.2	10.2	10.5	10.9	10.4	9.0	8.4	8.3	9.1	9.7	10.4	10.2	9.8
Minn.	Duluth	22	12.2	12.2	12.2	13.6	12.7	11.0	10.0	9.9	10.9	11.7	12.4	11.7	11.7
	Minneapolis–St. Paul	33	10.5	10.7	11.3	12.5	11.6	10.7	9.3	9.1	10.0	10.5	11.1	10.4	10.6
Miss.	Jackson	8	9.3	9.4	9.8	9.2	7.6	6.4	6.3	5.9	6.8	6.9	7.6	8.6	7.8
Mo.	Kansas City	32	9.9	10.3	11.6	11.5	10.4	10.0	8.9	8.8	9.1	9.1	10.2	9.9	10.0
	St. Louis	22	10.2	10.7	11.7	11.3	9.5	8.5	7.6	7.4	8.0	8.6	9.9	10.2	9.5
Mont.	Great Falls	30	15.7	14.9	13.4	13.2	11.4	11.3	10.3	10.6	11.7	13.8	15.1	16.0	13.1
Nebr.	Omaha	36	11.2	11.5	12.8	13.2	11.5	10.6	9.1	9.2	9.7	10.1	11.2	10.8	10.9
Nev.	Reno	29	6.1	6.0	7.5	7.9	7.6	7.1	6.5	6.1	5.4	5.3	5.1	5.1	6.3
N.H.	Concord	30	7.3	7.9	8.2	7.8	7.1	6.3	5.6	5.3	5.4	6.6	6.5	7.1	6.7
N.J.	Atlantic City	13	12.2	12.3	12.4	12.1	10.7	9.7	9.3	9.0	9.5	10.0	11.5	11.5	10.9
N. Mex.	Albuquerque	32	7.8	8.8	10.0	10.9	10.4	9.9	9.0	8.1	8.5	8.2	7.7	7.5	8.9
N.Y.	Albany	33	9.7	10.3	10.4	10.4	9.1	8.1	7.3	6.9	7.3	7.9	8.8	9.1	8.8
	Buffalo	32	14.4	14.2	13.9	13.1	12.0	11.4	10.7	10.2	10.7	11.5	13.1	13.6	12.4
	New York	52	10.9	11.0	11.1	10.6	8.9	8.2	7.7	7.7	8.2	9.0	10.0	10.4	9.5
N.C.	Charlotte	22	7.9	8.4	8.9	8.9	7.6	6.8	6.5	6.5	6.8	7.1	7.3	7.3	7.5
	Raleigh	22	8.7	9.3	9.8	9.5	7.9	7.1	6.9	6.7	7.1	7.4	7.9	8.1	8.0
N. Dak.	Bismarck	32	10.2	10.2	11.4	12.8	12.5	11.2	9.8	10.0	10.5	10.4	10.6	9.8	10.8
Ohio	Cincinnati	43	8.3	8.4	9.0	8.4	6.7	6.4	5.2	5.1	5.4	6.1	7.7	7.9	7.1
	Cleveland	30	12.4	12.4	12.5	11.9	10.5	9.5	8.8	8.4	9.1	10.1	12.3	12.4	10.9
	Columbus	22	10.1	10.3	10.7	10.0	8.5	7.3	6.6	6.2	6.7	7.6	9.5	9.7	8.6

Table E.8. (Continued)

State	Station	Length of Record (yr)	Jan.	Feb.	Mar.	Apr.	May	June	July	Aug.	Sept.	Oct.	Nov.	Dec.	Annual
Okla.	Oklahoma City	23	13.5	13.7	15.2	15.1	13.6	13.1	11.5	11.1	11.7	12.5	12.8	12.9	13.1
Oreg.	Portland	23	10.0	8.7	8.4	7.1	6.8	6.7	7.3	6.9	6.2	6.4	8.2	9.5	7.7
Pa.	Philadelphia	31	10.4	11.1	11.4	11.0	9.7	8.8	8.1	7.8	8.3	8.8	9.7	10.1	9.6
	Pittsburgh	19	10.8	11.1	11.1	10.8	9.4	8.1	7.5	7.3	7.7	8.5	10.2	10.7	9.4
R.I.	Providence	18	11.8	12.0	12.5	12.7	11.3	10.2	9.7	9.7	9.8	9.9	10.7	11.2	10.9
S.C.	Columbia	23	7.1	7.7	8.4	8.5	7.1	6.8	6.7	6.1	6.2	6.2	6.5	6.7	7.0
S. Dak.	Sioux Falls	23	11.0	11.0	12.5	13.4	12.1	10.7	9.6	9.7	10.3	10.8	11.7	10.7	11.1
Tenn.	Memphis	23	10.6	10.5	11.3	10.8	9.0	8.1	7.6	7.1	7.7	7.8	9.4	10.0	9.2
	Nashville	30	9.1	9.3	9.9	9.4	7.5	6.8	6.3	6.0	6.3	6.4	8.3	8.7	7.8
Tex.	Dallas	31	10.5	11.3	12.8	13.1	12.1	12.0	10.1	9.6	9.4	9.4	10.2	10.4	10.9
	El Paso	29	9.3	10.3	12.1	12.2	11.2	10.3	9.1	8.6	8.5	8.4	8.8	8.7	9.8
	Houston	20	11.9	12.1	12.7	13.0	11.7	10.2	8.8	8.4	9.1	9.8	11.2	11.3	10.8
Utah	Salt Lake City	42	7.6	8.2	9.1	9.4	9.3	9.2	9.3	9.5	9.0	8.4	7.7	7.4	8.7
Vt.	Burlington	28	9.6	9.4	9.2	9.4	8.9	8.2	7.8	7.5	8.0	8.6	9.5	9.7	8.8
Va.	Norfolk	23	11.7	12.0	12.4	11.8	10.2	9.4	8.7	8.7	9.6	10.4	10.7	11.0	10.5
	Richmond	23	8.1	8.6	9.0	8.9	7.9	7.3	6.8	6.5	6.7	7.0	7.5	7.5	7.6
Wash.	Seattle–Tacoma	23	10.4	10.2	10.4	10.1	9.4	9.1	8.6	8.3	8.4	9.2	9.5	10.2	9.5
	Spokane	24	8.6	8.9	9.4	9.6	8.6	8.6	8.0	7.9	7.9	8.0	8.0	8.5	8.5
W. Va.	Charleston	24	7.7	8.1	8.7	8.0	6.4	5.6	5.2	4.5	4.9	5.4	7.0	7.3	6.6
Wis.	Milwaukee	31	12.8	12.9	13.3	13.3	12.2	10.4	9.6	9.5	10.7	11.5	12.9	12.6	11.8
Wyo.	Cheyenne	36	14.9	14.9	14.6	14.1	12.3	11.3	10.0	10.1	10.9	11.8	13.8	14.6	12.8
P.R.	San Juan	16	8.7	9.2	9.5	9.1	8.6	8.8	9.7	9.0	7.4	6.6	7.4	8.5	8.5

Source. U.S. National Oceanic and Atmospheric Administration, *Local Climatological Data.* Monthly with annual summary.

490

INDEX